SOUNDS OF YOUR LIFE

The research and publication of this history has been supported by Ofcom

Cover photograph by Paul Carter [paulcarter-photographer.co.uk]
Cover design: Jonny Wood

SOUNDS OF YOUR LIFE

The history of Independent Radio in the UK

Tony Stoller

for Samuel, Louis, Abigail and Stanley

British Library Cataloguing in Publication Data

SOUNDS OF YOUR LIFE
The history of Independent Radio in the UK

A catalogue entry for this book is available from the British Library

ISBN: 9780 86196 690 5 (Paperback)

Published by
John Libbey Publishing Ltd, 3 Leicester Road, New Barnet, Herts EN5 5EW, United Kingdom e-mail: libbeyj@asianet.co.th; web site: www.johnlibbey.com

Direct orders (UK and Europe): direct.orders@marston.co.uk

Distributed in North America by **Indiana University Press**, 601 North Morton St, Bloomington, IN 47404, USA. www.iupress.indiana.edu

Distributed in Australasia by **Elsevier Australia**, Elsevier Australia, Tower 1, 475 Victoria Ave, Chatswood NSW 2067, Australia. www.elsevier.com.au

Printed and bound in China by 1010 Printing International Ltd.

Contents

	Introduction	1
Section I	**Prologue**	
Chapter 1	The long and winding road 1898 – 1970s	11
Section II	**Debate, design and implementation 1970 – 1979**	
Chapter 2	Paving the way for ILR 1970 – July 1972	27
Chapter 3	Making a start August 1972 – October 1973	41
Chapter 4	Turn your radio on The first year of ILR	54
Chapter 5	The pioneer years Summer 1974 – Summer 1976	68
Chapter 6	Is there anybody there? Audience research	84
Chapter 7	Now we are nineteen Autumn 1976 – 1979	96
Section III	**The independent radio experiment 1980 – 1989**	
Chapter 8	Victories and losses 1979 – 1985	115
Chapter 9	Doing well by doing good Secondary rental and programme sharing	131
Chapter 10	London Heathrow Calling The Heathrow Conference and its impact	142
Chapter 11	Left of the dial The failure of community radio: 1965 – 1989	154
Chapter 12	Changing the guard 1986 – 1989	162
Chapter 13	Copyright wars The long battle over music copyright	181

Section IV **Victory of the commercial model 1990 – 2003**

Chapter 14 Shadow and substance
1990 199

Chapter 15 Classic, Talk and Virgin
Independent National Radio: 1991 – 1994 210

Chapter 16 Glad confident morning
ILR 1991 – 1993 221

Chapter 17 Awards and re-awards
Licensing in London and beyond: 1993 – 1994 234

Chapter 18 High summer
1994 – 1996 244

Chapter 19 Challenging the regulator
1994 – 2000 260

Chapter 20 Radio by numbers
Digital radio 275

Chapter 21 Things can only get better
1997 and all that 292

Chapter 22 Weddings and wind-ups
1998 – 2000 299

Chapter 23 RSLs and Access Radio
The strange triumph of the social engineers: 1990 – 2006 313

Chapter 24 Breaking the mould
2000 – 2003 326

Section V **Postscripts**

Chapter 25 Epilogue 343

Acknowledgements, sources and bibliography 354

Annex A Radio advertising and sponsorship revenue 1972–2008 357

Annex B Independent radio licences in issue 1972–2008 358

Annex C Radio audiences 1972–2008 359

Index 360

Introduction

This is the history of a uniquely British experiment. After fifty years of BBC monopoly, *independent* radio was introduced into the UK in the early seventies. It was intended to be local, public service radio, funded by advertising but meeting a social as well as a market purpose. It was deliberately and avowedly not *commercial* radio. Across the eighties, as the old economic and political consensus came undone, what had been Independent Local Radio (ILR) gradually took on more of the characteristics of commercial radio as understood elsewhere in the world. Independent National Radio (INR) arrived in the nineties, and by the first decade of the new century it was clear that *independent* radio had been replaced by *commercial* radio. This book charts the history of the rise and fall of independent radio. It is also a fable, illustrating the failure of the social liberalism of the seventies, and what happened when that was replaced by the market liberalism of the nineties.

Set out west from Reading, along the A4, and you come to Calcott, once an almost rural suburb. Turn left, and at one time you would have driven along a poorly-surfaced road to a rather ugly Victorian detached house. This was the headquarters for Berkshire Civil Defence, part of a futile effort in the fifties and sixties to protect against the impact of nuclear war. In the mid-seventies, with the addition of a concrete studio block, it was the home of Radio 210. This was one of the smallest of the first batch of those Independent Local Radio stations which overthrew the BBC's monopoly of radio in the UK.

Radio 210 was an odd creature, but its story is a paradigm for independent radio. Enfranchised in March 1976 to serve just 200,000 people in and around Reading, and pleased to make a profit of £80,000 in a good year, it attracted substantial investments from such dominant media companies as News International and Thames Television. Its chairman was a scion of the Establishment, Sir John Colville, once Churchill's private secretary; its deputy chair was Charles Douro, the heir to the Duke of Wellington. Radio 210 employed four or five journalists in a serious local newsroom, and was able to mount election night specials with live reports from all the local counts. During daytime it ran mixed programming, with pop music, news, information and features, unpredictable interviews and rather too many in-jokes. In the evening it provided specialist music programmes of folk, jazz, big band, country and classical. Its music programming was eccentric, pretty much comprising whatever the presenters wanted to play.

And what presenters! Steve Wright, Mike Read – combining in a breakfast show called inevitably Read n' Wright – 'Whispering' Bob Harris from BBC television's *Old Grey Whistle Test*, Howard Pearce from Radio Luxembourg, and home favourites such as Mike Matthews, Gavin McCoy and Graham Ledger. Driven by its obligations to fund live music, it promoted Vladimir Ashkenazy and John Lill, and made the first ILR offer to the European Broadcasting Union network from a British commercial station of a live concert by the Eurythmics. Radio 210 was at the centre of local life, accessible to and courted by MPs and councillors, and for more than a third of local people it was the soundtrack to every day. When I was managing director there in the early eighties, our advertising slogan for the station tried to capture that aspiration: "Radio 210, sounds of your life". It now provides the title for this history.

By the nineties, the station had become Radio Two Ten, reflecting the increased significance of FM broadcasting and its position within the GWR group of stations. In its good years, it could aspire to profits not far short of £1m. The old buildings had been swept away in favour of a fine new studio and office complex, which also housed group functions, such as accounts and financial administration for the thirty or so GWR stations. The rather scruffy but leafy surroundings of earlier years had been replaced by modern housing estates, in keeping with Reading's growing prosperity and the radio station's own modernity. Two Ten's programming was strictly controlled according to the group policies laid down by group programming head Steve Orchard. The playlist of pop music was all carefully researched to appeal to 25–34 years olds, and that audience had risen accordingly. The specialist music shows were long gone, local advertisers often relegated to the margins. Presenters read their links between the records from pre-prepared liner cards, and were enjoined to say absolutely nothing else. When I visited the station, by then as its regulator, I was told in all seriousness by the Australian head of news that the station's policy was to carry only good, cheerful news which fitted the entertainment thrust of the station.

For the last 30 years of the twentieth century, mass audiences in the UK listened in their tens of millions to the unique experiment of ILR, and then later also to Independent National Radio (INR). Across the rest of the world, radio is either formal public service broadcasting, provided by a body more or less state-controlled, or private radio owned by commercial operators and run as a business. In the peculiar circumstances of the UK in the early seventies, an attempt was made to fuse the two approaches together. ILR was to be private radio, funded by advertising, but delivering public service output; populist for sure, but also a conscious contributor to engineering a better society. This book sets out to trace why private radio happened as it did in the UK, how independent radio flourished for a while against the odds, and how it still came to be displaced by its commercial avatar.

Before 1960, everyone over 30 knew what wartime experience and tradition had taught them. Radio in Britain was what the BBC did, and how the BBC did it. Then, as the post-war generation came into its own, deference died, authority lost its privileged position, and young adults were keen to reject the habits and the assumptions of their parents. These baby-boomers grew up amid an affluence their elders had never known, and with pop music as their *lingua franca*. "England Swings", sang Roger Miller in 1966. So what better than a new, popular music-based and demotic radio medium, to provide the soundtrack? Independent radio caught that mood, served those new audiences and, in what seemed no time at all, became an established part of the zeitgeist.

Then around 1990 the mood changed, as did the economics and the technology, and soon the 'market' triumphed. In that period, a medium which had been designed deliberately not to be unrestricted commercial radio, but instead a new approach to public service broadcasting, seemed as much out of fashion as flared trousers and flowered shirts. It would have taken a remarkable sense of self-assurance in independent radio to have stayed true to its mixed heritage, to have remained independent radio. There was hardly any will to do that anyway. The offspring of the seventies industry wanted to be commercial radio, for the rewards it would bring them and the place they believed it would give them in the brave new market-led broadcasting world. They gradually abandoned their home ground of localism, and with it their unique selling point to audiences and advertisers and among opinion-formers. Private radio – to use a neutral term – had struck a Faustian bargain in exchange for too little reward. It ceased to deliver all the sounds of their lives for the mass of listeners, and chose instead to be just another provider of music. Within the shortest time after the choice was made, the radio stations found themselves in a world where the new technologies permitted an infinite number of popular music providers, and the uniqueness was lost.

The tension between independent and commercial concepts of radio is the main theme of this book, with the rise of independent radio brought down in the end by its commercial doppelganger. Yet while it lasted, independent radio was the sound track of a generation. At its heyday, well over half of all radio listening in the UK was to independent radio. These stations were offering programming of continuous daily relevance, well above the lowest common denominator of output, with not a few programmes and social projects of enduring value and relevance. Independent radio practitioners broke new ground in radio, renewing the radio medium with improbable success sixty years after it first began, and having huge fun in the process. ILR stations defined their localities and their communities, both local and national. For that alone they deserve a proper history. Just as the fate of independent radio illustrates the failed hopes of the liberal seventies, so its history says a great deal about the state of Britain as the nation moved from a social to a market economy. Media reflect society in the sense that they make a living by showing their audiences a recognisable picture of the world in which they live. Any locally-based venture is an especially sensitive indicator, being that much closer to the people who really comprise society.

History moves seamlessly, but its changes are marked by milestones which are themselves sometimes the proximate cause of major shifts. On 16 October 1964, Harold Wilson won a general election by the narrowest of margins, and effectively frustrated the commercial radio lobby, which had been building up a momentum which it could not then recover. In a reversal of electoral fate, on 18 June 1970, Edward Heath's Conservatives won an unexpected victory, and their policies opened the way for the introduction of Independent Local Radio from 8 October 1973. Between those two dates – on 19 September 1970 – the new Conservative government shook off the commercial radio lobby and embraced independent radio instead. The publication of the Annan Report in March 1977 was the high water mark for the social market philosophy of broadcasting; the election of Margaret Thatcher on 3 May 1979 the surest indicator that the tide was changing.

The Heathrow Conference of 23 June 1984 was the point at which the independent radio industry broke with its regulated past and began to drive towards a resolutely commercial future. The 1990 Broadcasting Act made that future possible. The agreement by the Radio Authority on 6 July 1995 to extensive networking by most of the GWR Group's stations, and the major consolidations through that year, signalled that the commercial model could became a reality. The ownership consolidations permitted by the 1996 Broadcasting Act ensured that it would do so. Richard Park's invention of UK contemporary hits radio, launched at Capital on 28 September 1987, ensured commercial radio mastery over the heart of the popular radio audience for a decade. Jim Moir's re-creation of Radio Two, from April 1996, reversed that. The launch of the Access Radio experiment on 17 March 2001 signalled the end of the expectation that ILR would deliver community aspirations. The Communications Act of July 2003 was the final sign that legislation and regulation would henceforward care little for the previous distinctiveness of radio. The takeover on 29 September 2004 of Capital by GWR, and then on 22 June 2005 of Scottish Radio Holdings by Emap, finally extinguished independent radio in the UK in any substantial form. Global Radio's takeover of GCap Media in June 2008 swept away the lingering remnants of the ILR experiment. Channel Four's decision to pull the plug on its digital radio aspirations on 10 October 2008 seemed to have stymied the great hope for a future revival from that technology. Between the first and last of those two events lay four decades during which a unique experiment in broadcasting began, flourished, peaked and then fell away and was superseded.

The history falls neatly into three main sections, each covering roughly a decade: the establishment of independent radio in the seventies; its progression and the challenges to it in the eighties; and its replacement by commercial radio in the nineties, and then on into the new century. In telling that story I have tried to keep more or less to a chronological progression, and to use specific dates wherever possible, even at the risk of seeming repetitious. However, the narrative splits between 1979 and 1985 to consider in four separate chapters the licensing process and general matters; the use of secondary rental and programme sharing to enhance the output; the key Heathrow Conference of 1984; and the failed community radio experiment of the following year. The threads join up for the late eighties, before splitting again and between 1991 and 1995, in order to look separately at the progress of national radio and local commercial radio. This is intended to help avoid the confusion of too much information, although the stories really run along parallel tracks. Where particular themes make more sense when taken out from the strict chronology, though, they enjoy separate chapters: this applies to copyright, audience research, and community radio (which occupies two).

One cross-cutting issue provides a particular challenge, and that is popular music. This is the staple of independent radio, the figured bass of its output, the mechanism through which everything else was mediated. Yet the process of music programming occupied astonishingly little of the attention of those politicians, administrators, and regulators who are the subject of this book. The history of the relationship between popular radio and popular music needs a volume of its own. That is true also of the engineering and technical issues of these forty years, to which a lay historian cannot hope to do justice. In this political and administrative history, I have included only such references to key musical trends and milestones as are essential to place this perspective on independent radio's history into context. The same approach applies to engineering matters.

Where then to start, and where to end? However impatient we may be to get to the heart of the tale, it can only be understood in the longer historical context of radio. This study therefore begins with a prologue, reviewing the long pre-history of independent radio in the UK, from the very start of the last century, and then a detailed examination of the events of the three years up to its launch in 1973. Deciding where to end is more difficult. The arrival of Ofcom at the end of 2003 marked official and legislative confirmation that the notions of independent radio had been finally discarded, and is where I have chosen to end the full detail of this history. However, since that does not mark the completion of the themes with which this history will concern itself, I have added an epilogue tracing those strands either up to their completion, or as far as publisher's deadlines will allow. If that comes perilously close to being merely current affairs journalism – and will soon enough be overtaken by events – it still seems proper to update the story.

What to include, and what to leave out? Much of what independent radio did had its impact on localities outside London. However, the great majority of available source material is about the London companies – and other major city stations – and much of the attention of the press, political and opinion-leaders' attention was devoted to those. The London experience therefore has to serve by analogy for the local system as a whole. Certain stations also attract a disproportionate amount of attention. That applies especially to the early ILR stations, to Classic FM and a few others. Once again, where detailed information is available, I have quoted that to provide some depth to the narrative, and hope that the reader will see those instances for the examples they are of a much wider range of stations, rather than in some sense standing out from the general. If I give rather too many examples from the station I once ran, Radio 210, the academically respectable explanation is that it was indeed typical of the smaller, independent stations; the truth is that it still holds a special place in my memories.

As those at the Centre for Broadcasting History Research at Bournemouth University – who have made me so welcome as a visiting fellow – can attest, I am not an academic. Yet this book is determinedly not a memoir, despite its opening paragraphs. Even though these were my own times, this is, or attempts to be, a definitive administrative history of a medium, its societal impact and the extent to which it represents a revealing fable of its period. I have tried, therefore, to follow academic disciplines in writing it. All direct or indirect quotes of sources are referenced in footnotes, although for sequential quotes from the same source, the referencing is not repeated, unless clarity demands that. Footnotes on individual pages are also used where there are sidelights on a topic, which can be read alongside the main text, and also for essential cross-references. I have spoken with many of those who have lived through this history, and their quotes are referenced in footnotes. In a very few instances where other comments I have heard – but not in those interviews – are important or revealing, I have included them without attribution, asking the reader to take my word for it that they were said to me.

I use the three terms for this separate radio sector with deliberate intent, and not interchangeably: 'private' as the generic word; 'independent' specifically for the style of private radio which is the central subject of this history; and 'commercial' for what followed it, and what is more usual elsewhere in the world. I deploy italics for these words when that seems essential for emphasis or clarity.

I have relied extensively on files and other existing source materials, although I have assumed the historian's privilege – even duty – to comment on the relevance and implication of facts and trends. At times this approach has presented me with a dilemma. I was an active participant in radio between 1972 and 1984, and then again from 1995 until 2006. In a number of instances my own memory is the best or only source for events I am describing and analysing; or documents which I wrote are central texts. So in order to be clear, wherever I am describing incidents where I am the primary source, I switch from the third person to the first person.

I have also not forgotten E H Carr's maxim that history can never be written by an individual independent of his culture. Even Robinson Crusoe, marooned on his desert island, carried with him his cultural legacy. "Robinson is not an abstract individual, but an Englishman from York; he carries his Bible with him and prays to his tribal God."[1] I have been at different times a junior regulator of independent radio, its promoter, a radio station manager, its senior regulator, and always an obsessive radio listener. I was and am a believer in the ability of institutions and enterprises knowingly to shape society for the good. I admit to viewing with distaste the notion that all unregulated markets are naturally benign. That fashionable fallacy may have been exploded by the financial crisis of late 2008, which has been unfolding as I have been writing. I have made many friends among those whose tales are told in this history, and there are only a very few whom I have never managed to bring myself to like. I will try to treat them all fairly.

I am deeply grateful to Ofcom whose help and support made my research possible, which has opened its archives to me, and also supported this publication. Ofcom neither asked for nor had any say over its content. My thanks go also to all those who gave me generous access to their papers, memories and time. I also owe a considerable debt of gratitude to Professor Sean Street, and the Centre for Broadcasting History Research at Bournemouth University. I cannot hope to do justice here to the kindness and helpfulness I found at every stage in my research and writing, and although I try to include the most important acknowledgements together with a list of source materials and a bibliography at the end of the book. Emma Wray was both a researcher and a valued collaborator in managing the over-abundance of archive material. David Vick not only read and corrected my text in detail, but as so often before went out of his way to be helpful and supportive even where he fundamentally disagreed with me. Several friends and former colleagues also generously undertook the task of reading parts of the draft of my manuscript. I am hugely grateful to them all. The remaining mistakes of course remain mine alone. My family and friends have put up with me being increasingly pre-occupied with the task of researching, writing and revising over these past three years, and – as ever – I would be wholly lost without my wife Andy's support, encouragement and forbearance.

Independent Local Radio (ILR), and later Independent National Radio (INR) too, touched its tens of millions of listeners in a uniquely close relationship, at the best providing them all the sounds of their lives. Independent radio has been a lightning-conductor for both passionate approval and passionate dissent, which

1 E H Carr, *What is History*, Macmillan 1961.

has made being part of it a heightened experience. Writing about it demands care, to put trends and incidents back into the proportion they may have lacked at the time. Independent radio has wholly absorbed those of us who have worked in it, at whatever time over the past four decades. When we come together, whatever our differences of view, we share the bond of being independent radio people. I hope those passions will help illuminate this history, which is written most of all in celebration of those who made independent radio, and those who listened to it, knowing it was special.

Tony Stoller
Southampton
November 2009

I. Prologue

Chapter 1

The long and winding road

1898 – 1970

The history of commercially funded radio in Britain covers three separate centuries – just. It began in summer 1898, with Marconi's practical experiments with wireless transmissions along the south coast of England. On 12 December 1901 signals were sent successfully from Poldhu in Cornwall to Signal Hill in Newfoundland, and wireless technology was established as a commercial proposition. That added wireless transmission to the other necessary ingredient for radio, the technology to record and replay music and speech. Thomas Edison had intoned "Mary had a little lamb" into the new phonograph, built by his assistant John Kreusi in August 1877. Place sound recording and wireless transmission into the crucibles of post-Victorian England and gilded-age America, and radio became inevitable. Add the telephone from March 1876, and the processes for receiving, recording, replaying and transmitting words and music were all in place.

Efforts to transmit speech and music to a wider public, although hampered by wartime restrictions, continued both in Europe and the US at the start of the new century. The first recognisable radio programme was probably that transmitted by R A Fessenden in Branch Rock, Massachusetts, on Christmas Eve 1906. It was David Sarnoff, an employee of that same Marconi company, who pioneered transmissions on variable wavelength in 1916, and went on to found the Radio Corporation of America. The Marconi Company built an experimental transmitter at Chelmsford, Essex, in December 1919, using the call sign MZX and, on 15 January 1920, gave the first true broadcast in Britain – a programme of speech and music on gramophone record. At this point the government intervened to put a stop to this frivolous use of the radio airwaves, which the Postmaster General, Albert Illingworth, complained were interfering with "legitimate services".

The tide of enthusiasm was now too great, however, and in January 1922 – after meetings and petitions – a new Postmaster General, Frederick Kellaway, authorised a one hour programme of speech and music to be broadcast once a week. The Marconi company duly broadcast on station 2MT from Writtle in February, followed in May by 2LO in London.[2] Other companies showed a keen interest, and Kellaway announced in May the desirability of a venture jointly owned by all interested parties. Nominally designed to ward off the risk of monopoly, it actually provided for just such an outcome. At a meeting on 18 October 1922 of representatives of over 200 radio equipment manufacturers and

2 This summary draws extensively upon Sean Street's *A Concise History of British Radio*, Kelly Publications 2005.

others, at the Institution of Electrical Engineers on Savoy Hill, the British Broadcasting Company was formed.

Commercial companies would thus be allowed briefly to pioneer the spread of radio stations across Britain, but they would do so without the stimulus of commercial competition, and under the watchful gaze of the government. Radio in Britain was to be strictly on a British basis, rather than the American free-market model. When the Company's licence expired at the end of 1926, it was replaced by a Royal Charter establishing the British Broadcasting Corporation. The commercial owners were frozen out entirely, and the anti-commercial and evangelical spirit of its director general, John Reith, was all-pervasive.

Conditioned by the BBC-centric view of broadcasting history, it now seems to have been inevitable that British radio would not follow the American model. It did not seem quite so inevitable at the time. There were two main domestic driving forces encouraging a reluctant government to allow the newish technology to be used to provide a service to the public. On the one hand were the radio equipment manufacturers, who not unreasonably wanted to create a market for their products; on the other were the private 'experimenters' of the Wireless Society of London, with whose motives the official world felt instinctively more at ease.[3] The government department responsible was the Post Office, and it was as much as anything their fear of commercialism which made officials susceptible to the amateurs' arguments that radio needed to be available to the public on some basis other than as a business proposition.

The government's views were conditioned by what was happening across the Atlantic. In the United States, no such official scruple had got in the way of the first great wave of radio enthusiasm, a genuine new frontier with a Wild West style. As early as 1916, Dr Frank Conrad – an amateur operator and engineer for Westinghouse Electric – had begun broadcasting regular music and speech programmes from his garage in Wilkinsburg, Pennsylvania, as the amateur station 8XK. The warm response from other amateurs prompted Westinghouse[4] to build the first station specifically designed for broadcasting, KDKA. Its hurried launch (is any radio station launch ever anything else?) was in time to broadcast the election returns of the Harding/Cox Presidential race. This was the first programme to reach a sizeable audience, perhaps 1,000 amateur radio operators.[5] The local newspapers were so enamoured of the new medium that they printed daily broadcasting schedules. A huge rush of stations followed the Westinghouse innovation. KQW in San Jose, California, has a claim to have been on air before KDKA, having offered regular broadcasts of continuous music and information from 1912. There are other rivals for the title, including WOI at the Iowa State Fair in 1915. Whoever started it, the radio gold-rush had begun. By the end of 1922 there were more than 550 commercial radio stations in the US, owned among

3 Asa Briggs, *The History of Broadcasting in the United Kingdom*, OUP 1961 Volume 1 *The Birth of Broadcasting* p 50 ff.

4 It was Westinghouse who led the challenge to the European view that Marconi was the father of radio. The company had a relationship with Nikola Tesla, whose patents of 1898 pre-date those of Marconi by two years. The arguments got as far as the Supreme Court in 1943, whose ruling effectively anointed Tesla as the father of radio. However, if you visit a hundred miles or so west of Nashville, you'll find a town plaque honouring Nathan Stubblefield, who had demonstrated wireless back in 1892. That is inscribed with the words "Murray, Kentucky – Birthplace of Radio".

5 *Radio's First 75 Years* Ed. B Eric Rhoads Streamline Publishing Inc 1966 p 25 ff.

others by nearly 70 newspapers, and more that one and a half million radio receivers.

Across America, "utter chaos erupted",[6] of a kind to awake high moral dudgeon among responsible British civil servants. Many stations tried to broadcast on the same wavelength, 360 metres, and had to agree to shut down on what were termed 'silent nights' to allow other, more distant stations to be heard. There was no effective external regulation, so everything had to be done through bi-lateral deals. When the department store Bambergers opened its WOR service in 1922, it had to work out a schedule with WJZ to broadcast alternate daytime and evening hours every other day. Listeners were bemused. There was a dearth of programmes to feed the glut of stations until the arrival of American Telephone and Telegraph as a supplier of syndicated 'toll broadcasting', using revenue from the use of long distance telephone lines.

KDKA's announcers had given an on-air mention of a record store in exchange for being given records to play on-air, as did KQW. It used to be thought that it was WEAF which broadcast the first radio commercial as such. [7] However, it is now argued that there were other small stations that offered airtime to other businesses prior to WEAF.[8] Even if not the very first, though, the commercial aired at 5.15 in the afternoon of 28 August 1922, for the Queensboro Corporation, to sell tenant-owned apartments at Jackson Heights, New York, is an example of what was starting up. Similarly, WEAF then broadcast variety programming which attracted one of radio's earliest programme sponsors, the Happiness Candy Company. That deal required one of the show's acts, Billy Jones and Ernest Hare, to change their name to The Happiness Boys,[9] a level of control by sponsors over programme content which would have turned UK regulators weak-kneed.

It is small wonder, then, that the free-market model did not commend itself to the Post Office or its political masters. F J Brown, assistant secretary at the Post Office, had visited the United States in the winter of 1921, attended secretary of commerce Herbert Hoover's first radio conference, and reported back on the undesirability of following the American model. That was certainly the view of the British government. Briggs[10] notes that the Postmaster General, Kellaway, told the House of Commons that "it would be impossible to have a large number of firms broadcasting. It would only result in a sort of chaos, only in a much more aggravated form than that which arises in the United States."[11]

British newspaper proprietors were also much less sanguine than their American counterparts about the threat posed by the new medium. In 1922, press

6 *Radio's First 75 Years* op. cit. p 39.

7 Some recent academic work challenges the received wisdom that AT & T invented the concept of commercial radio, attributing that distortion to the first generation of radio historians, who favoured big business. If so, then just as the BBC view has dominated the conventional histories of UK radio, there was a similar effect at work in the USA, giving too much sole credit to the large corporations.

8 Clifford J Doerksen, *American Babel; Rogue Radio broadcasters of the Jazz Age*, University of Pennsylvania Press, 2005.

9 *Radio's First 75 Years* op. cit. p 45.

10 *Hansard* vol 152, col 1869, 3 April 1922 as quoted in *Briggs*, vol 1 p 68.

11 *Hansard* for 3 April 1922 (Vol 152, cc 1869w) however records Kellaway's reply as "I am entirely supportive of the idea of utilising wireless telegraphy for the broadcasting of messages of the kind referred to by the Hon Member". That was in reply to a question from Sir David Newton who asked him to "sanction and promote the daily broadcasting by established and suitably equipped radio stations of messages likely to prove of value to trade and industry or being of general public interest".

representatives had caught the ear of Sir Evelyn Murray, who, as secretary, was the chief permanent official at the Post Office. Murray then argued in a memorandum of his own to Kellaway that no news should be broadcast on radio which had not been previously published in a newspaper. That sounds fantastic to modern ears, but the Press Association had already gone further and made representations that no news items should be broadcast at all.[12] The industry which might have offset the innate tendency of British official opinion to oppose commercial radio was chiefly involved with trying to stifle it.

Within half a year of the reply to that parliamentary question, the British Broadcasting Company had been created. The first two of a long line of formal inquiries into broadcasting in Britain were the Sykes Committee in August 1923 and the more wide-ranging Crawford Committee in March 1926. Their reports, and the debates which surrounded them, established the principle of 'unified control' of radio broadcasting, which John Reith tried unavailingly to distinguish from 'monopoly'. Finance would come from a licence fee, charged to those who wished to operate radio receivers. Radio was to have a social purpose, never better characterised than in the Reithian triptych, 'to inform, educate and entertain'. The order of those tasks was far from accidental. On 1 January 1927, the British Broadcasting Corporation was constituted by Royal Charter, and an exclusively non-commercial, monopolistic model for British radio was enshrined for 45 years.

However, the official aversion to commercially-funded radio during five decades did not mean that it was wholly unavailable to the public between the wars. By the late 1930s, continental-based stations had attracted sizeable audiences and changed the nature of the BBC's offerings. Sean Street's ground-breaking study on commercial radio broadcast for a British audience between 1922 and 1945, *Crossing the Ether*, [13] describes the phenomenon of commercial radio stations transmitting from the near Continent to Britain. He traces their origins back to "the first commercial broadcast to Great Britain – a fashion talk sponsored by Selfridges"[14] in 1925, and sees the promoter of that fifteen-minute programme, Leonard F Plugge, as "central to the commercial challenge to the BBC ... the true founding father of British commercial radio". In 1930, his International Broadcasting Company (IBC), with a lease from French industrialist and radio enthusiast Fernand Legrand, established Radio Normandy. Its rival, Wireless Publicity Limited, swiftly established the English language service of Radio Luxembourg. They were soon to be followed by others – Radio Paris, Radio Toulouse, Radio Côte d'Azur – despite attempts by a BBC-dominated International Broadcasting Union meeting in Lucerne in May 1930 to forbid frequencies to stations broadcasting "the type of programme which is essentially based on the idea of commercial advertising in the international field".[15]

These stations on the near Continent had impact on audiences and advertisers, and also on the future development of radio as a whole. Inter-war commercial radio was an integral part of the British advertising mix. The advertising agency J Walter Thomson's estimate that it allocated around 3 per cent of its media spend

12 Both quoted in *Briggs* op. cit. Vol 1 p 130.

13 Sean Street, *Crossing the Ether – British Public Service Radio and Commercial Competition 1922–1945*, John Libbey Publishing 2006.

14 *This is the IBC* IBC 1939, quoted in *Crossing the Ether* p 53.

15 *Crossing the Ether* op. cit. p 151.

to radio, compares favourably with the ILR share of national advertising revenue right up until June 1993. The audience levels clearly give the lie to the carefully nurtured impression that it was the BBC, and the BBC only, that was the source of radio programmes for the mass audiences in Britain. The BBC itself did not even establish its first audience research panels until 1936, but the commercial stations needed credible audience research to sell their product to advertisers, and were able to draw upon a considerable body of expertise from America, not least through the JWT agency. By the mid 1930s, these commercial stations were showing audience shares for the commercial sector which refute the image created about this "golden age of wireless",[16] that the only radio provider of any significance was the BBC. By 1935 a survey showed that one out of two British listeners interviewed listened to Radio Luxembourg regularly on Sundays.[17] A larger study in 1938 estimated that over a million households were listening to Radio Luxembourg alone between 1 pm and 2 pm on Sundays[18], exploiting the effect of Reith's interdict on any entertainment broadcasts on the Sabbath. The 1938 audience figures for weekdays bear out commercial radio's more general impact. On weekdays, the commercial stations had 64 per cent of the available morning audience, with the share falling to 36 per cent across lunchtime but rising to a peak of 70 per cent by late afternoon. A social study by Quaker philanthropist Seebohm Rowntree, published in 1941, found the most popular single programme was the "Littlewood's Pool Programme from Luxembourg on Sunday at 1.30 pm".[19]

Thus, when the Ullswater Committee – which sat through 1935 to consider the renewal of the BBC Charter at the end of 1936 – was rejecting the concept of British commercial radio, British audiences had a very different idea. By 1938 "the commercial radio star was as its zenith",[20] and Leonard Plugge, who had been elected MP for Chatham in 1935, might well have begun an effective parliamentary challenge. But the drums of war were beating nearer and nearer, and the British Empire's 'finest hour' was also to be the BBC's. The outbreak of the Second World War, and notably the German occupation of northern France from which these commercial broadcasts were made, effectively put an end to the heyday of continental-based commercial radio aimed at Britain. It did so at the very point when it might have challenged even the domestic hegemony of the BBC. Sean Street has observed that "the Second World War saved the BBC from itself ... It was what a national public service broadcaster was born for. It also at a sweep wiped out everything except Radio Luxembourg".[21]

The BBC had a good war. The Home Service and the Empire Service, supported by Forces Broadcasting, were the sound-tracks of those years of fear, loss, striving and eventually success, the final and crowning achievement of the

16 *The Golden Age of Wireless* was the title of the second volume, published in 1965, of Asa Briggs' monumental history of UK broadcasting. There are 516 pages about radio in this volume, covering the period from 1927 to 1939, but the commercial radio stations merit significant mention on fewer than a dozen of them, so completely were they disregarded by official histories until Sean Street's work was published in its fullest form in *Crossing the Ether* in 2006.

17 Research study by the Institute of Incorporated Practitioners in Advertising (IPA) November and December 1935, quoted in *Briggs* op. cit. Vol 2 p 363-364.

18 IPA study 1938 in *Briggs* op. cit. vol 2 p 364.

19 Seebohm Rowntree, *Poverty and Progress*, Longmans 1941, p 409.

20 *Crossing the Ether,* op. cit. p 181.

21 Professor Sean Street in an interview with the author 3 January 2008.

doomed British Empire. The BBC's director general, the "austere didact" William Haley[22], was already in 1943 planning for the triple play of a Home Service (continuing the wartime domestic output), a Light Programme (following and replacing the Forces Programme) and a Third Programme of culture. The Light Programme began as early as 29 July 1945. The Third Programme, which came on air on 29 September 1946, was "the kind of cultural gem that could only have been produced in early post-war Britain under the condition of broadcasting monopoly",[23] but the social purpose of radio was re-stated as a central element of broadcasting policy, and that was to continue into independent radio in the seventies.

The historian of the Independent Television Authority, Bernard Sendall caught the contemporary official mood precisely. "In their use of the monopoly the BBC served Britain well, displaying an awareness of the obligations of privilege in accordance with the highest principles of *noblesse oblige*. Moreover, their wartime achievements – both in overseas and domestic broadcasting – earned them world wide admiration, not only for their programme as such, but also for the exemplary demonstration they offered of the workings of that unique British constitutional invention, the public corporation."[24] There was clearly no way in which an effort to break the BBC's monopoly of radio was going to get a hearing at this time.

Nevertheless, there were concerns about the monopoly position of that 'public corporation', an undercurrent of doubt and criticism which surfaced in parliament in 1946. That January, amid disquiet on all sides of the house, the new Labour government had made known its decision not to hold the traditional decennial enquiry into broadcasting, but to get ahead and renew the BBC's charter and licence. Despite a well-supported early day motion in the Commons, and a debate in the Lords which considered in detail the monopoly principle, and the arguments for commercial broadcasting as an alternative, July's white paper confirmed that things were to stay as they were for five years, with a fuller enquiry after that. The Beveridge Committee was duly constituted in June 1949, and its report rejected the notion of any commercial competition for the BBC outright. However, Selwyn Lloyd, an influential Conservative MP and later Chancellor of the Exchequer in Macmillan's government, produced a minority report. That opposed the principle of monopoly, and recommended establishing independent competing agencies for radio and for television, to be financed by carefully regulated advertising.

With the Conservative election victory in October 1951 over Attlee's exhausted Labour administration, broadcasting policy was re-cast. The Conservative Party's group on broadcasting policy now moved into office, with several of its members becoming ministers: Ian Orr-Ewing, John Rodgers and John Profumo. That group, however, soon conceded the radio argument, faced with "the achievements and prestige of the Corporation as an internationally admired radio service",[25] and influenced by their expectation that radio would decline in importance as television took centre stage. The radio debate never took place in those years, as the momentum built towards the introduction of independent television (ITV)

22 *Never Again, Britain 1945–51* Peter Hennessy Penguin Books edn. 2006 p 311.
23 Ibid. p 312.
24 Bernard Sendall, Independent Television in Britain, Volume 1 Origin and Foundation, Macmillan, 1982 p 4.
25 Ibid. Vol 1 p 13.

starting from September 1955. Uniquely in the developed world, therefore, the UK had a commercially funded service of television before such a radio service.

The main outlet for those who remained keen to make radio programmes other than for the BBC remained Radio Luxembourg, since there had been a tacit agreement between the Allied and Axis powers not to destroy the transmitter.[26] In the years after the war, Luxembourg reverted to offering a range of entertainment programming created in the UK, including shows such as *Take Your Pick, Double Your Money* and *Opportunity Knocks* which migrated across to ITV once the new television services were authorised. Ross Radio Productions had been formed by John Whitney and Joseph Sturge, fresh out of the Quaker Leighton Park School in 1951, joined later by Monty Bailey-Watson, to make programmes for Radio Luxembourg, thus providing a potent link between pre-war commercial radio and the eventual ILR stations. After 1955, Luxembourg increasingly became solely a record-based music station, with its unusual advertisers – such as Bulova Watches and Horace Batchelor's football pools predictions – etched in the memory of listeners who are now in late middle-age.

On 13 July 1960 Reginald Bevins as Postmaster General announced to the House of Commons that "the government have decided to set up a Committee of Enquiry into the future of sound and television broadcasting", to be chaired by Sir Harry Pilkington. This had comparatively little time for sound broadcasting, although its criticism of the ITV game shows inherited from pre-war commercial radio, and its championing of local radio, set an agenda for the natural social conservatives in official broadcasting circles.[27] In the event though, it was neither old style politics nor modern lobbying which moved the debate onwards. It took a combination of technology, youth culture and affluence – and rock 'n' roll – to overturn the complacency of the BBC and to usher in a radio revolution.

Transistors were being developed in both the US and Europe in the years after the war, with Bell Laboratories probably the first company to produce something suitable for manufacture. That company coined the name 'transistor' in 1948, and produced some prototype AM transistor radio receivers. The first all-British set was produced by Pye in 1956, but it took advances in battery technology to permit mass-production. In 1959 nearly 2m transistor radios were sold in Britain.[28] Radio was suddenly in its modern state: portable, personal, freed from electrical supply wires just as it had been wirelessly freed from telegraph wires by the twenties. The impact of the legendary 'tranny' on youth listening habits was colossal, equalling or exceeding that of the iPod in the next century. This was the first time that a widely available new technology, specifically for young people, had hit a generation which had new buying power and freedom from traditional authority as well. Taking your music with you was a badge of being young in the sixties.

The music was equally new and shocking. Elvis Presley recorded his first tracks for a major record label at RCA's studios in Nashville on 10 January 1956, including the ground-breaking (for mass popular music) *Heartbreak Hotel*. This was pretty much where rock 'n' roll began, and it took little time to catch on in Britain, where by the mid-fifties it had a growing following. Britain started to develop its

26 *Crossing the Ether* op. cit. p 197.
27 *Report of the Committee on Broadcasting 1960* (Pilkington Report). (Cmnd 1753). 1962.
28 *A Concise History of British Radio* op. cit. p 106.

own, home-based rock tradition, with skiffle being short-lived but influential. When mainstream UK performers started to make their mark they were building on a very successful UK adaptation of an imported musical culture. What Britain did not import, however, were the radio stations to play the new pop music. In the USA, top forty radio, first introduced in 1949 by the Mid-Continent Broadcasting Company, had met the case for American youth. That involved "round the clock playing of a carefully selected and programmed mixture of the highest-placed hits on *Billboard* magazine's Hot Hundred chart, newly released discs which were *expected* to reach the chart, and a sprinkling of past hits".[29] There was to be no legal UK equivalent.

Stephen Barnard's valuable book, *On The Radio; Music Radio in Britain*, shows how the BBC, immune to the changing social reality, decided to back the more wholesome skiffle, but to try to keep rock 'n' roll "at arm's length".[30] Green stickers marking play 'restrictions' appeared on a remarkable number of new pop records. *Heartbreak Hotel* wasn't exactly banned, it just didn't get played very much. This was in line with the mentality which required one obligatory piece of classical music in each Light Programme popular record request show – *Family Favourites, Children's Favourites, Housewives' Choice* – but decreed that it must only be played at the end of the show, where it would not be 'contaminated' by popular songs.[31]

Few 'teenagers' at the start of the sixties, therefore, thought of BBC Radio as the place to indulge their newly liberated taste for pop music. Radio Luxembourg via the transistor radio was the only broadcast source. Its AM broadcasts, afflicted by sky-waves, offered erratic reception, especially when you were trying to listen under the bedclothes. Yet those who wanted to hear particular programming settled for less than ideal reception; radio is far less technologically determinist than other media. There were a few attempts at that time to re-open the debate about alternative radio services. The *Times* carried two articles on the "future of sound broadcasting", citing "an evident demand for the continuous transmission of popular music".[32] Sidney Bernstein of Granada floated the idea of hiving off radio from the BBC to produce a renewed "vigour" in the medium.[33] The *Guardian* soon put that in its place with a concerted sneer at Luxembourg's output, and by implication a side-swipe at its Mancunian television rival. "The records provide no relief from the disc jockeys, nor the disc jockeys from the records; the whole is similar in effect to the uninterrupted ringing of a telephone bell. First it makes you jump, then it makes you forgetful, and finally it gets on your nerves and you put a stop to it."[34] Clearly there was to be no early provision from the official world for music-hungry teenagers.

That came from just outside the UK's territorial waters, in the remarkable phenomenon of 'offshore pirate radio'. Between 1964 and 1967, the pirates cocked a snook at authority, captivated the young music listeners, made money for some of their entrepreneurial backers, and briefly lived out a self-designed legend. The idea to fit out ships with radio studios and transmitters was not initially aimed at

29 Stephen Barnard, *On the Radio; music radio in Britain*, Open University Press 1989 p 41.
30 Ibid. p 38.
31 Ibid. p 26.
32 *The Times* 6 April 1960.
33 *The Times* 9 May 1962.
34 *Guardian* 7 August 1962.

the UK. Radio Mercur was broadcast from the vessel *Cheeta*, in international waters off the Danish coast in 1958; Radio Syd to Sweden in 1961. Radio Veronica broadcast a Dutch language service from a lightship off Scheveningen in 1960, and began English language transmissions only on 16 February 1961.[35]

The start of the three year offshore Bacchanalia is usually taken to be the first full broadcast of Ronan O'Rahilly's Radio Caroline, from the former Baltic ferryboat the *Frederica* on Easter Saturday 1964. The first record played by Caroline was the Rolling Stones' *Not Fade Away*. Allan Crawford's Radio Atlanta, broadcasting from the vessel now named *Mi Amigo*, began on 12 May, and Radio London from the *Galaxy* on 23 December. The staff list of the last of those, 'Big L', reads like a *Who's Who* of UK popular music radio from the sixties onwards, while Philip Birch, the skilled and urbane MD of Radio London, and Terry Bate, latterly MD of Radio Caroline, were later to be leading figures in ILR. This floating population of competitive stations broadcast until outlawed by the Marine &c Broadcasting Offences Act, in August 1967.

For young people with a new appetite for pop music, equipped with transistor radios and faced with official disregard, listening to offshore pirate radio was one of the crowning assertions of their new identity and potency. Dylan may have given those with intellectual pretensions their anthem, in January 1964, *The Times They Are A-Changin'*. For the mass audience of British teenagers, though, it was the 'top forty' sounds of the offshore stations starting a few months later which were the true sounds of their lives for those three years, and which created an expectation of what ought to be provided by legitimate stations.

Offshore pirate radio is relevant to this longer story at two levels: as an icon of pop culture, and as the business bridgehead of commercial radio. For the individual, as Paul Donovan has written, "if you remember Caroline, Shivering Sands, Big L and the Perfumed Garden, then you probably remember what it was like to be young in the sixties".[36] For the business community, and especially the commercial radio lobby, the extraordinary and immediate popularity of the ships provided an outstanding opportunity. Birch claimed in 1965 that UK advertisers were spending £2m a year on the radio stations.[37]

The UK government began work to find ways of sinking the new medium almost as soon as the first ships appeared. On 7 April 1964, Roy Mason MP (who was to become Postmaster General in April 1968) challenged the Conservative Postmaster General, Reginald Bevins, to stop them. The new Labour government instituted a series of diplomatic, policing and legislative efforts, culminating in the Marine &c., Broadcasting (Offences) Act in 1967, which effectively ended the era of wide-ranging offshore pirate radio. Public opinion was largely opposed to the forced closures. All but one of the ships went off-air the moment they were made illegal, as the shrewdness of Philip Birch and others told them of the importance of not being in a false position when and if legal commercial radio came to be permitted in the UK. Only Radio Caroline remained, as an intermittent rebel for the next 40 years, although Terry Bate, who had run it for its last two legal years, also took the road towards legal commercial radio.

35 Keith Skues, *Pop Went the Pirates*, Lambs Meadow Publications, Sheffield 1994 p 11.

36 Paul Donovan's *Radio Waves* in the *Sunday Times* 26 August 2007.

37 Local Radio Workshop, *Capital: local radio and private profit*, Commedia Publishing 1983 p 11.

Manx Radio on the Isle of Man was the first station to make a legitimate breach in the wall of resistance. The station began broadcasting on 29 June 1964, on Bevins' strict understanding that its signal would not generally reach the mainland, given fears that radio advertisements might actually be heard in Belfast or Liverpool.[38] Among those involved with Manx were some key names from the pirate ships, Rafferty of Radio Caroline, and Crawford of Radio Atlanta.[39] For all the government's insistence to the contrary, this seemed to the lobbyists to offer the prospect of re-opening the domestic debate.

Then, as so often in the history of radio in the UK, electoral politics took a hand. In the general election on 16 October, Harold Wilson won power by a whisker – an overall majority of just five seats – and the result might well have gone against him had the story of Nikita Khrushchev's 'retirement' as leader of the USSR not been delayed until the day after the poll. Most Labour politicians felt a visceral opposition to commercial radio. The new Postmaster General was Anthony Wedgwood Benn, and he became the scourge of the pirate radio ships.[40] Wilson was returned with a thumping majority of 97 seats at the election of 31 March 1966, and Benn's successor as PMG, Edward Short, proved no more amenable to the commercial radio lobby.

Despite this setback, the would-be radio entrepreneurs began to get organised. In 1965 John Whitney and Philip Waddilove took the initiative to set up the Local Radio Association (LRA). They appointed John Gorst – then working with Pye in public relations – to be its paid secretary, funded by Whitney's Ross Radio Productions. Gorst and Whitney convened the first meeting of the LRA at the Café Royal in London's Regent Street.[41] Prominent film-maker John Boulting of Watford Broadcasting was joint chairman with Whitney, whose company was Bournemouth Sound. The LRA published its first *Plan for Local Broadcasting* in May 1966, arguing that "the 'pirate' radio ships were reaching the peak of their popularity and were clearly indicating that, whatever the merits of the service provided by the BBC, there was also room for further choice in sound broadcasting".[42] The LRA did not have the arena all to itself. Hughie Green, famous as the ITV host of *Double Your Money* – which he had previously presented on Radio Luxembourg – and of *Opportunity Knocks*, co-founded Commercial Broadcasting Consultants with Tony Cadman in 1966.

At the BBC, Frank Gillard, who had been director of sound broadcasting in 1963, had been steadily leading BBC radio toward generic stations, arguing in 1965 that "a radio network nowadays is of the greatest value to its listeners if its output is consistent in character".[43] He was also pushing hard for local radio. Gillard was a visionary as far as local radio was concerned – he had been a regional controller before becoming radio's overall director – and his hope was nothing less than "to

38 *Sunday Times* 17 May 1964.

39 *Observer* 7 June 1964.

40 Benn, though, was no supporter of the status quo in radio. In 14 August 1970 he was writing in the New Statesman of the need "to establish a second public corporation – the Independent Broadcasting Corporation – which would develop sound broadcasting in competition with the BBC", although this model was not-for-profit community radio.

41 John Whitney in an interview with the author, 11 October 2007.

42 *The Shape of Independent Radio* Local Radio Association October 1970.

43 *Financial Times* 23 August 1965.

democratise the microphone".[44] Yet either he failed to see the societal changes and the consequent demand for pop music from the radio, or he was determined to forestall it. Within months, Radio Caroline and Radio London were each claiming more than eight million listeners, along with Radio Luxembourg.[45]

The Conservatives in opposition had been notably quick off the mark to respond to the *succes de scandal* of offshore pirate radio, and the renewed domestic lobby. In a pamphlet as early as February 1965, John Gorst and Captain L P S Orr MP (both members of the policy group whose lobbying had earlier paved the way for ITV) proposed to break the BBC's monopoly on legal radio in the UK. "We recommend that local commercial radio stations representing local social and commercial life be licenced *(sic)*; and that the only restrictions that are placed on them should be those that apply to the press ... The number of stations in any area should be governed by the availability of frequencies and the economic viability of the station, and not by arbitrary government decree."[46] At its Conference in Blackpool in October 1966, the Party confirmed its support for commercial radio, which in the fiercely partisan mood of the times would have strengthened the Labour government in its belief that it was right to reject it.

Sure enough, just before Christmas in 1966, the new white paper was published, and it was clear that "the BBC and its allies have won all along the line in their rivalry with commercial interests for the favours of the Postmaster General".[47] The BBC was handed the pop portfolio, as a new national radio service, and was also to be given three years to get ahead with experimental local radio, funded by local authorities and other local bodies. Anything "smacking of commercial gain" would be excluded. The *Times* was apoplectic about the local radio proposals – "this endorsement of propaganda and simultaneous condemnation of salesmanship is indefensible" – but did not notice the failure of the proposals to sate the pop music appetites of the nation. On 4 March 1967 John Lennon's claim for the Beatles that "we're more popular than Jesus now" made headlines across the world, but the UK government persisted in its view that pop music could be perfectly well supplied by a part-time channel from Auntie BBC.

The story of John Whitney's later meeting with Edward Short in 1968 makes clear the disdain felt by the Labour administration towards the commercial arguments. Whitney, as the Chairman of the Local Radio Association, had managed to get an appointment to see the Postmaster General. "Huge bronze doors, everything was huge. I was very impressed. In those days there were genuine flunkeys, who ushered me into the great man's presence. There he was sitting at a very ornate desk, quite theatrical. He didn't get up. I said 'I want you to consider the introduction of commercial radio'. He said 'what do you have to support this'. I had a letter from LRA members with me, which I produced and gave him. He said 'I can tell you we are not going to consider the introduction of commercial radio. It is not within our planning horizons. As far as I am concerned this issue is absolutely not a question of debate'. He looked at his watch, got up, and a flunkey

44 Michael Barton in an interview with the author, 14 February 2008.

45 Mike Baron, *Independent Radio*, Terence Dalton, Lavenham 1975 pp. 36-37.

46 *Conservative Policy; some thoughts and proposals,* February 1965, pp. 7 and 8.

47 *The Times* 21 December 1966.

who had been standing by the doors opened them. He went in one direction, I left with the flunkey in the other and found myself on Waterloo Bridge Road."[48]

The law outlawing the pirate radio ships came into effect on 15 August 1967. Radio One began broadcasting on medium wave on 30 September 1967, sharing its broadcast hours and its needletime allocation[49] with the Light Programme Radio One was not the full-time pop service which the offshore pirates had provided, for all their rough and readiness. Even four years later, Radio One's schedule ran from Tony Blackburn starting at 7 am to Alan Black's *Sounds of the 70s* at 6 pm. From 7 pm until closedown at 2 am, the offering was "Radio Two (with Radio One)".[50] If nothing else, the evenings were being left to Radio Luxembourg on 208, sky waves and all.

Gillard had won his opportunity to demonstrate local radio, nominally for a two-year period, and the first of the eight experimental BBC Local Radio stations got under way in November 1967; Radio Leicester, Radio Sheffield and Radio Merseyside in successive weeks. Gillard had established a funding system from local authorities and other community interests, and a supervisory pattern including Local Radio Councils. In Michael Barton's judgement "that was a very deft move ... it got local authorities engaged from the very beginning. If he was going to seek support after two years, how better to get them onside from day one."[51] They were to win the argument for the continuation of BBC local radio, but not because it filled the gap left by the pirates. Barton again: "BBC local radio was getting an audience of people who were probably 45, 50 plus. The youngsters were not listening to us, because we were not playing their type of music, and we were not using the type of dialogue they were comfortable, familiar with." Thus these new services found a role which did not satisfy the expectations which the widely popular pirate services had created, either among listeners or among the radio programme makers themselves.

Gillard retired in 1969, to be replaced by Ian Trethowan from ITN, and it fell to management consultants McKinsey to lead the work which led to the publication of *Broadcasting in the Seventies* on 10 July 1969. This proposed the reorganisation of programmes on the national networks into Radios One, Two, Three and Four, and the end of regional broadcasting, which would in turn pave the way for an expansion in BBC local radio. In part this was a response to the Conservative Opposition's plans for 100 local commercial stations. In May the *Sunday Times* Insight team reported that "Conservative Party leaders have given private assurances to the commercial radio lobby that, if and when the Conservatives are returned to power, they will open the door to local commercial radio". Jenny Abramsky thought that "*Broadcasting in the Seventies* was a visionary document", not least because the separation of Radios One and Two at last showed that "the BBC had realised that the music of Bing Crosby was on a different planet from that of the Rolling Stones". However, she concluded that a "narrowing of radio's ambition explains perhaps why there grew among some executives in the BBC a belief that

48 John Whitney interview op. cit. A slightly different account, a little more charitable to Short, is set out in *Street:A Concise History of Radio* op. cit. p 108.

49 For an explanation of 'needletime', and a detailed discussion of music copyright issues, see Chapter 13 below.

50 *Radio Times* programmes for Friday 5 February 1971.

51 Michael Barton interview, op. cit.

radio was somehow no longer central to the future of the BBC, and by the turn of the century would be a dead medium."[52] The report came in for fierce external criticism for its cost-cutting, yet a simultaneous rejection of any commercial funding. The *Economist* felt that "it has managed to think commercially about its output, but still baulks at financing itself commercially".[53] It certainly confirmed to those who argued for an alternative service of radio that, if radio were to be refreshed and re-invented to meet the expectations of the new youthful mass market, the BBC was not going to be the one to do it.

Rather like the BBC, recognising the new force of youth but unable to frame a response which caught the imagination of this uniquely empowered new generation, the Wilson government lowered the voting age in January 1970 from 21 to 18, reflecting its perception of the political power of youth. That belief had made it easy for the opposition Conservative Party to include in their 1970 manifesto, *A Better Tomorrow*, a specific commitment to introduce a new radio service, although the word "commercial" was carefully avoided. "We believe that people are as entitled to an alternative radio service as to an alternative television service. We will permit local private enterprise radio under the general supervision of an independent broadcasting authority. Local institutions, particularly local newspapers, will have the opportunity of a stake in local radio, which we want to see closely associated with the local community."[54]

It was commonly expected that Harold Wilson's Labour Party would win the general election, called for 18 June. The economy had recovered from the balance of payments crises of the mid-sixties, and all the opinion polls pointed that way, almost until polling day itself. The social agenda of the Wilson government – abortion and homosexual law reform, race relations legislation, the Open University – was part of the same broad trend towards a less constrained and less deferential society, which had produced pirate radio and would eventually permit ILR. However, there was also a subtler shift in public mood, away from social purpose and toward self-interest. Uncommitted voters, who wanted more scope for personal enrichment, in line with the social mood of the times, were starting to grow impatient with the fuss and bother which accompanied collectivist solutions. They were to reject Heath in 1974 and Callaghan in 1979 for much the same reasons, until Keith Joseph and Margaret Thatcher offered them a more individualistic approach.

There was more immediate trouble for Wilson on 14 June, four days before the election. With England leading West Germany by two goals to nil in the quarter final of the World Cup in Mexico, and only twenty minutes to play, manager Alf Ramsey substituted Bobby Charlton and Martin Peters to save them for the semi-finals. Goalkeeper Peter Bonetti allowed a soft goal, and England lost by two goals to three. Wilson's quip "have you noticed how we only win the World Cup under a Labour government?" had come back to haunt him. The feel-good factor he had hoped for from the soccer championship changed instantly to disappointment, and the election was lost. If Bonetti had not let that speculative shot from

52 James Cameron Memorial Lecture; *Public Service Radio, phoenix or albatross*, 25 November 2002.
53 *Economist* leader article, 27 July 1969.
54 *A Better Tomorrow*, Conservative Party, 1970.

Franz Beckenbauer under his diving body, how long might the UK have had to wait for an alternative radio service?

Heath's Conservative Party, with support from the Ulster Unionists, had an overall majority of 31 and that manifesto commitment to introduce commercial radio. Richard Park, later to be the outstanding music programmer in ILR, was a BBC DJ at the time, and his producer was Aidan Day, who was to be Capital Radio's programme director. Park recalls the moment. "I did Radio One Club live in the UK for the BBC on the morning when the result was announced. Round about twelve noon, one o'clock, we were absolutely certain that Ted Heath had been elected. Aidan and I said to each other 'so there you are then, that'll be commercial radio coming'. He wound up in London and I wound up in Glasgow."[55] At long last, the show could get on the road.

Was the outcome at that particular time mere accident, or the product of deeper forces? It is likely that, if Nikita Khrushchev had fallen one day sooner, the Conservatives would have won the 1964 election. In all probability, they would have legislated for commercial radio within a year or so, as a reaction to the offshore pirate radio stimulus, in accordance with the policy pressure from much the same people who had engendered ITV ten years before. In 1970, if England had not been beaten by West Germany in Mexico, Harold Wilson might well have won the 1970 election, and the private radio sector in the UK would have been again frustrated.

Yet there is logic in the arrival of Independent Radio when it came, and in the form which it then took. The social and political forces of the sixties, new transistor technology, youth affluence and empowerment, and most of all pop music, made the *status quo* unsupportable. That was evident by the late sixties. Once the balance had tipped, there was quite suddenly a huge but unspecific demand for something new on the radio, to match those underlying changes. Offshore pirate radio offered one response, but it was not a solution which the consensus of those times could contemplate. The ships were fun, they attracted notoriety for three years, but they were never seen widely as a long-term prospect.

Whatever the myths nowadays believed, the sixties were not a time when most people discarded all their prior assumptions. Rather, they contemplated vistas of radical change only through the frosted glass of tradition, caution and deference, which had characterised their society. Britain had rejected the American model of radio in the twenties, because it then represented a mismatch with the spirit of the times on this side of the Atlantic. Fully fifty years had passed between the definitive meeting at the IEE on Savoy Hill which created the BBC, and the Sound Broadcasting Act in June 1972, but the country was still not ready for full-on commercial radio. The arrival of ILR marked a moment when what ordinary people wanted triumphed over what their masters thought they ought to be allowed to have, but that still meant blending the pop revolution with continuing restrictions and public obligations, which were widely acknowledged to be essential. Therefore, the unexpected opportunity in 1970 to break the BBC's radio monopoly was bound to lead to a new form of public service radio funded by advertising, rather than a free market model of commercial radio. ILR was a child of the sixties in its restraint, just as much as in its novelty.

55 Richard Park in an interview with the author, 6 December 2007..

II. Debate, design and implementation 1970 – 1979

Chapter 2

Paving the way for ILR

1970 – July 1972

Britain in 1970 was an uncomfortable place. It was as if the people, the polity and the institutions were waiting in a state of vague anxiety for some unknown change. The sixties had promised social revolution, which for some parts of society – especially the newly affluent young – had begun to change their lives. For most people however, it had created a sense of dissatisfaction and anticipation, which the new present seemed unlikely to assuage. With politics still set in the post-Attlee consensus, the change of government from Wilson to Heath seemed to promise only a slight variant on the previous six years. Offshore pirate radio is a good example of how a sixties phenomenon raised expectations which remained unmet. The BBC had launched Radio One, flattering the pirate ships with extensive imitation. "DJs like Tony Blackburn, Dave Cash, John Peel, Keith Skues, Ed Stewart, Emperor Rosko, Stuart Henry and Duncan Johnson all came ashore to work for the Corporation's new pop radio station. Even the *Radio One is One-derful* jingles and station identifications were similar to the *Wonderful Radio London* set."[56] Yet the huge audiences for the ships still sensed that their high hopes were unfulfilled; this was just Auntie BBC lifting her skirts a little.

The general mood of social alienation was more brutal in the wider nation. Violence was building in Northern Ireland, with riots in Derry in June over the arrest of Bernadette Devlin. The next month, the government declared a state of emergency over a UK-wide dock strike. In May, a campaign of civil disobedience led by Peter Hain had caused the cancellation of the proposed South African cricket tour to England. The symbols of the sixties were falling: spaceship Apollo 13 came close to disaster in April, Jimi Hendrix died in September. The years of the Heath government were to be marked increasingly by public violence, industrial strife, and a growing gap between expectation and reality. In practice, this was a society with great potential. The long post-war economic boom was still running, and the first North Sea oil fields were about to come on stream. Social legislation in the sixties had eased the lot of many groups which had previously endured great discrimination, and Britain was freeing itself from the burden of Empire and some of its imperial mind-set. But it didn't feel that good.

Delivering commercial radio to an expectant public would have seemed a welcome and more gentle task for a government dominated by economic and

56 Mike Baron, *Independent radio; the story of commercial radio in the United Kingdom* Terence Dalton Limited, Suffolk 1975 pp 54-55.

industrial questions, Europe, and Northern Ireland. Christopher Chataway became the new Minster of Posts and Telecommunications (the Wilson government had re-shaped the old Postmaster General's department the previous year), and had already gained some familiarity with the issues in the debates within the Conservative Party in the late 1960s. Before the election, the opposition spokesman on broadcasting had been Paul Bryan, who was later to chair Piccadilly Radio. A director of Granada's television rental company, he seemed like many of his colleagues rather close to the more respectable of the offshore pirates. In 1966, he had argued against the closure of the ships "until some local sound broadcasting had been introduced", emphasising the point by a visit to Radio 270, a pirate ship off the Yorkshire coast.[57] Bryan set out Conservative policy on 3 March 1969 in a speech to the Monday Club, a grouping of notably right-wing MPs. That envisaged 100 or so stations, under the control of the Independent Television Authority (ITA), with levies on profits to "be used on social and cultural activities at present starved of funds, such as swimming baths".[58] This was not a conciliatory speech, either in its tone or in the chosen location. The BBC's local radio experiment had been "irresponsible" and "was heading for collapse".[59]

Harold Wilson and Edward Heath had clashed in the Commons in 1969 following a spat between Bryan and the then minister, John Stonehouse. Wilson asked Heath whether this was "another case of pressing what he thinks will be popular in the country, and half going along with it without committing himself".[60] The jibe had a ring of truth about it. The Conservatives were clearly keen to capitalise on the demand for pop music, evidently not satisfied by Radio One, but increasingly seemed to be shrinking away from a free market model. *Crossbow*, the publication of the moderate Bow Group of Conservative MPs, made clear in 1970 its view that "the Conservative Party is committed to local commercial radio, but we have not yet been explicit as to how it would be run. It would be a cultural disaster if we did no more than provide a pirates' charter, and a system which established what would be a private monopoly for local tycoons would be immensely damaging politically."[61]

Although the initial plans for a nationwide service of local commercial radio, which had been frustrated by the Wilson election victory in 1964, were what most Conservatives envisaged during the 1970 general election, once they were in office there was a significant shift. The campaigners for a new commercial network realised the risk of putting forward what seemed to be simply a money-making proposal. Even the free-market Hughie Green had begun to talk instead about providing "a necessary and much needed service which would be available to all at no cost to the taxpayer or the BBC licence payer", although he saw it happening in conjunction with "local councils and local newspapers".[62] For all that, when the Campaign for Independent Broadcasting met Chataway in mid July, they were arguing against an ITA-type organisation and for a free market model.[63]

57 *The Times* 29 August 1966.
58 *Financial Times* 4 March 1969.
59 *The Times* 4 March 1969.
60 *Hansard* 20 March 1969 col 742.
61 *'Mentor'* writing in *Crossbow* Bow Publications 14 July 1970.
62 Letter to *The Times* 26 June 1968.
63 *Television Today* 23 July 1970.

Chataway proved to be a more subtle thinker, and a more effective politician, than Bryan. He set about consulting over the introduction of commercial radio, and by September he was telling London Young Conservatives (a more appropriate choice of audience than Bryan's Monday Club) that his proposals would address the "increasing concern about the concentration of power over the communications media". While there was a need to protect local newspapers, "the major justification for introducing commercial radio is simply that it will provide, for those who listen and for those who work in radio, alternative sources of radio broadcasting".[64] He envisaged two objectives for the new service: to enable radio to cater sufficiently for local needs; and to provide a first-class news service, on the model of ITN's achievement in stimulating standards in television news. Chataway's vision was already far removed from the notion of commercial radio simply as a successor to the offshore pirate stations. "Radio is, in some way, unique in the services it can provide in a locality" he asserted. "But to provide these services the radio station has to cover an area to which most of its listeners can feel they belong. It is no use being given information about a traffic jam or a new swimming bath in places you never visit." On the same day, the Bow Group published *Home Town Radio.*[65] It argued for a central authority, owning the radio transmitters, regulating programming standards and advertising content, with an obligation to provide locally-originated programming and with protection against syndication and networking. If there was a single moment when it became clear that *commercial* radio was going to be displaced by *independent* radio, 19 September 1970 was the day.

John Gorst and John Whitney's Local Radio Association, which was close to Chataway, had also refined its own thinking. In October 1970 it published *The Shape of Local Radio*, a detailed set of proposals ranging from frequency allocation, through programmes and advertising to regulation. The pamphlet argued for local services to precede regional ones, in order to provide a genuine alternative to the county-wide coverage BBC services, and for them to be simulcast on VHF and medium wave. A report from the Marconi Company, annexed to the pamphlet, went into detail about "siting and frequencies for 150 local or area stations in 116 different places throughout the United Kingdom", seeking to cover "all localities where the population exceeds 75,000 within an 8 mile radius. Two stations are provided for large urban areas, and three or four for each conurbation."[66] It envisaged regulation by a new Independent Broadcasting Authority and "about nine Regional Controlling Authorities". However, this was not to be a re-organised ITA; "independent radio should be run by people who are exclusively interested in radio". The RCAs should lay down standards for transmission, but that would be the responsibility of the contractors. Those contractors would be chosen, taking into account seven factors: local connection of the applicants; shareholding arrangements; financial status, providing adequate financial backing; staff, to be suitably qualified; monopoly by other communications media to be avoided; local authorities to be excluded; and foreign investment to be limited to 25 per cent of capital. Local newspapers would be encouraged to participate. Programmes would offer a range of music, news and other speech, with an example given of a ratio of

64 *Financial Times* 10 September 1970.
65 *Home Town*, Bow Group publications, September 1970.
66 *The Shape of Independent Radio* Local Radio Association October 1970.

music 65 per cent, news 8.5 per cent, other speech 16.5 per cent and advertising 10 per cent, the last equating to the ITV statutory limit of 6 minutes in each hour. There would be unlimited 'needletime' for playing commercial recordings.

For what was still partly a polemic, the accuracy of prediction in *The Shape of Local Radio* is impressive. Hughie Green was vocal in his opposition, calling the LRA's proposals "claptrap"[67], but it was the broad thinking behind these recommendations which most closely coincided with the eventual outcome. The main differences are in the exact powers of the regulator, which was to cover both ITV and ILR, and which was itself to be the broadcaster, owning and operating the transmitters, in accordance with what was widely considered at the time to have been the very successful model of the Independent Television Authority for ITV.

The BBC was clearly worried about the possibility that Radio One might be moved across to the commercial sector. It embarked upon extensive press briefing at the end of December 1970, following a meeting with Chataway just before Christmas.[68] That might have been a diversionary tactic to protect its local radio services, but when the Corporation managed to 'discover' a new national medium wave frequency it suggested genuine fearfulness about the national channel. DG Ian Trethowan claimed that "we now believe that it will be possible to provide a high-power medium wave frequency for a national commercial network without interfering with any of the four existing BBC networks".[69] He need not have worried. John Whitney for the LRA stated that "we have never even considered the possibility that Radio One should be hived off to commercial interests. We should be very unhappy to see such a thing done."[70] By the end of January, Chataway had concluded that Radio One and the BBC local stations should remain within the BBC, and that ILR would have a separate and – in that sense at least – unthreatening existence.

The other public body in what was termed the broadcasting 'duopoly', the Independent Television Authority, was much less active. The Queen's Speech on 2 July 1970 had put forward the government's intention that ILR should be under the general supervision of an (unspecified) independent broadcasting authority. On 8 July the ITA director general, Sir Robert Fraser, discussed with the ITV companies the prospects for the new service. Tom Margerison for ITV observed that "there were a number of areas where the programme companies could make a valuable contribution", including news and cross-promotion.[71] Later that month, the ITA members[72] considered a paper written by the director general reflecting on the possibility that the Authority might "make representations to the minister suggesting that it should be the body which should be in charge of commercial local radio".[73]

67 *Financial Times* 15 October 1970.

68 *Financial Times* 28 December 1970.

69 *Guardian* 16 January 1971.

70 *Daily Telegraph* 29 December 1970.

71 Restricted circulation minute 113 from the Minutes of the ITA Standing Consultative Committee, 8 July 1970.

72 The ITA, the IBA and the Radio Authority each had boards of 'members', who were collectively the corporate executive for each the body. The employed executives, including the Directors General and Chief Executives, were referred to as the 'staff'. That nomenclature is used throughout this history.

73 ITA Paper 83(70) 10 July 1970.

The paper rehearsed the arguments supporting such a course of action: avoiding the establishment of a third public broadcasting authority; administrative cost savings; applying common content controls on advertising-supported broadcasting services; and some co-siting of transmitter masts. Crucially in hindsight, the ITA would only contemplate running ILR if it were itself the transmitting agency.

However, the arguments against taking on the radio duties have an elegance which probably betrays where Fraser's initial sympathies lay.[74] "Concentration of power and patronage is bad and diversity is good. Let another body have the responsibility for commercial sound." He noted that ITV and commercial local radio would be very different and in competition with each other, suggesting separate oversight. The administrative savings would probably not amount to much. With another television channel a possibility, the Authority would have "a wide jurisdiction for television and sound. Could it efficiently discharge it?" The Authority would need to adopt dual standards "in relation to the public service television side and the probably more 'pop' type of sound service", and it might have difficulty acting fairly in awarding franchises where the ITV companies were among the applicants. Weighing also with Fraser (and very telling for the later history of radio regulation) was a concern that "the organisation of local radio will be an immense and difficult task, which might divert the Authority from its television duties", a phrase which unintentionally but accurately predicts that radio would always come second in the IBA's affections.

The members of the ITA showed no enthusiasm to take on this potentially unruly new child, choosing the passive option "to wait and see … not oppose any suggestion in favour of ITA responsibility but not actively to foster one being made".[75] In the event, the white paper proposed that the ITA be renamed the Independent Broadcasting Authority. It would assume responsibility for what the ITA was now calling "a new local radio sound service, financed by advertising". The ITA said then that "the Authority stands ready to assume these responsibilities", but it was clearly not having any difficulty containing its enthusiasm for the task.[76]

The civil servants in the Ministry of Posts and Telecommunications had begun work on the radio question from early in the Heath government. John Thompson recalls that the task was led by D G C Lawrence, a former Post Office civil servant, who did not really like the idea of self-financing local radio. "He was a tough, able man. His great, private interest was the battlefields of the Duke of Wellington. His heart wasn't in it for ILR, but his head functioned well, and fortunately for the future of ILR, the permanent secretary, Frank Wood, was broadly supportive." [77] The early debates were over whether the new service was to be 'additional' to the BBC or 'alternative' to it; should it compete with, or merely complement what was already on offer? Through the policy discussions, the merits and de-merits of asking the ITA to take on the task seemed finely balanced, until

74 Robert Fraser retired as DG of the ITA in October 1970 at the age of 66. John Thompson recalls that, despite this implied initial reluctance, he was unfailingly helpful to those laying the groundwork for ILR. ""In my view" he says, "Bob Fraser was a skilled and shrewd operator".

75 ITA Paper 83(70) op. cit.

76 *Annual Report and Accounts 1970-1971* Independent Television Authority 23 November 1971 p 9.

77 John Thompson in an interview with the author, 8 November 2007.

settled by Prime Minister Edward Heath's opposition to the proliferation of public bodies.

The white paper, *An Alternative Service of Radio Broadcasting*, was published on 29 March 1971. It proposed the continuation of BBC radio largely unchanged, with the provision by the BBC of 20 local radio stations funded now by the licence fee. The Independent Broadcasting Authority (IBA) was to provide advertising-supported local radio from up to 60 stations, covering 65 per cent of the UK on VHF and 70 per cent of the UK in daytime on medium wave, with the availability of more than one service in London and perhaps elsewhere. The white paper saw the ILR stations competing directly with BBC Radios 1 and 2, and offering "a truly public service ... [government can see] no place for a system of broadcasting which did little more than offer a vehicle for advertisements".[78] There was to be a requirement for "high standards not least in the provision of news" and that "a major ingredient of the output of the stations will be local news and information". Above all, the services were to be local. "The IBA stations must be firmly rooted in their locality, and this should be reflected in the choice of station operators and subsequently in the output", which should combine "wide appeal with a first-class service of local news and local information". Chataway added to this in his House of Commons statement, saying that "with the possibility of up to 60 stations being set up, it is conceivable that cities with populations of between 100,000 and 150,000 will be served".[79]

The constitutional relationship between ILR companies and the IBA was to be the same as that between ITV and the ITA, which was widely thought to work well. The Authority would continue to be the transmission agency. Local newspapers would have a right to acquire an interest in a radio programme company, whether or not they were part of the successful applicant group, but not a controlling interest. ITV companies could invest in, but should not control, an ILR company operating in their television contract area. "Excessive profits must not be allowed", but "equally, the financial arrangements must be such as to attract broadcasters of ability, to support programmes of quality and to ensure that they are available as widely and as quickly as possible".[80] The white paper envisaged provision of national and international news, through a "central news company". It opposed a "sustaining service for any significant periods". ILR would be financed by spot advertising, not by sponsorship. There were to be local advisory committees for each station, and the IBA should experiment "in the early stages with the establishment of relatively small stations in order to determine the minimum size of community capable of sustaining a worthwhile local radio station".

This was all a very long way away from the free-market model of offshore pirate radio, and quite a distance also from the notions of local commercial radio favoured by the Conservatives in opposition. ILR was to be tightly regulated, public service local radio, laden with obligations and transmitted by a public authority, with a statutory localness and safeguards against undue profits. It was clear to commentators that the proposed arrangements "will not give the aspirant commercial radio companies the freedom they have been seeking".[81] It had become in the

78 *An Alternative Service of Radio Broadcasting*, 29 March 1971, Cmnd 4636 paragraph 2.
79 *Hansard col 1168* 29 March 1971.
80 *An Alternative Service of Radio Broadcasting* op. cit.
81 *Daily Telegraph* leader 30 March 1971.

end a very British compromise, one which was crafted to survive the challenges of a partisan political scene where the parties actually had little between them on most issues. Harold Wilson's response to the initial statement on the white paper was measured, hinting that ILR as envisaged might have the chance to survive a change of government. Yet once again, expectations raised by the now-fading image of the pirate ships were being thwarted for overtly commercial operators and broadcasters.

Much of the detail was to be left to the ITA to sort out, despite its lack of any radio experience. Indeed, because of legislative limitations in the Television Act[82] the ITA was actually precluded from hiring any staff or paying for any preparatory work. "I suppose we can think in our baths", one official was reported as saying.[83] As it turned out, they did a good deal more than that, although the formal constitutional position remained rather unclear. As early as the April 1971 meeting of the Authority, its director of engineering, Howard Steele, was reporting that "on the basis of the proposals made in the white paper, there appeared to be very serious technical problems in relation to frequency allocation".[84]

For the ITA's 27 May meeting, there were papers discussing how the establishment of ILR by the Authority was going to be funded, how less profitable companies might be assisted, and how excessive profits could be prevented. It is a measure of how vague were some of the ideas in the white paper that, right up until the passage of the eventual legislation in 1972, the financing of the embryonic radio work remained legally precarious. The ITA's legal experts thought that it could be justified only by the obligation on the Authority to "consider how the proposals [for radio] may affect it ... and be prepared to enter into discussions with the government about the proposals ... [which in turn] will involve the Authority's staff and members of the Authority devoting a good deal of time and thought to the problems involved".[85] Any detailed planning work was probably illicit, so that the provision of the first legal land-based commercial radio service for the UK was quite possibly based on technically illegal expenditure by the ITA.

At that meeting in May, the ITA members contemplated some financial support for smaller companies through a differential system of rentals – but opposed direct cross-subsidy – and continued to argue for complete separation of the future television and radio rentals. This is the first occasion when the possibility of additional rental was considered, which was later to become 'secondary rental'. The ITA envisaged at this early stage that the IBA might use the rentals charged to the larger companies, over and above the direct costs of their transmissions, to cover "the physical development, and improvement, of the sound network; assisting small companies ... [and] to provide employment for musicians, thus giving the independent sound system a bargaining counter for negotiation with the Musicians' Union about 'needletime'".[86]

This was a crucial meeting for ILR, and was being conducted by and between people with little or no background in radio. They were the *television* authority, and from the very first they were taking decisions which were inevitably informed by

82 Section 21 of the Television Act 1954 prevented the ITA from deploying its reserve funds "otherwise from the said purposes of the Authority", which did not include radio.

83 *Guardian* 31 March 1971.

84 ITA minutes, 22 April 1971.

85 ITA Paper 56(71) 20 May 1971.

86 ITA Paper 57(71) 20 May 1971.

their television experience. It would be wrong to impute any undue influence to the ITV companies in all this, but it is equally improbable that the ITA as it then was would have contemplated arrangements which would be damaging to their television contractors. That no one within the ITA had any knowledge or experience of radio was not their fault, but – taken together with their sense of a duty of care towards ITV – it goes a long way to explaining why ILR ended up even more regulated than its fiercest opponents could have hoped.

That is well illustrated by a key paper at the same meeting, which established the principle of advance approval of schedules for radio, just as required by the Television Act for ITV, and that "advertising practices and standards ... will remain as in the Television Act".[87] The strength of feeling in that paper is notable. "In television ... no programme may be broadcast unless it forms part of a schedule drawn up in consultation with the Authority and approved by it. We have said that this power cannot be regarded as a mere formality on the part of the Authority." The paper goes on to acknowledge the difficulty of "looking at" *(sic)* possibly 60 different sets of programmes, but argues that this may be done whenever necessary, "e.g. where one was failing to live up to the required standards". The later arguments over format regulation all flow from this initial over-structuring of radio regulation.

That same paper contains a proposal which is even more breathtaking. "It is the firmly held view of the director of engineering that consideration should be given to Authority ownership of sound studios." Concerned about the difficulty of "a uniform maintenance of high technical standards"[88], Howard Steele had been pressing hard for what now seems a barely credible extension of the Authority's powers. It was probably fortunate that the highly persuasive Steele was not able to be present at the May meeting. By the time the matter was discussed again in June, it had been established that "the Ministry ... would not be in favour of this extension to the Authority's powers". Even so, the minutes of the discussion make it clear that the Authority's staff engineers did not let the matter go lightly, even though it was opposed "by the staff outside Engineering".[89]

In June there was also detailed discussion of the frequency plans, which had long-term implications. For example, Oxford was excluded from the areas to be covered – and would remain so for 17 years – because "a station at Oxford might mean that the signal strength of the London and Birmingham stations would have to be considerably reduced".[90] The Authority's Policy Committee noted the need "during this preparatory period, [to] open up discussions with the Musicians' Union and Phonographic Performance Limited ... about the amount of time for which the new service will be able to use gramophone records ('needletime')". These television regulators also discussed the nature of the licensing system, and arrangements for the regulation of programming and advertising.[91]

ITV had by now begun to take a keener interest. Lord Thomson of Fleet of Scottish Television had met Chataway in September 1970, to urge that "the local press should be allowed a substantial share in the local stations. No one newspaper,

87 ITA Paper 58(71) 20 May 1971.
88 ITA Paper 58(71) 20 May 1971, paras 16 and 17.
89 ITA minutes for 24 June 1971, minute 25.
90 ITA minutes for 24 June 1971, minute 27.
91 ITA Policy Committee paper, 8 July 1971.

however, should hold a controlling interest."[92] 'Brum' Henderson of UTV had lobbied Chataway in October for a "virile broadcasting 'second force' in the private sector under public control of an independent authority", not omitting to mention that UTV's "programme record has – in this difficult area – given us a strong claim to obtaining the franchises in the major areas of population, such as Belfast and Londonderry".[93] In March 1971 he was writing to the ITA's director general, urging the Authority to make a bid to run commercial radio.[94] By June, Norman Collins of ATV (who had long been interested in commercial radio) was making detailed proposals to the ITA covering research arrangements, technical appraisal, how to start negotiating with the Musicians' Union, and the level of application fees. Collins envisaged stations owning their own transmitters, but in the view of new ITA director general, Brian Young, "the overall map would be complicated ... and I couldn't feel it was economical or sensible for some *(sic)* stations to run their own transmitters".[95] Collins even went so far as to offer the ITA advice on appointing their radio supremo, suggesting looking to Canada or Australia. Brian Young disagreed. This was to be a very British venture.

The government's planning continued, seeking sufficient detail to allow legislation to be introduced before the end of the year. Speaking to the Local Radio Association in June, Christopher Chataway explained the government's expectations and the rationale for the structures proposed.[96] Local radio needed to be regulated because of the scarcity of frequencies. That made it fundamentally different from newspapers, since "there is limited space for broadcasting, society has to be decide who shall have access to those frequencies ... and those with franchises must be held accountable for the service they provide". The ITA was the right body to take on this task, had the capacity and the enthusiasm, and had "already made it clear that there will be no question of their treating radio as a matter of secondary concern".

In that speech, Chataway also marked the ITA's card on a whole range of key decisions. The number of stations would be constrained by viability. The Authority would therefore want to begin with a mix of stations, in main cities, outside the conurbations and "a number which are catering to quite small populations". The West Midlands might well be ripe for dividing up further than BBC local radio had done (and it is surely not co-incidental therefore that – London apart – the greatest degree of overlap between any of the first 19 stations was between Birmingham and Wolverhampton). Local newspaper shareholdings should specifically be provided for after the winning consortium had been selected, since "it seems reasonable that those newspapers that may be affected should be given some priority to participate in the new medium". ILR might well carry a more advertising each hour ('minutage') than ITV. Chataway there and then mooted the nine-minute-per-hour maximum which was to be adopted. ILR programmes should compete with Radios One, Two and Four, but must be "firmly rooted in their locality and carry local information and local news". It would be for the ITA to negotiate needletime with the Musicians' Union, and "it will be necessary ... to

92 Meeting note 10 September 1970.
93 Letter from R B Henderson to Christopher Chataway 8 October 1970.
94 Letter from R B Henderson to Brian Young 12 March 1971.
95 Memorandum Brian Young to Lord Aylestone 14 June 1971.
96 Christopher Chataway speech to the Local Radio Association lunch on 22 June 1971.

provide opportunities for live music and employment of musicians; and contributions to that end will have to be made by the companies according to their capacity to pay". For national and international news, "an IRN certainly has attractions".

Chataway's creation, the IBA, would be expected to do as directed. The shape of ILR as a tightly regulated, locally-based public radio service was being set from the very start. The government had also sorted out a way of funding the ITA in its preliminary radio work, but the advertisement for a Senior Adviser for Local Independent Radio was for someone to be employed initially by the ministry. On 23 September the new ITA director general, Brian Young, was able to tell the Authority that Mr J B Thompson was to be offered this appointment.

John Thompson is the father of Independent Local Radio. He designed it, implemented it, and sustained it through periods when its survival seemed quite improbable. Subtle, personally generous and politically far-sighted, with an almost limitless capacity for the details of the new service, and taking great delight in manoeuvring people on the ILR chessboard, Thompson's strong opinions could be concealed at times by a Byzantine courtesy and a fog of cigarette smoke. Good chance played its part in his application. "We were going on holiday in Corfu. Reading the *Economist* on the flight, I was caught by one of the very few recruitment advertisements, and I thought – this all looks quite interesting. I then forgot about it for two weeks – but by coincidence, flying back from Corfu, I again saw a copy of the *Economist* on the aeroplane, and there was another reference to this. So when I got back to London I wrote off."[97]

Working somewhat in limbo in the ITA premises opposite Harrods, at 70 Brompton Road in London's Knightsbridge, Thompson was nominally employed by the Ministry from October 1971, until the passing of the Sound Broadcasting Act allowed his formal employers to become the new Independent Broadcasting Authority in July 1972. By that time, some key decisions had already been taken by government and the ITA; those apart, ILR was Thompson's to establish, with an eye always on the political fragility of the concept. Even within the ITA, Thompson observed that the relationship was complex. "In the circumstances of 1970, 1971, it just seemed a very strange idea. This topsy-turvy introduction of radio after commercially-funded television was peculiar. I always found my ITA colleagues very friendly and helpful, but rather puzzled."[98]

The Sound Broadcasting Bill was introduced in the House of Commons on 11 November 1971, with a surprising amount of detail left until the committee stages of the bill. Chataway's statement in the Commons talked about the competitive challenges for ILR, but also the opportunities. "We do not seem to be a country in which radio listening figures are yet particularly high. In some other countries radio has staged quite a recovery over the past decade. Its immediacy and flexibility still give it great advantages as a medium of communication." Radio, he said, was "a totally different animal"[99] from television.[100]

97 John Thompson interview op. cit.

98 Ibid.

99 The line about ILR as a different animal surfaced again in the Annan Report in 1977, which asserted that "local radio is a different animal and requires a different keeper". For those of us who worked for John Thompson, it is hard to suppress the sense that both phrases came from his skilful pen.

100 *Hansard*, House of Commons 11 November 1971 cols 1250–1380 Second Reading of the Sound Broadcasting Bill, col 1276.

The parliamentary debate that day confirmed that ILR was at least to have a chance to establish itself and demonstrate its potential, whatever the electoral future. Ivor Richard, the opposition spokesman, told the Commons that a future Labour government would freeze the expansion of ILR, not abolish it, pending a Royal Commission into broadcasting.[101] Commentators calculated that such a report could not have any effect until the late 1970s. A key topic for discussion in the press and elsewhere was the commercial potential of ILR. The government's estimate was £10m (foresightedly around 2 per cent of UK display advertising). The *FT* vacillated, predicting £10m in May but envisaging as much as £20m by the time the bill was published. Bill MacDonald of the Evening Newspaper Advertising Bureau, and soon to be the first MD of Radio Hallam in Sheffield, thought that higher figure might be reached by the end of the decade.[102] This was not going to be a bonanza, even before the realities of the 1973 oil shock and industrial strife could be anticipated. The ITA's assumption was a gloomy £6.5m[103] compared with ITV revenues that year of £143m.[104]

1972 opened with the National Union of Mineworkers striking against the Heath government, cutting coal supplies to power stations. Electricity to homes and businesses was turned off for up to nine hours a day. Although a deal with the miners was agreed by the end of February, the mood of the year was established. British troops killed 13 protestors on the Derry Bogside in January, and the IRA bombed the British Embassy in Dublin, and Aldershot Army barracks in February. Against that bleak background, fundamental decisions about independent radio had to be confirmed, and they must be understood in the context of the times. Take, as a prime example, the discussions over who should own and operate the local radio transmitters. John Thompson's recollection is that "the general mood was against separate arrangements. Extremely skilled as some of the ITA engineers were, in an ideal world it might have started as a separate activity. But in the political and social mood of 1972, it simply wasn't on. The IBA Engineering Division had a lot of experience in dealing with these things and – a crucial point – they knew how to work 'with' the BBC. Although the larger question of how much sense it made for the radio transmissions – or the studio codes – to be the responsibility of a central authority is an arguable point, granted the political and social realities of the time I think there was no alternative."[105]

That was indeed how it worked out. Throughout the year, the ITA's engineers worked on arrangements for transmission, studio codes, and the lines linking the two. Other preparations proceeded as well, so that late in May the ITA members could consider a long "list of subjects concerning which the Authority has already taken firm policy decisions".[106] Technical planning had advanced to the stage where it was now confirmed that there should be about 60 stations, all transmitting on both VHF and MF, with two stations in London and one or more small-population stations included in the initial programme. The first ten stations were to be London I, London II, Manchester, Glasgow, Birmingham, Liverpool, Tyneside,

101 *Ibid.* col 1272.
102 *Advertising Weekly* 2 April 1971.
103 ITA Paper 71(71) 17 June 1971.
104 Advertising Association.
105 John Thompson interview op. cit.
106 ITA Paper 74(72) 19 May 1972.

Swansea, Sheffield and Plymouth. For advertising, it was confirmed that up to nine minutes should be allowed in any 'clock hour', although that might be reviewed in the light of experience. The ITA's Advertising Code would be used for radio also. In John Thompson's view, this conferred legitimacy on the whole system. "In my experience, the advertising code was mighty useful. Here and there it might have been a bit heavy, but by and large to have had a code in the political circumstances was very helpful."[107]

On financial issues, the Authority "has accepted the desirability of ensuring that the larger radio companies do not make excessive profits", but "does not regard the direct subsidisation of some radio companies out of excess rental paid by others as being the course it would favour". Part of the ITA's planning was based on a concern not to have profits siphoned off to the Treasury, which was what happened to the ITV levy. Their aim was "to secure as far as possible that if there is any excess profitability anywhere in the system some part of it, at least, should flow to the Authority for re-deployment by it in the interests of the system as a whole".[108] A second tier of rentals was to be levied, on a sliding scale on profits above a certain percentage of a company's net assets. That was partly intended to correct any imbalances in the primary rental arrangements, given that "it is possible – indeed highly likely – that the initial apportionment between companies of basic rental could prove inequitable". Such money could help repay more quickly the government loan which the IBA would receive to set up ILR, but "perhaps more importantly, there are many positive uses to which the funds from the second-tier rental could be applied to the advantage of the ILR system as a whole ... Each year a list of needs would be considered, and the best of those would be approved by the Authority. These might include, for example, the commissioning of special programmes that would then be made available to the stations, the institution of special research projects, the support of training ..." There was also the opportunity to deploy second-tier rentals for the employment of musicians, to meet the minister's expectation, and to "help towards an agreement with the Musicians' Union".

Thus was born 'secondary rental', one of the most characteristic features distinguishing *independent* from *commercial* radio. In the context of 1972, and the design of the ILR system, it offered the prospect of linking together political realities with a highly practical way of enhancing the new services. Over the years which followed, that prospect came to fruition and underpinned many of the most notable achievements of ILR in the years up to 1987. However, it came at the cost of increasing still further the marginal tax rate paid by the ILR companies, and aggravating – in their corporate thinking at least – their sense of being hard done by.[109]

Across the UK, groups were beginning to coalesce with the intention of applying for the new ILR franchises. In Glasgow, for example, Radio Clyde announced its formation on 30 March, with a formidable list of backers including Collins publishers, Rangers Football Club, the Scottish CWS and Scottish Television. The group's spokesman, Esmond Wright, former Conservative MP for

107 John Thompson interview op. cit.
108 ITA Paper 79(72) 19 May 1972.
109 Secondary rental is discussed more fully in chapter 9.

Pollock, accurately catching the political mood of the times, said "it is not a money-making exercise. Of course it has to be viable, but commercial radio is not comparable with the hey-day of commercial television. It has to be seen as a public service."[110] Among the backers of Radio Clyde were also three substantial newspaper publishing groups, between them offering the *Glasgow Herald* and *Evening Times* (Outrams); the *Paisley Daily Express* (Scottish and Universal); and the *Scottish Daily Express*, *Evening Citizen* (Beaverbrook); plus the *Greenock Telegraph*. They had been brought together by Hugh Stenhouse, in many ways the prime mover in the Radio Clyde consortium. Jimmy Gordon, who joined late in the process, recalls that "Beaverbook's Max Aitken and Hugh Fraser, who ran the *Express* and *Herald* were well on the road to forming separate groups. Stenhouse banged heads together and said 'look, if you two guys go against each other somebody else will come in … if you get together, you're unbeatable."[111]

The issue of newspaper shareholdings within ILR dominated parliamentary discussion of the new system. The report stage of the Sound Broadcasting Bill on 11 April was not debated wholly on party lines. John Gorst argued that to let newspapers buy into ILR would set a dangerous precedent for television. However, the main challenge to the government's position came from Labour's Ivor Richard and others, who were concerned about the concentration of media ownership power in too few hands. Nevertheless, Chataway decided that the IBA should be required to allow newspapers in a local area to acquire shareholdings in the new ILR companies if they wished to do so. Thompson concluded that the IBA would "need to distinguish between newspapers who come forward as active, positive participants in a consortium, and those newspapers which may prefer to rely (rather more negatively) on their entitlement under the Act" and should be prepared to allow those in the first category higher shareholdings.[112] The Newspaper Society met the IBA on 27 July, pushing the shareholding entitlement for an individual newspaper group up from the IBA's proposed 12.5 per cent, to the eventual 15 per cent.

The parliamentary process through the Commons was fairly smooth. An amendment to remove the notion of a London News station attached to the national radio news service had been carried at committee stage on 18 January, where John Gorst and two other Conservative MPs sided with the Labour opposition, but it was reversed at the report stage. Two Labour peers tried unsuccessfully in the Lords to delay the introduction of ILR until 1976, describing the bill as "squalid and horrid", to prove that some of the old passions still remained.[113] Despite the occasional skirmish, the Sound Broadcasting Act received the Royal Assent on 12 June 1972, and on 19 June Chataway as Postmaster General announced the location of the first 26 ILR stations.

What were the expectations of those who had, one way or another, fought the long battle to get an alternative service of local radio up and running in the UK? They seem to have been modest. Philip Birch was clear that there was not going to be another bonanza like the pirate ships. "At Radio London we had a straight Top 40 format, low overheads, and the advantage of being an experimental station

110 *Glasgow Herald* 31 March 1972.
111 Jimmy Gordon in an interview with the author, 25 April 2008.
112 ITA Paper 79(72) 19 May 1972.
113 *Hansard*, House of Lords, col 488.

– which advertisers were eager to try quickly, before it was suppressed."[114] John Gorst, who had been for so long the promoter of commercial radio, was disappointed. "I have heard estimates of a potential audience for one major station of between half and one million people. That isn't what I meant by local radio." John Thompson was keen to stress the unknown issues which would have to be coped with in a quintessentially British experiment. "This is a curious and fascinating country that we live in, and this seems to be an opportunity to do things our own way."[115]

And indeed that is the key point. The eventual form of private radio in the UK – *independent* radio – was a quintessentially British compromise, in which the vulgar vigour of the free market was to be held in check by the cautious wisdom of an established system of licensing and oversight. This was wholly at one with the official mood of the times, where social engineering and management was seen as a prime duty of the state. Introducing an alternative system of radio involved taking a significant risk; all the more reason, then, to hold that initiative within tried and trusted structures. To understand how Britain worked in the seventies, and where it failed to work, you do well to look at the history of independent radio.

The form which ILR eventually took served the official mindset, but did it meet the proper expectations of listeners? Sean Street has argued that "the appetite that was fuelled by the pirates also created an expectation that when ILR was born in 1973 it didn't meet – not initially, anyway".[116] However, at this precise moment the task was to implement *independent* radio. The nature of that service was determined by the zeitgeist of the times, which led to the decision broadly to replicate the ITV model of licensing and regulation for ILR. Without the extremely careful management of the political realities achieved by Chataway and Thompson, there would have been no new radio system at all – or perhaps one which the incoming Labour government would have shown no qualms about dismantling immediately. The broader public interest was well served in the end. Listeners gained a whole swathe of new radio stations, doing different things, which clearly had great audience appeal, and the checks and balances held firm. To echo Thompson's phrase, this was the UK doing things in its own way.

114 *Sunday Telegraph* 18 June 1972.
115 *Campaign* 4 August 1972.
116 Sean Street in an interview with the author, 3 January 2008.

Chapter 3

Making a start

August 1972 – October 1973

There was to be little more than a year from the passage of the Sound Broadcasting Act in June 1972 to the first station broadcast in October 1973. A great deal of administrative and licensing detail needed to be settled in a short time, as well as some smart work in building transmitters and arranging lines and links. For the potential programme companies, there was the business of getting together both a group and an application, and also finding time to lobby the new Independent Broadcasting Authority for favourable terms and rules. The membership of the IBA, established when the Sound Broadcasting Act came into effect on 12 July 1972, hardly differed at all from that of the ITA. The only change was Christopher Bland replacing Sir Ronald Gould as deputy chairman. Bland was then known chiefly as a merchant banker and former Conservative councillor, and brought no radio expertise to the IBA, nor would he have found any in the board on his arrival.

However, there was technical expertise aplenty among the ITA's engineers. They were to implement the new radio network, and they would need to re-use medium wave frequencies in multiple locations across the UK. There is debate about who first developed the idea of using highly directional aerials to squirt a powerful signal across an area, while leaving the space behind it to be re-used on the same frequency for another locality. Hughie Green's group were making such proposals to the BBC in the late sixties, but it was the ITA's team at their Crawley Court engineering headquarters who brought this into practical use. The advantage was that relatively low power transmitters could be deployed, reducing capital, operating, and maintenance costs. Directionality enabled frequency re-use, and also allowed relatively strong signal levels to be delivered into the centre of metropolitan areas. Furthermore, it made it possible to match more closely the VHF and daytime medium wave coverage eras. The disadvantage was the need for large and elaborate aerial arrays, which made demands on finding sites, and then getting planning permission to build on them.

The first shock for the IBA came around quickly, and it was as a direct consequence of using directional aerials. The ITA's engineers had struggled to find a suitable site for the London MF services. On 14 July, two days after it came into formal existence, the IBA board was told that there was no prospect of permanent medium wave transmitters for the two London stations being ready for commissioning by autumn 1973. A solution was available, but it was not attractive. It involved "building a temporary MF station which would give reasonable coverage of the London area during the hours of daylight only … using non-UK shared frequencies at fairly low power levels".[117] Consequently, London stations would have to begin the ILR adventure with their medium wave transmissions, on temporary frequencies, coming from a wire slung between two chimneys of Lots Road power station in Chelsea. It took almost a year and a half from the temporary launch before the permanent transmitters came on air.

The chief shock for the would-be radio stations surfaced around the same time, when the outcome of the IBA's negotiations with the copyright bodies was announced. For the Authority, however, this huge issue had been around since the start of the year, when John Thompson had begun contact with the Musicians' Union and the two principal copyright societies. Music copyright was (and remains) a critical issue for radio in the UK. The long story is covered in detail in Chapter 13. Here, it is enough to note the outcome of challenging negotiations. The programme companies were allowed enough 'needletime' to allow them to play half of their hours as gramophone records issued by the major manufacturers, up to a maximum of nine hours for an eighteen hour day. The shock came chiefly from the cost thus incurred. ILR stations would have to pay a rising percentage of their net advertising revenue on a sliding scale to the copyright bodies. Since each ILR company was also required to spend at least 3 per cent of its net advertising revenue on the employment of musicians, the total cost of music, as a percentage of net advertising receipts, would be 10.5 per cent in year one, 12 per cent in year two and 14 per cent in year three.[118]

The ITA had already discussed its attitude to the composition of local radio companies privately back in May, although this was not made public until July. For almost the first time, we see an attempt to introduce the primacy of radio considerations into a licensing process otherwise heavily influenced by ITV precedents. John Thompson argued in a paper to Authority members that "the outstanding positive factor guiding the IBA's policy toward the composition of companies will need to be (and be seen to be) the relevance of the contractors to the needs and opportunities of radio in the UK in the mid 70s".[119] This led him on to assert that the IBA would need to aim to select companies capable of providing "a truly public service" on radio, comprising diversity of ownership, control and influence among the programme companies selected. Diversity of nature and ownership between different stations, as well as within each individual company, was the aim from the start.

The details of the IBA's requirements and expectations of franchise applicants were set out in a series of "Notes on Independent Local Radio",[120] published in July. They serve as a summary of how ILR was expected to operate, and are therefore central to understanding the early stage of the new medium. Significantly, the Notes begin with the technical arrangements. The IBA was the transmission agency; literally, the 'broadcaster'. It wanted to be sure that this was understood, and that its standards would be met. As well as the transmission arrangements and coverage areas, there were specifications for the links between the studios and transmitters, a detailed code of practice for studio construction and

117 IBA Paper 106(72).
118 *Notes on Independent Local Radio* IBA 12 July 1972.
119 ITA Paper 65(72) 17 May 1972.
120 *Notes*, op. cit.

performance, and requirements on the stations to monitor their output; chiefly to ensure no breaks in transmission, although a side product was monitoring tapes for content regulation.

The general statement, preceding the detailed requirements regarding the composition of radio companies, summarises the IBA's radio approach. "In every area the Authority will be looking, above all, for men and women capable of running a lively and responsible station; one which will provide, in the words of the white paper, 'a truly public service ... combining popular programming with fostering a greater awareness of local affairs and involvement with the community'." Reluctant to allow, in the early stages of ILR at least, any one investor to have a major role in more than one area, anyone with more than 20 per cent of one company was told not to expect to have a significant holding in any other. The Authority would "examine with special care applications where there is a significant representation by overseas interests". It was prescient, and just as well, that the Notes did not rule them out. ILR was to be much indebted to Canadian investment in its early years.

Programming would be expected to provide popularity, local awareness and community involvement. There would need to be a range of content, specialist music output (although the amounts were not specified) and programming "with a particular relevance to the needs and opportunities of local life". Local and national news output was required, and must be in a "fresh and distinctive radio style". The national and international news service was not mandated, but there was a strong enough steer in the Notes that applicants proposing other arrangements would have an uphill task to win a franchise.

Confirmation of the primary rentals and of the levels of secondary rental came with the advertisement on 4 October, inviting applications for the first five franchises. The London General station, with an anticipated coverage (on medium wave during daytime) of 9.2 million people, would pay an annual rental to the IBA of £315,000, rising to £380,000 at the end of three years, when there was to be a break in the contract. The London News franchise, with the same coverage area, would pay £185,000, rising to £230,000. In Manchester, a coverage area of 2.4 million would cost £108,000 to £132,000; Glasgow, at 1.9 million, £85,000 to £105,000; and Birmingham, with 1.7 million, £75,000 to £95,000. Secondary rental rules were that once a contractor's pre-tax profit rose above 5 per cent of net advertising revenue, a quarter of those profits would go to the IBA, rising to half when the profit on turnover was 10 per cent or more. The closing date for applications was 8 December, and the services were expected to begin broadcasting simultaneously at a date not yet set, perhaps in the spring of 1974. The Local Radio Association met John Thompson on 9 November to plead for reduced rentals, citing mostly "unspecified withdrawals" among potential applicants, but with no success.[121]

There had been a number of loose confederations of potential applicants, coalescing and separating during the run-up to the application deadline. Among the radio entrepreneurs eyeing the London General franchise, for example, John Whitney and Phillip Waddilove, the original progenitors of the Local Radio Association, were active with one group. Barclay Barclay-White, a Weybridge

121 *The Times* 10 November 1972.

dentist, had brought together in another a range of show-business talent including Bryan Forbes and Richard Attenborough. Among the communicators who were prominent in the horse-trading were Michael Kustow, who had once run the Institute of Contemporary Arts, literary agent Michael Sissons, broadcaster-tuned-Oxford-don Tony Smith, and former BBC television producer Neil Ffrench Blake, who had brought in Lord (Ted) Willis, famous at that time as the writer of the long-running *Dixon of Dock Green* television series. Hughie Green had high hopes for his Wigmore Radio group, chaired by Conservative Peer Lord Mancroft.

Based on their respective London evening newspapers, Associated Press (proprietor of the *Evening News*) and Beaverbrook Newspapers (the *Evening Standard*) were potential players. Associated had brought in Philip Birch, formerly the head of offshore pirate Radio London. Beaverbrook had Terry Bate, whose strong Canadian accent, and an entrepreneurial style learned in North American commercial radio, belied his English roots. The *Evening Standard's* Jocelyn Stevens had been Bate's chairman at Radio Caroline and instrumental in bringing him back to the UK from Canada in1965. Bate himself ended up as a substantial shareholder in nine of the first 19 ILR companies – either personally or through his joint venture with Beaverbook – and as the owner of Broadcast Marketing Sales Ltd.[122]

The Dimbleby family's south-west London newspapers were part of Bate's group. Outside London, the big and relatively independent provincial newspapers such as the *Liverpool Daily Post & Echo*, the *Birmingham Post* and the *Wolverhampton Express & Star* made their dispositions, seeking to play an earlier part in applications rather than only taking up their statutory guarantee. The general feeling among commentators was that the IBA would seek to broker marriages between some of the more disparate groups.

Anecdote suggests that there were, at one time, perhaps fifty bidders for the London franchises. However, after what the *Economist* referred to as "the shake-up"[123] that number was reduced to thirteen by the closing date of 8 December 1972. The IBA received a total of 26 applications for these first five franchises: eight for London General; five for London News; six for Glasgow; three for Manchester; and four for Birmingham. Each applicant was required to submit detailed written proposals, which were then analysed by the IBA staff. For the non-metropolitan stations, a group of Authority members and senior staff visited the cities concerned, to carry out preliminary interviews with all the applicant groups. As a result, three Glasgow groups were invited back to attend a further interview with the full Authority in London, two groups from Manchester, and two from Birmingham. For London, the full Authority interviewed all groups, and then conducted second interviews with two groups for each of the General and News franchises.

Those final interviews were remarkable set-piece occasions of a type rarely seen nowadays. The Authority members and senior staff lined two thirds of an immense oval table on the ninth floor of 70 Brompton Road. All the IBA people had an extensive brief on each applicant, including detailed analysis, and the list of questions determined by the staff which the members were to ask, each allocated to a specific member. In my experience they usually kept to their brief, but not invariably. The applicant group was then ushered in, having been kept quarantined

122 Email from Terry Bate to the author, 28 December 2008.
123 *Economist* 10 February 1973.

in another room until the requisite time, to sit at named places agreed with them in advance. After a brief opening statement from the consortium's chairman, the questioning proceeded for an hour or more, with a stenographer taking a verbatim record. At the end, signalled by the IBA chairman, the group would depart, rather shaken, to take coffee in *Richoux's* or somewhere similar along the Old Brompton Road. Robert Stiby, who was part of the Capital Radio team for the London interview, recalls that "Lord Aylestone [the IBA chairman] terrified me ... it felt like an Admiralty board court martial ... "[124] The Authority members would discuss their views in closed session, aided by the senior staff only, and reach their decision. There might well then be further private discussion between John Thompson and the provisional winner to resolve points of detail, before an announcement was made.

The Authority conducted the final interviews for the two London franchises, and announced the offers of contracts on 8 February 1973. The London Broadcasting Company (LBC) won the news franchise, and Capital Radio the general and entertainment franchise. The press journalists had fun with what were unexpected selections. "Radio choice – muzak or newzak", scorned the *Sunday Times*.[125] Lewis Chester's *Sunday Times* piece set the tone for the inevitable purveying of conspiracy theories from unsuccessful applicants, which dogged the radio regulators ever afterwards. The IBA at once learned that awarding two franchises from thirteen applications earns you two uncertain friends but eleven immediate enemies, whose *amour-propre* cannot allow them to believe they lost fairly.

Among the losers was Hughie Green. A long-time campaigner for commercial radio, and one whose group had done a great deal of practical work, it was claimed that Green could not resist using his very popular ITV talent show *Opportunity Knocks* to push his ILR application. *Private Eye* alleged that a Mr Frank Avis was invited onto the show to support an "abysmal" singing group named *The Kinsmen*, to provide an excuse for showing "an interesting film all about the public service aspects of commercial radio".[126] Mr Avis, its story said, "represented Melbourne's commercial radio stations". Inconclusive correspondence followed between the IBA and London Weekend Television. When the matter came before the full Authority in September, it was invited simply "to draw its own conclusion from this correspondence",[127] but the episode can scarcely have helped Green's cause.

LBC's largest shareholder was Charterhouse Securities, closely followed by Associated Newspapers. The mix was flavoured by the presence of IBC, in an echo of the days of pre-war commercial radio,[128] and the Canadian newspaper and broadcasting group Selkirk Communications. LBC's chairman, Sir Charles Trinder, had been Lord Mayor of London. His deputy chair, and very much the driving force of the application, was Sir Gordon Newton. He was among the most successful and perhaps the most idiosyncratic of British newspaper editors since the second world war, and had transformed the *Financial Times* during his time as editor up to 1972. With marketing man Michael Levete as managing director, and Michael Cudlipp as chief editor, this was a company created according to the model

124 Robert Stiby in an interview with the author 21 January 2007.
125 *Sunday Times* 11 February 1973.
126 *Private Eye* May 1972.
127 *IBA Paper* 123(72) 19 September 1972.
128 See chapter 1.

of a broadsheet newspaper – an error of judgment from which it never really recovered.

If LBC was Fleet Street establishment, then the original Capital Radio was showbiz, and its success allowed the *Evening Standard* to deploy its familiar tone of amused condescension, with the headline "Capital's 'radio dentist' wins it".[129] Barclay Barclay-White had trained as a dentist, and owned several practices in the Weybridge area of Surrey. A chance conversation with his son-in-law interested him in the commercial potential of radio, and "drawing on his wide range of connections among his patients"[130] he put together a consortium, with Richard Attenborough (later Sir Richard and then Lord Attenborough) as its chairman. Barclay-White's Dominfast group owned 30 per cent of Capital Radio, and Graham Binns' Rediffusion 15 per cent. Beaverbrook's *Evening Standard* and the *Observer* each held 8 per cent, thus setting the two London evening newspaper giants side by side in the two London ILR stations.

The other major shareholder in Capital, with 15 per cent, was Local News of London, which had been created by local newspaperman Robert Stiby. He had been in Canada in 1972 to buy a new press for his *Croydon Advertiser*, and hearing radio there, came back enthused with the idea of playing a part in the new UK service.[131] He brought together a consortium of many local weekly papers in London, and put the group into play. LBC showed a keen interest, hoping for access to all the local newsrooms. However, after a conversation with Binns and then with Attenborough, Stiby decided instead to go with Capital. His 20 investors subscribed a total of £100,000 between them, an investment that was to be worth £43m when they sold in 1998.

LBC had its senior management already in place, but Capital Radio – deliberately, according to Stiby – had not done so. John Thompson indulged in some marriage-broking. Capital was told that it was the favoured candidate, but that there was another group with more experience and better management and "we needed to talk with each other as that would be their [the IBA's] favoured choice".[132] Attenborough was not persuaded about all the members of the rival group, but made contact with John Whitney on his car phone.[133] The process of bringing London Independent Broadcasting Company's management across to operate Capital Radio's station was underway. Together with Attenborough, they were to be the backbone of Capital Radio – and, to a degree, of ILR as a whole – through its tumultuous first decade.

The three provincial companies had more diverse shareholdings, once the various elements of marriage broking had been completed. Birmingham Broadcasting (BRMB) included all the region's main newspaper groups, with the *Birmingham Post* and *Mail*, the largest single shareholder, owning 10 per cent of voting shares and 15 per cent of non-voting. Of the other shareholders, including

129 *Evening Standard* 9 February 1973.

130 *The Times* obituary of Barclay Barclay-White 22 January 2007.

131 Stiby interview op. cit.

132 Stiby interview op. cit.

133 Whitney had installed a car phone in his Mini, in conscious imitation of his hero, Leonard Plugge, the commercial radio pioneer of the inter-war years. On that phone he received a call from Attenborough, whom he had never met, inviting to him to a meeting to discuss becoming the new station's MD. Plugge's company, IBC, also owned 13.33% of the shares of LBC, the first ILR company, providing remarkable continuity from 1925 to 1973 and beyond.

local department stores, engineering companies, two trade unions and a wide range of individuals – including senior managers of the station – none held more than 6 per cent. Greater Manchester Independent Radio, which was to operate as Piccadilly Radio, showed a similar range – although not quite as broad – with its only two large shareholders being the *Bolton Evening News* group and the *Manchester Evening News*, with 11 per cent and 10 per cent respectively. In Glasgow, Radio Clyde outstripped the others in the range and diversity of its shareholding, but again four newspaper groups were among the seven largest shareholders.

Looking only at the London awards, the *Financial Times* reported that national newspapers were "notably absent" from the successful groups, attributing their absence to the IBA in the run-up to the selection "making it known that newspapers might be taking too much for granted in assuming a prime place in radio almost as of right".[134] However, the local papers did indeed have that right, granted under statute, and in every one of the first five franchises newspaper holdings dominated the shareholding lists, even if falling short of controlling the companies. The regional ITV companies also had a toe-hold. ATV owned 2.5 per cent of BRMB, Granada TV 8 per cent of Piccadilly, and Scottish TV 7 per cent of Radio Clyde. The established commercial media thus ended up with what seemed to be potentially blocking holdings in the new medium. That represented the political reality of the times, which had obliged the government to grant shareholding rights to local papers – even if they were not part of the winning consortium – and also reflected the IBA's nervousness about the commercial climate into which ILR was to be launched. The new companies would need support from those with deep pockets, but the intense suspicions of the anti-commercial groups in the radical left of UK politics were fuelled by the outcome.

It has been important to dwell at some length on the composition of these first companies, and not just because so many of those involved went on to play a significant role more widely in ILR. The nature of these companies set the tone for ILR, right up until the consolidation of ownership and the arrival of stock-market quotes in the late eighties. Companies would have a broad base, a strong local flavour, and shareholders often as much interested in the kudos which came from being part of the local station as in early commercial returns. The existing commercial media would be a little too prominent, but there would be some improbably successful blending of apparently opposed interests, such as the industrial and trade union shareholdings in both BRMB and Radio Clyde. These were companies brought into being by the aspirations of the social liberal consensus of the 1970s, and they reflected those aspirations in their own composition.

The first nine months of 1973 were a period of preparation, and of making practical arrangements. For the new companies, recruitment and planning were to the fore. The new chief editor of LBC, Michael Cudlipp, who at the time was senior deputy editor of the *Times*, was characterised as "radio's first newspaperman", which was a pessimistic observation.[135] His experience lay chiefly in print, and the two weeks spent in the US and Canada to "learn all about radio" could only do a little of that. LBC also hoped that its Canadian investor, Selkirk Communications, would bring the necessary expertise. LBC (and the crucially

134 *Financial Times* 9 February 1973
135 Roger de Freitas in *Broadcast* 30 March 1973.

important IRN service) was to be run from premises dubbed 'Communications House' in Gough Square, one of the Dickensian spaces which open up from the maze of alleys and small streets round the back of Fleet Street. The location was, in many ways, a good one – central for political and other guests, and in the traditional heart of the UK newspaper industry – but it reinforced the company's identification with the press rather than with broadcasting. It was a long way from the newer broadcasting hubs, and when Rupert Murdoch's News International Group made its seminal move to London's Docklands in 1986, starting a flight from Fleet Street by all the national papers, LBC/IRN found itself alone.

For Capital Radio, the merger with the London Independent Broadcasting consortium at the IBA's instigation gave it most of its top team. As well as John Whitney as MD, Tony Salisbury became general manager, Tony Vickers sales director, and Gerry O'Reilly chief engineer. Capital acquired offices in Euston Tower, just beyond the acceptable edge of the West End, and set about building studios.

This rather unprepossessing building had the advantages of space and easy access, and become an unlikely icon for the new medium. The large foyer in particular was to enable Capital to offer access to its listeners and the whole London community, and to stage events such as the public debates run during the early years. The company also set about recruiting the rest of its management and staff. For such a new venture this was a challenging task, as it involved finding people to do several senior jobs which were quite new within the traditional world of UK broadcasting, and the net had to be cast wide.

As programme director, Capital appointed Michael Bukht, who under another name was television's Michael Barry, the 'Crafty Cook'. Attenborough described Bukht as "fizzing with ideas",[136] a description which characterised his huge contribution to independent radio over the next thirty years. In the late sixties he had been programme director of the commercial Jamaica Broadcasting Corporation, then joining BBC Radio where he had been one of the pioneers of UK phone ins. One of the first appointments Bukht made to Capital Radio, in late June, was that of BBC TV's Ron Onions as news editor. Onions was later to carry much of the burden for driving LBC through its successive crises.

The commercial aspects of independent radio also began to take shape. Capital issued its first rate card at the end of May, and its close resemblance to a television rate card was an early reminder that this was not a print medium. Capital was to sell its own advertising, but LBC – advised by its major shareholder and board member Kenneth Baker, vice-president of Selkirk Communications – decided at the start of July to let its sales be handled by Radio Sales & Marketing, an independent sales house owned by Selkirk which would bid also for other radio stations' national advertising contracts.

Both Capital and LBC recognised the need for a presence in the national advertising scene, and the crucial importance of educating advertisers and agencies in the new medium, a view shared by the provincial stations. However, competing advertising sales agencies were destructive. Selkirk's RS&M was to be joined in the marketplace by Broadcast Marketing Services, BMS – linked with the other Canadian investor in ILR, Standard Broadcasting – and by Associated Newspapers'

136 *Broadcast* 8 June 1973.

Air Services. With Capital Radio Sales, that meant that there were four different sales teams trying to get appointments with advertising agency media buyers. John Thompson, for one, was "baffled" that so much energy should be spent, and so much revenue diverted, into internecine competition.[137]

In London, the presence of two new radio stations meant that they were competitors for audiences and revenue. This brought Capital Radio into conflict with the IBA. The Authority had decided not to mandate formally the taking of the national and international news service of Independent Radio News, IRN, for all the other ILR stations, and Capital Radio was wary of being dependent on a service provided by LBC, its commercial rival in London. The station intended to gather news itself – hence the recruitment of the heavyweight Onions as news editor – and bridled at paying the fees demanded by LBC/IRN. After some serious arm-twisting, Capital conceded, John Whitney confirming in late August that Capital would be taking "most of its news from the main news station" with its own news operation being supplementary to IRN, and concentrating on London interest stories.[138] Nevertheless, Capital's board was already becoming alarmed at its high level of fixed costs, of which the IRN fee was effectively part.

The IBA was also moving ahead with its planning. London VHF was to come from the Croydon mast, already used by the IBA for the London ITV services (one of the clear benefits of a single regulator for both media). Radio Clyde's VHF service was to come from a site shared with ITV at Black Hill, as was BRMB's at Lichfield, but the MF services needed new transmitter sites, as also did Saddleworth for Piccadilly's VHF output. The new local radio planning group of the Authority, based at its sylvan engineering headquarters at Crawley Court, in the Hampshire countryside, had "considerable difficulty in finding suitable sites for the medium-frequency stations which need to be located close to their urban target areas, and which, by virtue of their complex aerial systems, take up a large area of land".[139] Directional transmission for medium wave meant deploying a three or four mast radiating aerial, but that required a good deal of land for both the aerials and the ground-mat in each location. Adopting circular polarisation for the VHF transmissions was aimed at improving car radio reception and portable out-of-doors signals as a whole.

The IBA's engineering standards were too high, being largely copied from the BBC's principles for regional radio centres. Chris Daubney, who was responsible for testing and commissioning the first phase of ILR stations, recalls that "we had to rationalise some aspects of the acoustic standards".[140] The adoption from the start of stereo output for all the ILR stations had taken the BBC aback, and conferred a technical legitimacy on the fledgling system. This was the other side of the argument about whether or not companies should own and operate their own transmitters. John Thompson points out, in the context of VHF stereo, that "what is not always appreciated ... by a number of the radio companies is that there were some real advantages in having these very skilled blokes [the ITA/IBA engineers] involved".[141]

137 John Thompson in an interview with the author, 8 November 2007.

138 *Daily Telegraph* 28 August 1973.

139 Independent Broadcasting Authority *Annual Report and Accounts* 1972-3 p 47.

140 Chris Daubney in an interview with the author, 8 November 2007.

141 John Thompson interview op. cit.

It was now clear that neither in programming, nor in transmission and technical standards, was the IBA going to opt for the lowest-cost or lowest common denominator option. Its ambition for ILR was set much higher than that, for better or worse. Even so it is legitimate to wonder how the stations, dominated as they were by recorded pop music, reacted to the IBA Handbook's description of stereo broadcasting for ILR. "The conductor taps the podium twice. Sounds begin to swell and fill the auditorium. The movement builds up to a tremendous crescendo. An evening at the concert hall … When you have purchased your stereo recording system, take care in positioning your loudspeakers and listeners ... near the centre line between the two loudspeakers. Now you can take your seat."[142] It probably raised a wry smile or two.

The IBA's Codes of Practice for the technical standards required of their studios amused the ILR operators a good deal less. These made stringent demands on acoustic isolation and audio quality. The first five stations encountered what was to be one of the key rituals before a new ILR service came on air; the formal inspection by the IBA's engineers. They would arrive at the studios, often to what resembled a building site rather than a finished studio complex, to measure all the programme origination equipment, studio signal paths and recording equipment, both for mono and stereo; to examine the arrangements for logging tapes, studio acoustics – including reverberation time and ambient noise level; to look at radio cars and other outside broadcast equipment; and to check that the studio-to-transmitter paths delivered the correct channel bandwidth, field strength, polarisations, modulation level, limiting characteristics, compression (on MF), pre-emphasis (VHF) and stereophonic pilot tone. All this was to ensure "the establishment of a high-quality stereophonic VHF broadcasting service, backed up by an MF service which makes the best possible use of this congested band".[143]

Unless the engineers said OK – and they were never without *some* major reservations – the station could not go on the air. The engineering teams actually did all they could to help the infant stations, while not abdicating their duties. Chris Daubney recalls approaching the task with genuine enthusiasm. These younger staffers were keen to be " … enablers. The stations just wanted to get on air. We were a bit of a pain, and had fairly demanding standards, but in the end I think they recognised the worth of it … We were keen to work with the stations, not just say 'this is how it has to be'."[144] This was as far as could be imagined from the days of the offshore pirate radio ships, where the hum of the generator and the jumping of needles on the record turntable in high seas were all part of the 'charm'. Yet it was less than six years from the day the pirates were scuppered.

Regulatory arrangements within the IBA comprised a small Radio Division headed by John Thompson, plus a few ITA staff adding radio duties to their television ones. Advertising control for radio – another added-value feature which having a single regulator brought to the new medium – was managed by the IBA's deputy head of Advertising Control, the genial Harry Theobalds, whose sympathetic flexibility within what was otherwise an inflexible structure was a huge help to the new service. The Radio Division had some autonomy over its financial

142 *ITV '75* guide to Independent Television with Independent Local Radio supplement p 222.

143 Ibid. p 48.

144 Chris Daubney interview op. cit.

management, since the finances of the two media within the IBA had, by statute, to be kept separate, and Christopher Lucas was recruited to run that side of the work as senior assistant, radio. Bob Kennedy, who had been the first programme organiser of the experimental BBC Radio Leicester in 1967, became the other senior adviser, with specific responsibilities for programming. When I joined the Radio Division in the summer of 1974, to the entirely undefined job of senior officer, radio, it was clear that in military terms we were essentially staff officers for 'General' Thompson.

The IBA's radio team was as much of a start-up in style, tone and experience as were the ILR stations themselves. One of Bob Kennedy's earliest issues was how to prevent the broadcasting of offensive records. Early in September, he had written to the first five companies to seek their views. Clyde, Piccadilly and BRMB had replied, LBC said it did not worry them, and by the end of the month Capital still "were thinking about it".[145] The pragmatic conclusion was that there would indeed be records that should not be broadcast, but that the stations would be likely to know of these first. There needed to be a mechanism for a station to consult the Authority about an unacceptable record, and then a return loop for advising and alerting all the other companies. That consultative style was to characterise the early years.

One quintessentially public service element imposed by the Sound Broadcasting Act onto ILR was the Local Advisory Committee (LAC). Mirroring the arrangements for BBC Local Radio Councils, the IBA was required to appoint a committee for each franchise, going beyond being merely a local version of 'the great and the good', with one third of the members nominated by local authorities. The LACs' statutory purpose was to give the Authority "such advice as in the opinion of the Committee would be appropriate for reflecting, so far as is reasonably practical, the range of tastes and interest of persons residing in the area".[146] The committees could have been interfering busybodies; it is to the credit of those who served on them, and those who managed the process from Brompton Road, that they were overwhelmingly helpful to ILR. Whilst they never became cheerleaders for their local station like their BBC counterparts, they were generally supportive, rarely disruptive, and a genuine source of information for the regulator. The LACs were not to survive the 1990 Broadcasting Act, and its shift to *commercial* radio philosophies. Over their decade and a half, however, almost 2,000 local people[147] served diligently and with enthusiasm on their local committees, and were a real part of the public service apparatus underpinning *independent* radio. The modest amount of involvement of the IBA regional television offices with local radio worked very much with and through the LACs.

Technical regulation was also mostly worked out on a co-operative basis, although it could veer into the overly formal, reflecting the traditional practices of the original ITA. The Authority's decision to require all stations to have a delay mechanism, capable of providing at least seven seconds between a comment being made and it being broadcast, was non-negotiable. Similarly, the new-fangled radiophones were not under any circumstances to be used for broadcasting

145 Memorandum Bob Kennedy to HR (John Thompson) 24 September 1973.
146 Sound Broadcasting Act 1972, section 3(5).
147 Based on a manual count of IBA card records.

purposes, although they were permitted for contacting outside broadcast crews. That caused a spat at a meeting in September 1973, as the minutes relate. The IBA's head of local radio planning, Baron Sewter, pointed out "that this matter has now gone to the Minister who has refused to grant permission and the Authority can go no further. Mr White [Piccadilly Radio] pointed out that since the local post office in Manchester had granted permission, he was dissatisfied with the minister's ruling. Programme companies were in general dissatisfied that the Post Office Headquarters' decisions were taken to affect the whole country."[148] Jon Snow was to break new ground by defying this rule two years later at the Balcombe Street siege.[149]

Joint advertising supervision had been one of the key arguments for keeping radio within the then ITA. The Authority said that it understood "the special need for speed and flexibility in dealing with the requirements of radio advertisers".[150] With the anticipated predominance of local advertising, it guided the stations towards establishing their own copy clearance capacity locally, working to a very detailed Code of Advertising Standards and Practice. There was, at that stage, very little difference between what was permitted on ITV and what was to be allowed on ILR, not least as the aim was "to secure consistency of standards throughout the Independent Broadcasting services". The ITA had delegated the detail of copy clearance for television commercials to the Independent Television Companies Association (ITCA), and radio was accommodated within what became a joint advertising copy clearance operation, which was well regarded throughout the advertising industry. It was slower than radio would have liked, and showed an infinite capacity for taking pains over the smallest detail. Still, in the social and political mood of 1973 it was exactly what was required, and lent further authenticity to ILR.

The IBA was also moving forward with its plans for further stations. When the Authority met in May 1973, it agreed a list of another eleven stations to bring the total to eighteen on air by July 1976. In addition to the five already awarded, franchises were to be advertised for Swansea, Tyne/Wear, Sheffield, Liverpool, Edinburgh, Plymouth, Nottingham, Portsmouth, Bradford, Bristol, Wolverhampton and Teesside, plus either Ipswich or Reading. IBA chairman Lord Aylestone explained to the minister that "the inclusion of three 'unprofitable' stations ... means taking something of a financial risk; but the Authority feels it is a risk worth running".[151] By that, he meant that the IBA could afford to build and equip the transmitters, the issue which had dominated their correspondence for the past year. The risk for the operating companies was not a factor in the decision. When I went to run Radio 210, the Reading station, in 1981, it was modestly profitable, as were the other small stations. None of us was ever told that we were in charge of stations which the IBA's chairman had thought "unprofitable" from the outset.

The question of whether to include Ipswich or Reading had been finely balanced, as Ayleston wrote to Chataway. "The argument for Ipswich is that it would be the one station in a large area (East Anglia) where an ILR station is much

148 3rd Joint IBA/AIRC Technical Committee Meeting 14 September 1973.

149 See chapter 5.

150 *IBA Annual Report* 1973-4 p 38.

151 Letter from Lord Ayleston to Christopher Chataway, 30 May 1972.

desired, and where there is no BBC Local Radio station. The argument for Reading is that it would be the London peripheral station, which we undertook last summer should be brought into being as soon as possible."[152] The IBA favoured Ipswich, but wanted to check that it had the minister's agreement. Chataway was "most reluctant to see Reading omitted from the next batch of stations".[153] He advised leaving Bristol until later in order to accommodate both, so ILR would have to manage without the western city, which hurt its attempts to sell itself to advertisers as a national network. In the event, the number of franchises was brought up to nineteen in order to include Belfast.

The IBA continued to be very vexed over its own costs, with its director of finance noting that the government's proposals would increase the required capital expenditure by £11,500, and decrease the rental income by £20,000. "These amounts may not sound important in relation to the whole. But the Authority is already taking a gamble to please the minister ... We do not yet know where the break-even point will be. We do not yet know that even the Swanseas can swim, much less the Plymouths, Readings and Ipswiches ... Suppose it turns out that Swansea [and smaller stations], and conceivably even Portsmouth and Bradford, cannot even stand their present rentals, small though they are. It is, I think, most unlikely we should be allowed to let them go broke. Who then will pay the piper? Clearly not the minister."[154]

Evidently, not everyone in the IBA was happy. This was a marginal activity when compared with the core work of television regulation, but hugely demanding on time and attention. Meanwhile, the radio companies were working feverishly to get their studios ready, their staff appointed, and their programme schedules in place. The IBA's own Radio Division was making up new regulations as it went along, firmly in the context of the relative rigidity of the IBA, but equally excited by the unruly prospects which ILR had to offer. By October 1973 everyone had to believe that they were prepared to launch a unique broadcasting venture.

152 Ibid.
153 Letter from Christopher Chataway to Lord Aylestone 25 June 1973.
154 Memorandum Anthony Curbishley to Brian Young 26 June 1973.

Chapter 4

Turn your radio on

The first year of ILR

LBC began broadcasting at 6 am on Monday 8 October 1973, David Jessel following a 30-second station ident with: "It is six o'clock in the morning, Monday October the eighth. Welcome to London Broadcasting, the news and information voice of independent radio." Actually, Jessel had recorded that announcement two afternoons before, mistrusting his constitution at 6 am in the morning.[155] The first advertisement to be broadcast was for Birds Eye fish fingers. The first news bulletin was read by an Australian, Ken Guy. LBC had the luck it would have wished for; it was a 'good' week for news. Two days before, on the holiest Jewish day of Yom Kippur, Egypt and Syria launched a surprise attack on Israel. On LBC's first afternoon, Prime Minister Heath announced the next stage of the government's prices and incomes policy, limiting wage increases to a maximum of £350 per annum, and maintaining statutory controls on prices. LBC expected to be able to do well with 'hard' news, given its overwhelmingly Fleet Street background, and it had its chance on day one. It carried welcome statements from each of the three UK party leaders. Although many Labour politicians harboured a doctrinaire distaste for commercial radio, Harold Wilson was notably generous. "It is well known that the Labour Party have opposed the introduction of commercial broadcasting, but that, in no way, prevents me from offering my best wishes to the Independent Broadcasting News service." Through all the disasters which followed, LBC and its news service IRN were to build – and keep – good relations with politicians of all parties.

ILR, and LBC/IRN within it, was intended to be, from the start, much more accessible than the rather hidebound BBC. The previous month, the revolutionary fashion retailer of the late sixties and early seventies, Biba, had opened its own department store in the seven-floor Derry and Toms building on Kensington High Street, West London. When Janet Street-Porter followed David Jessel on to the airwaves at 9 am that first LBC morning, in a programme presented with Old Etonian Paul Callan, the equivalent demotic and classless nature of ILR seemed to be confirmed. An astute radio critic observed that "as for that controversial Twiggy of the air, the Cockney-accented Janet Street-Porter, well she'll either be fired for impudence or become a national cult. I rather suspect the latter."[156]

For all that, LBC did not start well. Media launches are notoriously fraught. The station could have weathered the links that failed to work, and more than its

155 *New Statesman* Memoirs of an LBC Broadcaster 25 January 1974.
156 Jeremy Rundall in the *Sunday Times* 14 October 1973.

fair share of technical glitches. However, the inadequacy of its management quickly became manifest, in minor and major matters. Anecdote has it that the new Uher tape recorders, neatly boxed in a corner of the newsroom, proved irresistible to the occasional staffers who populated the Gough Square studios. The double stack had disappeared by the end of the second day on air. The broader problems were even greater. LBC was finding it difficult to invent a way of operating an all-news and talk radio station. It had plenty of newspaper experience among its editors, but too many of its radio skills came from mostly Antipodean freelancers. The company was also severely under-capitalised, and its failure to attract anything like enough advertising, even at the start, was to leave it needing a new injection of capital within little more than a month.

Its natural critics had a field day. In the Commons, Labour's Gerald Kaufman maintained that "as the London Broadcasting Company, in addition to being in clear breach of the prospectus on which it gained its franchise, is grossly overworking and underpaying its staff, has hardly any listeners, gets hardly any commercials and seems to be on the verge of financial disintegration, would it not be kinder to put it out of its misery?"[157] He was taking aim at what was by then seen as an easy target, as a proxy for Labour's distaste for ILR in general. Even Conservative maverick John Gorst thought him "absolutely right" in describing LBC as offering an "expensive and totally inadequate news service". [158]

The IBA's Annual report for that year, normally a model of understatement, thought the early mistakes "embarrassing, if hardly unexpected".[159] John Thompson says now that his heart sank at LBC's first 48 hours, adding that if LBC had not been so fragile, IRN would have been so much easier to establish. It was "a pity LBC's management and directors in the early years were seemingly not fully focused on the job in hand", at least for the first eighteen months or so, by which time the damage had been done.[160] The most worrying label of all, applied instantly by the *Economist*, and one which stuck to the station even during its years of success, was "amateurism ... it lacks the sort of authority essential to a news and current affairs station [competing with the BBC]. That is not helped by ... the station's political correspondent, in the second week of operations, popping up with a report on a beekeeping exhibition."[161]

Capital Radio began eight days later, with the daytime mix of popular music and speech that characterised all the other ILR stations; arguably, 16 October 1973 was the real start of ILR. This also started a launch-day tradition, with everyone associated with the station – board members just as much as engineers – working frantically with screwdrivers and paintbrushes all though the previous night in their Euston Road premises, right up to and beyond that first on-air moment. Just before 5 am on 16 October 1973, the station's General Manager Tony Salisbury had to wake up DJ David Symonds, who had fallen asleep in the toilets amid all

157 David Jessel claimed in the *New Statesman* that this resonant rhetorical question was first posed not by Kaufman "but by an indiscreet member of the IBA at Nuffield College High table, an ungracious remark since the member concerned had been one of those responsible for granting LBC the franchise".

158 *Hansard*, House of Commons, 26 November 1973, cols 6/7.

159 *IBA Annual Report 1973-4* p 31.

160 John Thompson in an interview with the author, 8 November 2007.

161 *Economist* 20 October 1973.

the frantic last-minute wiring and soldering of the new studios. After Chairman Sir Richard Attenborough's "Good morning, this for the very first time is Capital Radio", the opening song played on the first of the new popular music-based stations was *God Save The Queen*,[162] in a new version recorded by the London Philharmonic Orchestra conducted by Sir Arthur Bliss, Master of the Queen's Musick. Symonds then played Simon and Garfunkel's *Bridge Over Troubled Water*. In that fusion of the traditional and the fairly new, the medium which was to provide the sound track of their lives for those who lived in the towns and cities of the UK, for the next 30 years, spoke for the first time to London. Or at least to as many Londoners as could then find the newly launching Capital Radio on its temporary wavelength. Capital's start seemed altogether more encouraging than LBC's. It shared the fish fingers commercial, and some of the hesitancy which comes from the challenge of live broadcasting in unfamiliar studios, but it seemed to be quickly into its stride. Contemporary pop music pleased the *Daily Telegraph*. "It's music to drive by. It is characterised by songs like the Bee Gees' *Massachusetts*, the Mamas and Papas' *Monday Monday* … it is Schmilsson, Carly Simon, Elton John, Don Maclean. Nothing to jangle the nerves of a parent."[163] Music director Aidan Day, recruited only at the end of August from his Scorpio Sound studios, pitched the music to match Michael Bukht's intention to provide popular contemporary music for adults during daytime, with rock and then specialist music feature programmes for jazz, country and western, folk and classical in the evenings, and space for live pop and even classical concerts at weekends. This was the archetypal pattern for ILR, and it was getting an early test with the London audience. With presenters including Roger Scott, Tommy Vance and Joan Shenton, Dave Cash and Nicky Horne it was being given every chance.

In pratice, the initial music selection was not a success. Five years later John Whitney was to admit "we got it wrong".[164] Capital listened to Radio One's output of pop/MOR and went "too far left of that". The solution was a fundamental change in music policy, which was neatly accomplished at the end of December. "We went into Christmas playing one kind of music and came out playing another, and no one knew we had done it." Tim Blackmore, who was Head of Music and Head of Programmes at Capital from 1977, explains that "the launch policy had been very biased towards the 'hippier' side of contemporary music eg: Joni Mitchell et al. The change was simply to recognise more of the top 40 tracks that had been initially eschewed, but without ever owning up to being a top 40 station … [later on] our most listened to programme was the Saturday morning *Capital Countdown* show which was the 'top 40' we made up ourselves on Friday afternoons and which served as the station playlist. No research, just professional judgement."[165]

Tensions between LBC and Capital were a feature of the early weeks behind the scenes. Capital was then warned off by the IBA from covering the wedding of Princess Anne and Mark Phillips on 14 November, despite having booked its seat and interviews with the organist, the head flower arranger, and possibly the Archbishop of Canterbury too. John Whitney's teeth were clearly gritted as he

162 *Now We're 5*, Capital Radio's 5th anniversary programme.
163 *Evening Standard* 16 October 1973.
164 *Broadcast* 16 October 1978.
165 Email from Tim Blackmore to the author, 29 December 2008.

commented "we believe we have established our own broadcasting style … sufficiently different from London Broadcasting … however, we understand the IBA's point of view and of course we accept".[166] Capital's key competitor, Radio Two, broadcast the wedding, along with Radio Four and fourteen of the nineteen BBC Local Radio stations.

The BBC had been little disturbed by the launch of LBC, which it also stigmatised as "amateurish".[167] ILR seemed more threatening when – in David Hendy's exposition – local stations "started opening on medium wave frequencies the government had snatched from Radio Four … Over the next few years it was commercial stations such as Radio Clyde, Radio Forth, Swansea Sound and those in Liverpool and Manchester which succeeded best in stealing listeners from the BBC, playing as they did on strong local identities and popular suspicion that the Corporation was still too London orientated." The BBC tied its leading disc-jockeys into two year exclusive contracts.[168] These included a notable roll-call of those drawn from the offshore pirate stations – such as John Peel, Ed Stewart, Johnnie Walker, Tony Blackburn – emphasising that Radio One was, if anything, more the inheritor of the pirates than Michael Bukht's "station for adults".[169]

Meanwhile, the IBA processes continued, apparently undisturbed. On the day LBC began broadcasting, the Authority published the programme plans from its application for the franchise, as it did with Capital and the subsequent stations. That was, for its time, a significant piece of openness. The advertisement, analysis, interviewing and award of new franchises saw new stations announced in short order: Metro Radio in Newcastle; Swansea Sound; Radio City in Liverpool; Radio Forth in Edinburgh; Plymouth Sound; and Radio Hallam in Sheffield. Between them, by the middle of May 1975, they would cover 16.4 million people, nearly 30 per cent of the UK population. Two more franchises were to be advertised in February 1974, for Nottingham and Teesside. This progress was too slow for the members of the Local Radio Association, who were "astonished and bitterly disappointed"[170] that the IBA aimed at an annual roll-out of only six stations a year after the first batch. The Authority, however, was hamstrung by financial and capital expenditure constraints, and also by a growing awareness that all was far from well with the commercial prospects of the first stations.

LBC had faced financial problems from the start. Even if its original capitalisation had been adequate, and its cost management had been scrupulous – and both were far from the case – income was elusive. In mid-November, 1973, advertising pundits were forecasting revenue for the station of barely £150,000 in its first three months, hardly enough to pay even the IBA transmitter rental. On 21 November, with the IBA's agreement, the company called up £622,500 of its capital reserves. That in itself almost tripled the working capital, which had begun at a derisory £375,000, but left available only around a quarter of a million pounds out of the original investment.[171] LBC was not helped by having to bear so much of the cost of the IRN service, with only one other station on air, but its real

166 *Daily Telegraph* 3 November 1973.
167 David Hendy, *Life on Air, the History of Radio Four*, Macmillan, 2007.
168 *Financial Times* 27 November 1973.
169 Michael Bukht presentation at the Café Royal, London 23 August 1973.
170 Local Radio Association announcement, 12 July 1973.
171 *Daily Telegraph* 22 November 1973.

problem was lack of credibility among advertisers in its challenging early weeks. Capital Radio was also coming under fire from advertisers for witholding publication of audience research, leading to assertions – accurate as it turned out[172] – that "Capital must have something to hide, and they cannot buy advertising as an act of faith for ever".[173]

On 10 December it was leaked that LBC was planning to cut ten posts from its remarkably high complement of 150 staff, and the first of a seemingly endless round of labour disputes started. These were to dog the company – and to a degree the rest of the industry – for years afterwards. The National Union of Journalists, entirely in keeping with the industrial mood of that year, refused to contemplate any redundancies until Chief Executive Michael Levete and chief editor Michael Cudlipp were sent packing. They demanded a meeting with the IBA and summoned the spectre of support from the Capital NUJ Chapel.[174] As with any industrial dispute within a media company, the briefing of journalists who were 'objectively' covering the story, by those journalists involved in the dispute, put LBC at once into a false position. Its initial lack of credibility was made worse by damaging press coverage, which had negative effects too on Capital Radio and ILR as a whole. That weekend saw a non-coincidental series of stories: "London's ailing broadcasters"[175], scorned the *Guardian*, "a massive indictment of the entire system", challenged the *New Statesman*,[176] while the *Sunday Times* scoffed at "why the jewel in the crown of commercial radio looks tarnished".[177]

A response in the House of Commons by Postmaster General Sir John Eden the following Tuesday, to questions from Philip Whitehead, John Gorst and John Grant[178], drew cheers from Conservatives hearing from him that "one should not conduct a campaign against something which has only recently been launched, which is still going through fairly early difficulties and troubles, but which none the less offers fairly good prospects for a bright future".[179] Yet Eden was clearly hedging his bets, noting that "the awarding of franchises is the responsibility of the IBA, not the minister".

The NUJ got its wish within days. There were extensive discussions behind the scenes involving LBC and Lord Aylestone, Brian Young and John Thompson. Following a formal meeting of the IBA on 20 December, which gave its approval, Michael Levete resigned as chief executive and Michael Cudlipp as chief editor. Chairman Sir Charles Trinder resigned the same day, and was replaced by Sir Gordon Newton, distinguished from his time as editor of the *Financial Times*, but without any broadcasting pedigree. The station offered an unspecified aspiration about bringing in an experienced broadcaster from Canada to act as chief executive. A statement from the IBA emphasised that "at no time has the Authority been asked, nor would it consider, any proposal to change the basic format of the company as an all-news and information station ... discussions will continue, and

172 See chapter 6.
173 *Financial Times* 24 November 1973.
174 *Financial Times* 10 and 11 December 1973.
175 *Guardian* 14 December 1973.
176 *New Statesman* 14 December 1973.
177 *Sunday Times* 16 December 1973.
178 Grant was later to become a member of the Radio Authority.
179 *Hansard* 17 December 1973 Sir John Eden response col 934.

in January the Authority will consider the company's plans for entering 1974 on the basis of long-term stability".[180] The Authority's own credibility was also on the line.

ILR was not well-placed to cope with any new storms. Yet a tempest was breaking throughout the developed world, and reserving its especial fury for Edward Heath's struggling British economy. The day after Capital Radio had begun broadcasting in October, the Organisation of Petroleum Exporting States (OPEC), plus Egypt and Syria, announced an embargo on shipments of oil to nations which had supported Israel in the Yom Kippur war; that is to say the USA, and its allies in Europe and Japan. Petrol quickly became scarce, and prices rose from around $3 a barrel to a previously unprecedented $11 by January 1974. Restrictions were imposed across the west, with long queues and even fights breaking out at filling stations. Stock markets in the US and Western Europe dropped dramatically, ending the long bull market which had run largely uninterrupted since the Korean War in the early 1950s. Two decades of fairly steady economic growth turned into a shuddering recession, with fiercely rising inflation.

In Britain, the National Union of Mineworkers (NUM) chose this moment to renew its challenge to the Heath government. On 13 November 1973, the miners banned overtime working, severely reducing the output of coal from British pits. With coal stocks dwindling, and under pressure from oil prices, the country faced a major energy crisis. To reduce electricity consumption, and thus conserve coal stocks, Heath introduced a series of measures including the Three-Day Work Order, more commonly known simply as the 'three day week', to come into force at midnight on 31 December. Commercial consumption of electricity would be limited to three consecutive days each week. Essential services, which included radio stations along with restaurants, food shops and newspapers, were exempt from the energy restriction, but not from the swingeing blow dealt to the economy. Advertising confidence and expenditure were early casualties.

On 4 February 1974, the miners rejected offered wage increases of 16.5 per cent and voted overwhelmingly for a national strike. Three days later, the railway unions also went on strike. Heath called a snap General Election for 28 February, in what he hoped would be a referendum on 'who governs Britain' – the government or the unions? The outcome was inconclusive, with no party winning an overall majority. On 4 March, despite polling 200,000 fewer votes than Heath, Harold Wilson formed a minority administration, returning to office for a third term. His emollient words to LBC five months before suddenly took on more significance. The country was reeling from recession, industrial and political strife, and assaulted by IRA terrorism. No one seemed likely to have any time to attend to the troubles of independent radio. At least on 6 March the miners called off their strike, settling for a 35 per cent pay increase, but inflation was set to climb to 17.2 per cent, and in that climate there was to be no easing in the industrial relations pressure within ILR.[181]

180 IBA statement 20 December 1973.

181 The rise in retail prices for ILR's first years was shattering for the UK economy and for most ordinary people. The annual percentage increases year on year were: 1973: 8.4 per cent; 1974: 17.2 per cent; 1975: 24.2 per cent; 1976: 16.5 per cent; 1977: 15.8 per cent; 1978: 8.3 per cent; 1979: 13.4 per cent; 1980: 18.0 per cent; 1981: 11.9 per cent; 1982: 8.6 per cent. Except in wartime, there had been nothing comparable in the modern era.

What a time to be starting a new and commercially marginal medium! Radio Clyde began broadcasting in Glasgow on Hogmanay, 31 December 1973, and brought some relief to the proponents of ILR. The company under its Chairman Ian Chapman – the head of publishers Collins – had brought in STV political editor Jimmy Gordon to be its MD at quite a late stage. Clyde was well-managed behind the scenes, and well-led on air by Andy Park as head of entertainment, and Alex Dickson as head of news and current affairs. Clyde set the gold standard for ILR, at a time when the currency elsewhere seemed at great risk of being debased. It offered distinctly high-grade popular radio, mixing a skilfully chosen selection of music with ambitious news and features programmes and programming.[182] It played the local card to perfection, becoming a statement of ambition for Glasgow in its drive to self-esteem and regeneration. Clyde's consortium included many Scottish opinion-leaders, so it was not going to face the easy condemnation which lay in wait for the two London stations if it made mistakes.

It did not make many. Gordon had noticed the talent of Richard Park, once an offshore pirate DJ. He joined Clyde from BBC Radio Scotland, and recalls the pleasure of being involved in well-managed success. "Every day was so exciting, I cannot tell you. For the first year my heart beat faster every time I approached Anderston Cross [Radio Clyde's studio location]."[183] That was the mood all the subsequent ILR stations wanted to create, for their staff and their listeners – and advertisers too – and Radio Clyde was continuing proof through the dark days that it could be done. Its sure-footedness, and Gordon's close links with the Labour Scottish establishment – notably John Smith and Donald Dewar – were to be crucial supports to ILR as a whole, when Wilson's fourth administration was able to turn its mind back to the question of the future of independent radio. Without both, it is entirely imaginable that ILR might not have been able to muster enough political support to keep the system going.

BRMB followed on 19 February 1974, broadcasting to Birmingham and the West Midlands, an area badly damaged by the collapse of the motor industry and increasing racial tensions. Run by the experienced trio of managing director David Pinnell, sales director Reg Davies, and programme director John Russell, it had from the start a distinctively Brummie style which was to endear it to its locality. Piccadilly Radio in Manchester began broadcasting on 2 April, and was more obviously metropolitan in style. Ex-pirate Philip Birch as managing director provided probably the closest personal link to the offshore stations, but with programme controller Colin Walters bringing strong BBC experience, the station combined a disciplined approach to popular music programming with a strong range of the speech features which were the hallmark of an ambitious ILR service. Piccadilly was the first station out of London to broadcast round the clock. The IBA made much of its powers to determine the number of hours a station could be on air, and frequently denied requests for extended hours if those were longer than the Authority thought a company could readily manage.

The Tyne/Wear station, Metro Radio, launched on 15 July. It began with its

182 It became the convention to refer to the fluid mixed broadcasting output, characteristic of ILR's informal daytime broadcasts, as 'programming', to distinguish it from the more structured and scripted 'programmes' which made up more formal radio.

183 Richard Park in an interview with the author, 6 December 2007.

own pop music chart, to which it added five hours a week of classical music, plus jazz, country and western, folk and underground music, to live up to the requirement for a broad range of music. Metro had one of the hardest times at the start of all the early ILR companies, as the recession hit the north-east with particular ferocity. Its initial technical and administration director, Neil Robinson, was quickly to become managing director. The first engineer to lead an ILR company, he carried out that role – and played a leading part in the radio industry – with considerable distinction. Swansea Sound, from 30 September, was a genuinely bi-lingual station, with its alternative name Sain Abertawe prominent in publicity. The smallest of the first batch of ILRs, it mixed together English and Welsh output, employing a head of Welsh programmes, Wyn Thomas, to work with programme controller Colin Mason, who was widely known as 'the Brigadier' for his army connections. Managing director Charles Braham, a local newspaper owner and well-connected with the local *crachach*, had been simply a non-executive director until a falling out within the original group left Swansea Sound desperate for an MD at short notice. Braham stepped in on a temporary basis and remained firmly in place, as one of the founding figures of ILR.

Bill MacDonald, another Briton with extensive experience of Canadian radio – and a noticeable Canadian accent – led Radio Hallam, which opened to serve Sheffield & Rotherham on 1 October 1984. He had attracted a number of known radio voices from offshore pirate radio and the BBC, including Keith Skues, Roger Moffat, Johnny Moran and Bill Crozier, and a talented management team to match. Launching just beyond the first anniversary of ILR, but completing the 1974 line-up, was Radio City in Liverpool. Strongly reflecting the energy and character of its founder and managing director Terry Smith, a Yorkshire journalist and news agency man, it was nevertheless a firmly Scouse effort. Smith recruited star talent, most notably Gillian Reynolds as programme controller and David Maker as news editor. City had high aspirations, especially for live football coverage from Smith's beloved Anfield, but also in drama and features. Alan Bleasdale's *Scully*, and Scaffold member John Gorman's *PC Plod*, were ambitious early efforts at a new style for radio drama.

ILR's prospects were looking especially grim at the start of 1974, as the recession bit into advertising across all media. ITV revenues dipped over Christmas by 15 to 20 per cent, but for the two ILR stations the fall was more in the order of 80 to 90 per cent. One saving grace from the 'three day week' was that ITV and BBC television were required to shut down at 10.30 pm to help the nation save power; radio continued to broadcast into the night, and thus gained exposure to new listeners in large numbers. To try to weather the storm, Michael Bukht at Capital re-arranged his schedule. He featured two maverick DJs, Kenny Everett and Dave Cash, in what became the legendary *Kenny & Cash* show, from 6.30 to 9 am. Actor and mature heart-throb Gerald Harper hosted a mixture of sweet music and gifts of champagne and roses on Sunday lunchtime. With Tony Myatt at midday and Roger Scott in the afternoons, the station began to take on the sound which would distinguish it for the next 13 years.

At LBC, the main immediate response was at the top, with Canadian broadcaster Bill Hutton becoming managing director. Previously Selkirk Holdings' director on the LBC board, Hutton at last brought some real radio experience to the floundering Gough Square operation. David Jessel was one of the early casualties, his voice thought too light, his show uninteresting.[184] LBC raised a further £600,000 from its major shareholders, Charterhouse Securities, Associated Newspapers and Selkirk. After a further spat with its NUJ Chapel, the station compromised on proposed cuts in the middle of January, and looked for some stability in an unstable world.

Some good news arrived in the form of an LBC-commissioned set of audience figures, which indicated a weekly audience for LBC of around 1 million adults, and for Capital Radio some half a million more. Capital was doing particularly well in the evenings, but still trailed Radios One, Two and Four by some margin. It was not triumphant, but it was defensible.[185] The two London ILR stations were still sparring, though, and fell out again over a plan to cover jointly the February 1974 general election, providing input also to Clyde and BRMB. Again, militancy played its part. LBC's NUJ Chapel protested when five journalists arrived from Capital, leading to "heated telephone calls between Bill Hutton and John Whitney" before the Capital team was withdrawn.[186]

Hutton had advertised for someone to fill the "toughest job in broadcasting", that of editor for LBC and IRN,[187] perhaps more in hope than expectation. The eventual appointment exceeded his fondest hopes. Marshall Stewart, who had for the past four years been editor of Radio 4's flagship *Today Programme*, decided in February to jump ship to LBC, in a move of profound importance for both the station and, through the IRN network, for the ILR system as a whole. By the following month, he had brought across Douglas Cameron from *Today* in a striking coup, and then Ron Onions from Capital to be editor of IRN, under Stewart. It was no longer going to be so easy to deride LBC, with talks of amateurism or lack of radio expertise. Crucially, with Stewart and Onions in charge of IRN, the national and international news service for all the ILR stations now had the chance to become a force in its own right. That did little to alleviate NUJ militancy, however. January 1974 had ended with another week of demands, offers and threats, until the NUJ eventually called off a threat to black out LBC and IRN's coverage of Denis Healey's first budget speech as Labour chancellor, and potentially put the station off the air permanently.

To find a way of handling issues where joint work was desirable, the first five companies had met on 30 March 1973 and agreed to form an Association of Independent Radio Contractors (AIRC), which all those gaining radio franchises would be invited to join. They told the IBA that they expected it to operate in areas where the ILR companies needed to negotiate jointly; including labour relations, dealings with copyright bodies, joint purchases, taxation matters, advertising terms and conditions, joint sales operations and audience research. They did not add negotiations with the Authority itself, although that later became central. The IBA worried unavailingly about the acronym, AIRC, preferring the more euphonious formulation Association of Independent Radio Stations, AIRS, as "a possibly more elegant and serviceable alternative".[188] However, AIRC it remained until Paul

184 *Broadcast* 11 January 1974.
185 See chapter 6.
186 *Financial Times* 13 February 1974.
187 *The Times* 18 January 1974.
188 IBA Radio Consultative Committee minutes 1(73) 12 April 1973.

Brown renamed it the Commercial Radio Companies Association (CRCA) in the mid nineties.

AIRC provided an important forum for discussing a joint approach to radio rate card terms and conditions, and to engage with the IPA and ISBA to discuss audience research, although in June "there were ... some individual differences of approach",[189] a state of affairs which was to characterise the trade body ever afterwards. By September, there was a small AIRC secretariat headed by Michael Flint, who had been acting managing director for Capital Radio until the arrival of John Whitney, and was a distinguished copyright lawyer. AIRC represented the companies in discussions with the British Radio Equipment Manufacturers Association (BREMA), the Musicians' Union, the copyright societies, and the trade unions.[190] The last of these led with grim inevitability to national union agreements governing pay and conditions. AIRC engaged a labour relations expert, W J Skinner, and an audience research consultant, Tony Twyman. At the end of 1973, Flint was joined by Cecilia Garnett, who was to take over from him the following summer as AIRC secretary until 1979.

The National Sales Agencies grew in significance within these councils: Broadcasting Marketing Sales, run by Terry Bate and Mike Vanderkar; Air Services, run by Eddie Blackwell; and Radio Sales and Marketing, under Terry Williams and then Dick Seabright. These were to split and then coalesce, joined for a long while by Capital's own sales house, with the larger-than-life figures of Tony Vickers and Philip Pinnegar. The radio companies tended to follow the interests of their own major shareholders in their choice of representation. When the Canadian companies began playing a more major role in ILR as a whole, their influence predominated within each of the sales houses and their client stations. As early as July 1974, Air Services was taking over RS&M and trying to sell LBC alongside BRMB, Piccadilly, Hallam and Plymouth. "Can anyone sell LBC?" asked *Broadcast*, pointing out that "RSM had the most talented, most experienced radio sales crew in town before its members left to revive their sales talents by selling a product people wanted".[191] John Thompson simply despaired that the advertising agencies were being confused by the turf wars between the sales houses. "It was wholly baffling to me. I could understand why they [the sales agencies] wanted to make a go of things, and good luck to them, but it only needed five minutes' thought to realise that if they went into a national advertising agency in London or Manchester or Glasgow, it was virtually inconceivable that any placer of ads was going to listen."[192]

In the political world, Wilson's government, still a minority administration, had the courage to take on the established media interests. Labour had been planning a Committee of Enquiry, to be chaired by Lord Annan, when it was unexpectedly turfed out in 1970. Back in office early in 1974, it resumed its intention. On 10 April, home secretary Roy Jenkins announced the setting up of a Committee "to consider the future of the broadcasting services in the UK". On 12 July, he appointed the other fifteen members who were to work with Noel Annan, and set the process fully underway. On 31 July, Jenkins announced that

189 RCC minutes 2(73) 11 July 1974.
190 RCC minutes 3(73) 11 October 1974.
191 *Broadcast* 8 July 1974.
192 John Thompson interview op. cit.

pending the report, expected towards the beginning of 1977, ILR would be limited to nineteen stations, offering rough parity with the BBC's twenty local stations in England. Paradoxically, that limit actually meant that the IBA had to accelerate its plans. At the time of the announcement, only thirteen companies had been appointed, with Radio Forth in Edinburgh, Plymouth Sound, Radio Trent in Nottingham, and what was to become Radio Tees on Teesside joining the stations already on air by the end of 1974. Six more franchise awards would need to be completed quickly.

Early in August, LBC/IRN hit another union crisis. It followed what was to become a repeated pattern. While the company was negotiating for a further cash injection, this time from Associated and Selkirk, the NUJ Chapel took offence at something fairly minor, in this instance a promotional trail on air which they claimed was "knocking the BBC". The union threatened to strike; Hutton threatened to close down the station; and John Thompson played peacemaker while denying doing any such thing. "It was", he said, "an incident I could well have done without".[193]

By the time of ILR's first anniversary in October, the national economic outlook was as bleak as it could be. Businesses generally cut all discretionary expenditure, and far too often for the comfort of ILR that meant advertising. The ILR stations responded as best they could, but as new ventures they had hardly had time to accumulate any fat to shed. Locked into negotiating a national pay deal, the AIRC's position was undermined by the BBC agreeing a 22.5 per cent pay increase across the board in the summer of 1974. When I started at work at the IBA that autumn, it was rumoured that David Pinnell at BRMB had walked along the corridors of its Birmingham offices and studios, taking out every second light bulb. Hutton had more to cut at LBC/IRN, starting with the monthly taxi bill of £3,100. By dint of extreme efforts to control costs, Radio Clyde, BRMB and Piccadilly were by October in sight of making a profit, or actually doing so, but the pressure on the smaller stations was acute, and the two London stations were in real peril. Max Hastings characterised the period perfectly as "a year of sleepless nights".[194]

This was the moment for Capital Radio's own cash crisis. After earning around £1m up to the end of 1973, the steep recession of 1974 had left the station some £40,000 short of covering its monthly running costs of £175,000. Capital, like LBC, was undercapitalised at the start, and Attenborough admitted that it had overspent its set-up budget of £400,000 by £125,000.[195] Out of an initial capitalisation of £650,000, nothing was left. Whitney set out to negotiate redundancies, which involved closing down the much vaunted newsroom. It is a measure of the scale of the problem which the company faced, that they actually secured agreement with the unions at the end of November. However, even this did not solve the problem, which was soon to be made worse by the need to promote the move to a permanent medium wavelength of 194 metres.

An oft-told story has it that at this point Richard Attenborough sold two Cezanne paintings to provide the extra money to keep Capital going. John Whitney says, "I think he did sell, or at least he put them up as collateral".[196] Capital's bankers

193 *Adweek* 9 August 1974.

194 *Evening Standard* 3 October 1974.

195 *Daily Telegraph* 10 December 1974.

196 John Whitney in an interview with the author, 11 October 2007.

were threatening to foreclose, and Attenborough managed to persuade some of the directors to stump up a loan. Robert Stiby recalls that "there was a board meeting at Capital where all the directors actually had to pledge personal money to get through the week. We were insolvent otherwise."[197] Eventually, it was the arrival of Canadian money and expertise in 1975 that bailed Capital out. Standard Broadcasting took around a quarter of the equity, and, like their opposite numbers at LBC, Selkirk, brought real expertise in running commercial radio. That also closed one of the longer-running stories from early commercial radio; Capital chose to take the Canadian money in the face of an offer from Hughie Green[198], who was thus finally frustrated in his long efforts to become a major player in the new UK radio sector.

Nationally, the political picture became a little clearer. After running a minority government for eight months, Harold Wilson called another general election for 11 October 1974. The Labour party scraped a victory. Their three seat overall majority – at least achieved with a 2 per cent plurality of the votes cast – allowed them to govern for the next 5 years, but only with difficulty. On Christmas Eve, former broadcasting minister John Stonehouse – who had been presumed drowned a month before when his clothes were found on a Miami beach – surfaced in Melbourne, Australia, where he had been living under the assumed name of Donald Clive Mildoon. As Labour postmaster general and minister of posts and telecommunications, he had sustained the party's opposition to commercial radio between 1968 and 1970, but now found himself merely an outré news report on those very stations he had opposed.

IBA policy had made Independent Radio News central to the fortunes of ILR in the early years. It was therefore the greatest of relief when it came of age in July 1975. The late spring of the year was notable nationally for a referendum over continued UK membership of the European Economic Community. Between the announcement of the referendum on 19 May, and polling on 5 June, there was intense political activity for the radio news broadcasters to report, and IRN acquitted itself creditably. However, its real challenge came four days after the result, when a four-week experiment began with the live and recorded radio broadcasting of the proceedings of parliament. This was to be the acid test, for IRN was to be judged directly against BBC Radio. Could IRN measure up? Failure might well have jeopardised the whole ILR project, being a test in the very foreground of political attention.

The UK parliament had always guarded access to anything other than its public galleries with a remarkable ferocity. The idea of broadcasting the proceedings of parliament was first suggested by the BBC in the 1920s, but permission was refused. Richard Crossman, when leader of the house, was keen to bring in both microphones and cameras, but in the crucial cabinet discussion in November 1966, Harold Wilson realised that snippets of coverage might be used in magazine style programmes and said, "that couldn't possibly be allowed ... At this point there was a general clamour, making it clear that nobody wanted the television experiment".[199] There was a closed circuit radio and television experiment in 1968, after

197 Robert Stiby in an interview with the author, 21 January 2008.

198 *Music Week* 22 February 1975.

199 Richard Crossman, *Diaries of a Cabinet Minister* Hamish Hamilton and Jonathan Cape p 35.

which the select committee concluded that "radio broadcasting was both feasible and a most effective method of bringing parliament to the public",[200] but still no progress was made.

Parliament debated the issue frequently, with support varying, until finally on 24 February 1975 the Commons voted by 354 to 132 in a free vote to allow a four-week experiment. Its house services committee agreed with the BBC and the IBA that this would run from 9 June to 4 July, although it could allot only a single commentary position each for just two reporters in very close proximity. So it was that IRN's political editor Ed Boyle, and his colleagues Peter Allen and Julian Manyon, ended up sharing a cramped, sound-proofed, wood-and-glass box inside the chamber with BBC political editor David Holmes and his team.

The practical arrangements involved taking a live feed from the Commons' own sound system, and mixing that with commentary from the reporters on the floor of the Commons, in a caravan (described rather heroically as a 'mobile studio') situated in a courtyard outside. The resulting transmissions, recording and news extracts were then relayed to the ILR stations up and down the UK, and also to IRN. A daily programme, *Parliament from the Inside*, was broadcast by LBC and made available to all stations through IRN. During those four weeks, IRN produced nineteen live relays from inside the house, gave 50 live reports on the day's business, produced twelve and a half hours of complete programmes and provided some 500 recorded extracts for the ILR companies. There were also eleven special programmes covering matters of interest to local stations, notably the proceedings of the Scottish grand committee for Forth and Clyde.[201] Advertisements were not allowed to be broadcast within ILR's relays of parliamentary proceedings, although there were no restrictions at the start or end of broadcasts. This was a huge undertaking for a company such as LBC/IRN, with plenty of troubles of its own, scant resources, and with its staff working in unsuitable and cramped conditions.

In a neat historical circularity, the first voice heard on air was that of the speaker, Selwyn Lloyd, calling the house to order.[202] It had been Lloyd's minority report, dissenting from the Beveridge Report, presented to parliament in January 1951, which had paved the way for the breaking of the BBC's broadcasting monopoly. The first question heard in the radio experiment was on shipbuilding and aircraft nationalisation, from Geoffrey Pattie; the first matter of substance was a challenge from Ian Gilmour to Tony Benn, then secretary of state for trade and industry, about his known anti-Common Market views in the wake of the 67 per cent support for continued membership in the referendum. All were surely aware that they were being broadcast live, but even the arch populist Benn was talking in a parliamentary language that sounded incongruous on radio. Politicians would learn quickly the need for crisp short sentences, when they were speaking more for the benefit of radio listeners than for their colleagues in the house, and in doing so moved society one step further down the road to 'politics by sound-bite'.

200 *Second Report from the Joint Committee on Sound broadcasting* March 1977 p v.

201 *IBA Annual Report 1975-6* pps 35-36.

202 The cry "order, order" from the start of each day's parliamentary proceedings, used by radio as its own starting sound, came to be known best in the voice of the subsequent speaker of the house, George Thomas. From 1976 until 1981, his ringing Welsh tone caught the demotic intentions of such broadcasting far better than Lloyd's upper class accent.

Parliamentary writers bemoaned the yah-boo sounds, but for most listeners what they heard was a revelation. They relished the set-to at Prime Minister's Question Time. Most of the ILR stations broadcast that live on 9 June, and several continued thereafter, with Forth and Clyde also carrying Scottish Question Time live. Listeners heard Benn called a "crackpot" for his proposals for nationalisation, and prime minister Wilson say that he would need time "for a little eclairissement" to work out exactly what Margaret Thatcher had said about the European Referendum result. Audience research conducted by the IBA showed that of those who had heard the radio broadcasts, nearly nine out of ten wanted them to continue.[203] It indicated also that the public found parliament, as offered to them on the radio, to be "remote from their own affairs, as well as a noisy place".[204]

The IBA strongly recommended that radio broadcasting of parliament should be made permanent. It would require better facilities, and an improved basic live feed. The costs of that would be shared between parliament and the broadcasters; selecting, editing and transmitting recordings and live broadcasts for ILR would be paid for by the companies.[205] This was eventually agreed in 1978, with IRN receiving support from secondary rental funds.[206] Television was largely held at bay by MPs for another twelve years, although sound feeds were regularly used by ITN and BBC Television. After a television experiment limited to the House of Lords only, starting in November 1984, full live television coverage became a permanent feature of UK political life only in July 1990.

For IRN, and by extension for ILR as a whole, the radio experiment was a much needed triumph. Independent radio, for all its uncertain commercial fortunes at that time, had shown the political classes and ordinary listeners that it could match the BBC in a major new broadcasting development. The timing could hardly have been better. In the period following the referendum there was a renewed level of parliamentary interest in the independent radio experiment, with the Annan Committee investigations underway. IRN's unquestioned success in holding its own, within the overall achievements of the experiment, legitimised independent radio news coverage in political circles and among ILR's audiences. The IBA's Brian Young considered that "broadcasting from parliament had been extremely successful and ... had helped put ILR, and in particular IRN, 'on the map'".[207] Chris Daubney, who was prominent in the IBA's engineering team making the technical arrangements, puts it succinctly: "You realised, in a small way, that this was a moment in history".[208]

203 *Memorandum to the House of Commons Select Committee Broadcasting Sub-Committee*, IBA (undated, apparently late July 1975).

204 *The Sound Broadcasting of Parliament* James M Cross in *Independent Broadcasting* June 1982.

205 Ibid.

206 See chapter 9.

207 RCC Minutes 10(75) of a meeting on 16 July 1975 p 6.

208 Chris Daubney in an interview with the author, 8 November 2007.

Chapter 5

The pioneer years

Summer 1974 – Summer 1976

The years between the announcement in the summer of 1974 that there were to be nineteen ILR stations, and the start of the expansion which followed the Annan Report at the end of the decade, were the formative period of independent radio in the UK. The freeze on any future development of ILR, or BBC Local Radio, offered a strange kind of stability. Once all the first nineteen stations were on air, they became the explorers and early settlers along independent radio's own version of the Oregon Trail. They had new ideas about how to get things done, and displayed a hardy survival instinct which offered a pattern to those who were to follow them in easier times. These years vindicated the notion of public service local radio, financed by advertising and reflecting back to listeners the sounds of their lives through a local, accessible medium. For all the problems and challenges, it felt surprisingly good to be among the pioneers.

They were, though, hard years for the UK. Under the stress of a failing industrial base and an embattled currency, the British electorate was no longer satisfied by the centrist similarity between the major political parties. That consensus had been mugged by the militancy of the trade unions, and left almost for dead. Inflation was rampant and apparently out of control, accompanied by stagnating output. When revolution came to the UK in 1979, it came from the right rather than the left, yet the uncompromising form which Thatcherism took was in its own way a mirror image of the bloody-mindedness and sheer unpleasantness of the militant industrial left. Business and industry were pretty feeble. They had largely failed to modernise, in part at least because of the extent to which they were protected from competition – first by Empire, and then by the comforts of nationalisation – and could offer little of the dynamism needed to confront a harsher world climate.

Discontent in Northern Ireland spilled over in savage fashion onto the streets of mainland UK, including bombs at Harrods (opposite the IBA's offices in Brompton Road), the London Hilton, Parliament, Tower Bridge, and the home of prime minister Heath. Vicious bomb attacks also lacerated Birmingham, and an armed siege closed London's Balcombe Street. There were political assassinations, unprecedented in modern Britain: Earl Mountbatten, Tory grandee Airey Neave, British ambassador Sir Richard Sykes, and right-wing TV presenter Ross McWhirter. Ordinary people went in fear of being caught in bombings, or – for those of Irish background – of reprisals which went largely unreported. The situation in Northern Ireland itself was even more extreme, as communal violence flared.

ILR was ideally placed to apply some salve to the wounds of a damaged society, in those areas where there were to be radio franchises. Almost all the stations quickly formed a bond with surprisingly large audiences. ILR was friendly and local, demotic in tone and accessible in practice, playing approachable music well seasoned with good humour and useful local information. For almost the first time, mass-audience broadcasting was available in a demystified form, where you had every chance of meeting the station DJs, and quite possibly knowing someone who worked in the station. The studio premises were mostly as idiosyncratic as the city-centre and out-of-town locations they occupied. There were some converted large houses, either in town or outside, and a number of fifties office buildings, often just on the wrong side of the tracks; or, in Metro's case, deep inside a failing industrial estate in Swalwell, much further from the centre of Newcastle than the nominal distance in miles. Downtown was behind high barbed wire fences in County Down, Plymouth Sound was housed in a former organ factory. Victory occupied the ground floor of a converted church hall, with the problems of noise from Boy Scouts jumping around upstairs in the evenings.

The requirement for a broad and local ownership base also meant that many local businesses and individuals had a stake in the companies. For listeners, ILR's appeal was broad, fairly classless, and covering a good age range – at least up to those who were aged 50 or so. Although ILR gave them little credit, the BBC local radio stations were doing much the same job for the older groups, often rather well; but they were all part-time in their hours, and associated with the monolith that was the BBC. The society of which the BBC was such a pillar was out of favour; it had left the mass of people feeling, at best, ill at ease. If ILR could survive, for ordinary people it would be one of the few positive developments of a grim decade.

Taken all in all, these were good times for independent radio people, as well as bad. They could not fail to be influenced by national events, which were depressing the whole society, but there was a vibrancy and an energy among the radio pioneers which no amount of doomsaying could suppress. For some companies, the dire economic circumstances, combined with the pressures of starting a new untried medium, brought them to the very brink of failure, yet there was a sense of belief – and even mission – which kept the individuals going beyond the bounds of exhaustion. Above all, they were having fun, and sensed that their listeners were sharing in that, too, in increasing numbers.

As 1975 began there were nine stations broadcasting: LBC, Capital, Clyde, BRMB, Piccadilly, Metro, Swansea Sound, Hallam, and Radio City. They were joined on 22 January 1975 by Radio Forth, for the Edinburgh area. Along with the IBA radio folk, these early stations carried the entire responsibility for the survival of independent radio in the UK. By and large, the individuals rubbed along pretty well together, aware that they were participants in a shared venture which each knew to be special, although there were some genuine rivalries between stations which competed within overlapping areas, and some synthetic antagonisms generated by the rival advertising sales houses. The characteristics of the strongest – both commercially, and in terms of their wider ambitions – came to be those of ILR as a whole, when added to John Thompson's founding vision.

Radio Forth faced the problem which confronted all the media of Scotland's capital city, of how to reconcile the expectations of the Scottish middle class establishment with the genuine working class character of much of the eastern Central Lowlands, and the unmistakably rural hinterland. It lacked the huge instant appeal of Radio Clyde in Glasgow, and often seemed overshadowed by the prominent success of ILR in its great rival city. However, it started with a notably strong newsroom team and offered serious material alongside the usual popular music, information and entertainment. The company had a wide ownership, with no single investor holding as much as seven per cent of the shares – but newspaper, television and banking interests from Edinburgh predominated.

Some mild controversy surrounded the appointment of its chief executive. Christopher Lucas had been one of the two senior officers within the IBA's new Radio Division, working to John Thompson. His move from regulator to franchise operator raised cries of 'foul'. Brian Young wrote swiftly to the Ministry of Posts and Telecommunications that "we can claim, I think, to be as sensitive as any branch of the public service to the risk of securing (or seeming to secure) favours by our association with patronage. There is no question of that in this case."[209] Lucas ran Radio Forth for three years, before departing to become for seventeen years secretary/director of the Royal Society of Arts, Manufactures and Commerce, where he seemed much more at home. The strength of his senior team at Forth, notably programme controller Richard Findlay and head of news Tom Steele, provided the on-air substance of the station.

Plymouth Sound, the first of the four 'small' stations among the original nineteen, was the opposite of Radio Forth in many ways. Despite two large newspaper groups and a local ITV company dominating the shareholding, the wide spread of the remaining equity ensured a very local and individual character.[210] Strongly aware of the limited commercial possibilities, MD Bob Hussell doubled up as sales director. Starting broadcasting on 19 May 1975, Plymouth Sound was an archetype for what seemed then to be the unfeasibly small ILRs, with multi-tasking and cost-cutting which seemed revolutionary at the time. The station announced that it was "not a 'pop' station ... building one of the largest libraries in the system and [playing] on average 1,100 records per week".[211] It featured phone-ins extensively, with open-line discussions for two hours each morning and afternoon. The company made much of the fact that 50 per cent of the station staff were women, as were two of the five on-air presenters, including Louise Churchill who was to become a Devonian institution.

Sound Broadcasting Teesside was the first company to be unopposed in its application for an ILR franchise. It launched as Radio Tees on 24 June 1975 under MD John Bradford, who had been one of the most junior members of the Local Radio Association when it was originally lobbying for the introduction of independent radio. Bradford himself appears at many critical moments in the history of independent radio, including chairing the pivotal Heathrow Conference ten

209 Letter from Brian Young to Jolyon Dromgoole MPT 1 April 1984.

210 Information on shareholdings, directors etc, comes from successive *IBA Annual Reports* for 1973-74, 1974-75, 1975-76, 1977-78 and 1979-80, The Reports show the initial shareholdings for each company when it commenced broadcasting, and changes in its Board and senior management thereafter. To avoid a forest of footnotes, the summaries here are not individually referenced.

211 *TV & Radio 1976* IBA yearbook p 156.

years later – with toothache[212] – participating in the first major ILR merger between Wiltshire Sound and GWR in 1985, and later as director of the Radio Academy. When I went to visit the Victorian premises in Stockton-on-Tees on behalf of the IBA, to sign off the pre-operational checks, Bradford, programme controller Bob Hopton, and news editor Bill Hamilton were as much installation engineers as anything else, as the open wiring ducts and unfinished studio desks showed the pressures of getting on air.

Radio Trent in Nottingham came on air on 3 July 1975. It had a balanced shareholding between corporate and individual owners, and notably three major trade unions among its investors. With the Greater Nottingham Co-operative Society, these interests jointly represented over 20 per cent of the equity. The first broadcaster on Radio Trent was John Peters,[213] a former GPO engineer who had graduated through the United Biscuits Network.[214] The station faced a turbulent early year through industrial action – ironically, in view of its shareholding – and soon lost star DJ David Jensen to the BBC. Managing director Denis Maitland had formerly managed the offshore pirate Radio London, Big L, in the 1960s.

Pennine Radio in Bradford was the second of the small population coverage stations. The government had required the IBA to include four smaller franchise areas among the first nineteen, but a station in Bradford rather than Leeds was both commercially challenging in itself and diminished ILR's national sales pitch. This was conceived very much as a community station, with broadcasters such as Austin Mitchell, who has been MP for Grimsby since 1977, and Steve Harris, who still leads the campaign for local television, thereby embracing a clear populist agenda. Here, as elsewhere, BBC local radio, firmly established in a traditional BBC regional location in Leeds, enjoyed first-mover advantage.

In Portsmouth, Radio Victory faced some of the same pressures as Bradford. Its locality was the poor relation in the Solent area when compared with Southampton, the base for the BBC. Once again the largest single shareholder was the newspaper group, Portsmouth and Sunderland Newspapers, but a range of corporate and individual investors provided a wide and local base. It started broadcasting on 14 October 1975 – for some strange reason at 1 o'clock in the afternoon[215] – and it did not make a success of its early years. Mild-mannered MD Guy Paine had to contend with some of the wilder spirits in ILR, and oversaw what was seen locally as "a bit of a Mickey Mouse operation, where the sales team turned up in kaftans".[216]

The smallest of the first nineteen ILR stations was Radio Orwell in Ipswich, serving a population of just 200,000 in its primary service area. It boasted a distinguished pedigree. Its chairman, Commander John Jacob, was the brother of Ian Jacob, who had been director general of the BBC from 1952 to 1960. Its MD, Donald Brooks, had run the national radio service in Hong Kong. When the station came on air on 28 October 1975, it broadcast only between 6 am and 10 pm, in

212 John Bradford in an interview with the author, 27 February 2006.

213 *Radio Trent, the Castle Gate Years 1975 – 2005* anniversary DVD.

214 See chapter 9.

215 *Farewell to Victoryland*, documentary about the loss of the franchise broadcast on Radio Victory 28 June 1986.

216 Chris Carnegy in an interview with the author, 18 December 2007.

line with the IBA's controls on broadcasting hours. The need to take its name from the local river, the Orwell, also arose at least in part from the strictures that the Authority placed on station names, all of which required its advance approval.

The other little station was the one serving Reading, which has featured already in the initial introduction to this history. This was also a deliberate experiment, negotiated by the IBA with the government, to judge the viability of a station in the hinterland of Greater London. It broadcast as Radio 210 from 8 March 1976, taking its name from its medium wavelength, and was quite the strangest ILR animal. It had been brought together by Neil Ffrench Blake, who had been one of the unsuccessful London general applicants. He had assembled both Thames Television and News International into an application that promised and delivered extensive public access to programme making, including a group of local children making their own regular programmes. When I joined Radio 210 as its MD five years later, I was told that Ffrench Blake used to form his playlist by looking at the deepness of the grooves on single records, and throwing those which threatened to be too noisy from the open mezzanine onto the floor below. The station began without a managing director, having Ffrench Blake running programming and Michael Moore sales. However, some inspiration from Ffrench Blake's contact list had brought in as chairman Sir John Colville, erudite patrician among patricians, and the gods would never have allowed anything headed by 'Jock' Colville to fail.[217]

Downtown Radio in Belfast had, in so many ways, the hardest task of all the first nineteen stations. The circumstances in Northern Ireland were ferocious, with daily violence and consequent social and economic disruption. Yet, as a benign consequence for local radio, there was a tremendous hunger for broadcasting which understood more than the 'national' television and radio correspondents. Downtown launched on 16 March 1976, the day Harold Wilson announced his shock resignation. The day before, in London, a tube train driver had been shot dead pursuing a bomber presumed to be from the Provisional IRA. The day after, four Catholic civilians were killed by a bomb planted by the Ulster Volunteer Force outside the Hillcrest Bar in County Tyrone.

Everyone was willing Downtown to succeed. A typically broad shareholding mix meant that most interests in the narrow commercial community of Belfast were part of the company. The huge news focus meant that Downtown had to be at least as credible as its better-resourced BBC competition, and news editor David Sloan (later to become the station's MD) oversaw over 200 news bulletins each week. The nature of the challenge is illustrated by the degree of attention which the station had to give to the access it allowed to views on air. This was controlled by "four simple ground rules – no proselytisation *(sic)*, no overt contradiction of other denominations, all material pre-recorded and no participation by those

217 Sir John "Jock" Colville had been Churchill's private secretary during the Second World War and afterwards, and one of the great diarists of the age. His two volumes *The Fringes of Power, Downing Street Diaries* tell how the government was and was not run between 1939 and 1955. He organised the founding of Churchill College for his great mentor, and has a hall there named after him. He could summon ambassadors and open the route to Royal interviews with ease. When I was MD at Radio 210, we would retire after Board meetings to the Calcott Inn where, over a steak and bottle of red wine, he would tell of what it was like to draft the first letter Churchill ever sent to Stalin. He was the kindest and most considerate of men to his managing directors. But then all who knew him averred that "no one can ever say 'no' to Jock".

directly involved in electoral politics".[218] It was all quite a challenge for a new station, in an area where advertising was even harder to come by than on the mainland. Within eight days the *Belfast Telegraph*, a major shareholder but also a fierce competitor, was reporting on "an atmosphere of carefully calculated intelligent mindlessness. It assures you that happiness is around every corner ... yet behind this all-pervading happiness is the harsh reality of the news, though it is presented so simply that one ear alone will do for listening."[219] MD David Hannon will have settled for that initial verdict.

The last, and unquestionably the wildest, of the first nineteen stations was Beacon Radio, serving Wolverhampton and the Black Country from 12 April 1976. It was dominated by its larger than life station manager, Jay Oliver, who drove a dune buggy around the uncomprehending streets of Wolverhampton, and brought more than a dash of US West Coast style to Dudley and West Bromwich, where it was not normally to be found. Beacon was kept just on the right side of the rules – most of the time – by Oliver's programme controller and assistant station manager, Allen Mackenzie. They had little time for the more homespun efforts of British local radio, aiming for "a bright and commercial sound, recognising that a bumbling amateurish approach would hardly be suitable".[220] BRMB and Beacon's areas overlapped to a considerable degree, and there existed from the start something more than border skirmishes but just less than full-scale warfare.

Beacon, and to a lesser degree Plymouth Sound, had faced problems raising their necessary capital. The only applicant for its franchise, the Wolverhampton company had proposed to the IBA capitalisation of £450,000, to be raised by public subscription. The public offer failed, and it took a massive injection of cash from the Canadian radio group Selkirk Communications to rescue the situation. Given the precarious political situation into which ILR had entered, the IBA simply could not allow one of its precious nineteen ILR stations not to get on the air.

By August 1976, Canadian investment of £1,360,000 represented 17 per cent of the total capitalisation of around £8m of the first nineteen stations. The Standard Broadcasting Corporation had invested directly in Metro, Pennine, Trent and Plymouth, with holdings of between 5 and 9 per cent. In 1975, the company had come to the rescue of Capital Radio itself, with a £750,000 investment, half of which was a loan, the rest acquiring a 29 per cent shareholding, and – in an echo of Selkirk's role at LBC – the provision of a full-time Standard executive as part of the rescue operation. The same year, when Metro Radio found itself close to bankruptcy, Standard lifted its stake in that company to 32.5 per cent. Bill Hall, Standard's resident executive in the UK, also had links with Terry Bate's Broadcast Marketing Sales (BMS). He commented shrewdly on the management weaknesses of the early radio companies, "it's half show business, but it's half business too".[221] Selkirk had increased it shareholding in LBC to 49.9 per cent in 1975, and also owned a modest 4 per cent of Radio Forth.

Lacking the opportunity to invest further in Canada, these two companies welcomed the opportunity to move into the UK, and were happy to take the longer view at a time when short-term profitability looked elusive and few other investors

218 IBA yearbook p 141.
219 *Belfast Telegraph* 24 March 1976.
220 *TV & Radio 1977 op cit* p 159 – however did that get past the IBA's editors?
221 Quoted in the *Daily Telegraph* 30 August 1976.

were available. Robert Stiby on the Capital board, while respecting the skills that were brought in and the restraint shown, believed that Standard was simply waiting for things to get so bad that they would be able to "put more money in and take control" of the company,[222] and that an upturn in sales fortunes in late 1976 came just in time. He recalls Standard's Bill Hall, who was by that stage deputy chairman of the Capital board, coming to a meeting of Capital's finance committee to say the business can't go on any longer, and there needed to be a rights issue. But Hall was then shown the sales figures for the next 3 months, which had suddenly taken off, and was reduced to observing that "the cavalry just came over the hill".

The competitive spirit the Canadians engendered in their rival sales houses was damaging to ILR too, but overall they brought experience and professionalism. Without that, and their initial investment and later rescues, ILR's prospects in the awful economic circumstances of the mid-seventies would have been bleak. Gillian Reynolds, departed from Radio City to become the doyenne of radio critics, observed in August 1976 that "whatever the future of Canadian interests in British broadcasting, there is no doubt at all that their faith and confidence in commercial radio has so far been the saving of the system as a whole".[223]

Nevertheless, on 9 April 1976, close to four years from the date of Royal Assent to the Sound Broadcasting Act, and despite topsy-turvy political fortunes and all manner of practical obstacles to overcome, the IBA was able to announce the completion of ILR's first phase of development. "In the thirty-one months between the first company going on air in October 1973, and the opening of Beacon Radio in Wolverhampton, the Authority has, on average, brought a company on to air every six weeks. In this time, the total number of people in England, Scotland, Wales and Northern Ireland able to receive ILR has reached over 25 million on VHF, and probably over 30 million on medium wave at some times of day."[224] Benchmarking against the requirements of the white paper, the IBA confirmed that "it has ... been the Authority's policy to achieve the widest possible geographical spread of stations, offering experience of local radio in all major regions, and the first ever local radio stations in Scotland, Wales and Northern Ireland ... The nineteen ILR areas also include substantial rural coverage".

John Thompson had set out his stall; now he was pointing out that he had delivered what had been promised, more or less on the original terms. In a lecture in January 1976, he had allowed himself some rare publicly-expressed satisfaction. "On the independent broadcasting side, there is common ground for believing that one of the most fruitful areas for broadcasting development is for local radio throughout the UK on an independent and self-financing basis." For the benefit of the Annan Committee's deliberations he added that "at the IBA we now consider it highly desirable that there should be an expansion of the new radio system as soon as possible".[225]

The IBA had instituted the Radio Consultative Committee (RCC) in 1973, and this was at the heart of its regulatory engagement with the ILR companies, or 'our temporary contractors' as some mischievously used to dub them. The companies were required to meet together quarterly with the IBA's senior executives

222 Robert Stiby in an interview with the author, 21 January 2008.
223 *Daily Telegraph,* op. cit.
224 IBA News Release 9 April 1976.
225 John Thompson in the first of the 1976 IBA Lectures, 21 January 1976.

at the RCC, with the Radio Division staff in attendance. These highly formal meetings, chaired by the director general, were where policies were handed down and where the companies reported on their progress. They were grand and often rather solemn sessions, and you usually felt that you broke the solemnity at your peril. However, one RCC meeting took place on 21 July 1981, when Bob Willis was bowling out the Australian cricket team at Headingley, following Ian Botham's heroics with the bat. The then managing director of Beacon Radio, Peter Tomlinson, had sneaked a transistor radio into the RCC meeting. There were subdued cheers at the fall of each wicket, shared among the company people present, to the mystification of the 'top table'.[226]

At the first three meetings in 1973, and for a good time thereafter, the conduct was much more decorous. The IBA's Director General (Sir Brian Young was knighted in 1976), working from a detailed written brief, would run though a full agenda, supported by papers circulated in advance, and for two to three hours the companies would learn about how the IBA was to carry out its duties under the act. As each new company came on air, it would be invited to report on how things were going. That led to some useful sharing of information – and a bit of self-promotion, too. LBC's short-lived MD, Geoffrey Wansell, reported in January 1974 that "LBC had been faced with a tumultuous rush to get on air ... LBC had very nearly achieved their targets in advertising for the first months, but many costs were higher than anticipated and the immediate future seemed bleak".[227] At the same meeting, Jimmy Gordon said that "the audience reaction to Radio Clyde at the outset seemed to be good", but warned that "Clyde's experience was that all four of their cartridge machines had broken down in the first week".

The IBA was concerned to establish a body of regulations within which the ILR companies would operate, particularly – but not exclusively – concerning broadcast content, programming, advertising and promotions – and to do that at least partly in a consultative way, although still within a paternalistic relationship. A typical 1974 meeting covered IBA rules for programmes and advertisements with prizes; the Gaming Act and publicity for charities; the Code of Violence and the use of 'bad language' in radio programmes; standby generators; stereo transmissions; broadcasts by the Queen and other members of the royal family; conventions in station promotions; and a common year-end accounting date for financial accounts.[228] At times, the subjects and the language now seem rather ludicrous. The first paper I wrote for the RCC, as a member of the IBA staff, followed a complaint from the BBC about ILR advertising banners at football grounds. "The BBC expressed their concern because a sign was so positioned that the Match of the Day cameras were obliged to televise ILR station promotion. The Authority could not agree to the BBC's suggestion that LBC should be asked to keep the sign

226 Following that 1981 Ashes triumph, England captain Mike Brearley wrote thoughts on captaincy which were directly relevant to what was to become within a few years an anachronistic relationship between the IBA and its franchise operators. "Social changes ... have over the past fifteen or twenty years made the captain's job more, rather than less, difficult. Social hierarchies have become flatter; authority figures are taken for granted less and criticised more ... The aristocratic tyrant has given way to the collaborative foreman." (*Wisden Cricketers' Almanac* 1982 p 109) With only minor changes to make it apply to radio, that helps explain the frustrations which spilled out in the Heathrow Conference four years later (see chapter 10).

227 Radio Consultative Committee minutes 4(74) of a meeting on 9 January 1974 p 8.

228 Radio Minutes 5(74) of a meeting on 3 May 1974.

covered during any BBC transmissions, for in practice a contractual agreement had already been made for this season. Radio companies are asked to note that this is a sensitive subject, and if they are planning to purchase any similar advertising signs to discuss their plans first with Mr. Stoller."[229] I cannot recall that any did, and who can blame them?

An early concern had been how the very new, untried ILR newsrooms would cope with a bitterly contested general election on 28 February 1974, and the raft of legislation controlling broadcasting at election-times. Two papers were hurriedly produced late in January 1974, just ahead of the spring general election, covering balance and impartiality in programming, and the appearance of candidates in radio programmes at the time of parliamentary and local government election. By the time of the second general election that year in October, there had been relative leisure to provide direction and guidance on Party Political and Party Election Broadcasts. Section 9 of the Representation of the People Act presented a particular challenge to the fledgling newsrooms in the new stations, as it was then thought to make it compulsory that all candidates for a particular constituency must appear on air if any one of them did, during the pending period for an election. Although the interpretation was slightly modified later, this presented a headache in organisational terms for the stations, and also allowed one candidate to 'gag' all the others simply by refusing to appear. To guide the ILR news editors through the minefield – and the political parties were keenly aware of their rights – the IBA produced an algorithm which proved a superb guide.[230] It was an excellent example of regulation working well for the good of the stations, and offering the type of support which individually they would have found difficult to obtain.

During those years, almost every ILR station expected to broadcast discussion programmes with local candidates, and to run a full 'election night special', with live reports from the local counts and close engagement with politicians. Those offered a strong way of identifying with the locality, and of gaining credibility among local politicians; but they had to be got right. In 1974 ILR's future hung by a political thread. In the event – both events, February and October – the ILR stations did better than could possibly have been expected, given that some had been on air for barely a few weeks before the elections. Quietly, they were showing that there were to be a force be reckoned with, even in the unlikely area of political coverage, at least when there was a major occasion.

In its role as promoter and protector of ILR, the IBA had also been busy twisting the arms of the ITV companies to get some radio programme listings into *TV Times*, which was then the biggest selling magazine in the UK. The television companies were reluctant to promote their potential commercial rivals, but the closeness of the relationship between the IBA and its television contractors meant that the Authority could ask this sort of favour, which it thought would be a significant promotional tool for ILR. I was sent off to conduct 'shuttle diplomacy' between the Tottenham Court Road offices of a sympathetic *TV Times* and the suspicious radio companies – gaining from Brian Young the doubtful soubriquet of "our Henry Kissinger" – but it was a fruitless effort. *TV Times* confirmed to

229 Radio Paper 20(74) 7 October 1974.
230 Radio Paper 4(74) 24 January 1974.

AIRC on 6 November 1974 that it agreed in principle to including ILR programme details in the magazine, but the cost had risen from £50 to £100 per station per issue, and that gave the companies the excuse to refuse. There was quite a heated discussion at the RCC meeting that day,[231] but the companies decided to put the matter to an AIRC sub-committee as a way of setting it aside, and the opportunity was lost.

Independent Television Publications, the publishers of *TV Times*, launched *Radio Guide* with AIRC as a separate publication in May 1976. That followed the relevant AIRC committee's conclusion that "it seemed to ... all concerned that there was very little hope, at this stage at least, of the radio programme listings appearing in *TV Times* due to copyright problems and cost" and that this would be "the next best thing".[232] It did not last long. The forecast of monthly sales of 80,000–100,000 proved hugely over-optimistic, with only 20,000 sold in a good month. *Radio Guide* closed in June 1977, unlamented by the radio companies who had anyway given it very limited co-operation, and who were finding that gaining audiences was not a huge problem.

With the exception of the announcement on Thames TV and London Weekend of new London frequencies, the *TV Times* notion was the only time such close promotional co-operation between ITV and ILR was to be within reach. In later years, commercial radio was to complain bitterly about the BBC's exploitation of its ability to cross-promote between its television and radio services. Yet it had discarded its chance to set an equivalent precedent for ILR, at this very early stage. This was also the first instance where the radio companies, despite being "not unaware of the Authority's wishes in respect of *TV Times*",[233] decided to go down a different road.

As the ground rules were increasingly seen to be in place, the IBA began to drift into a more *dirigiste* approach than the legislation strictly required, partly out of concern for the political credibility of ILR, which seemed likely to be judged against the expectations created by the BBC. A paper was produced concerning "pronunciation on Independent Local Radio",[234] which recommended four standard books, plus grammars for Welsh and Irish Gaelic. A few incoming DJs, with perhaps 'mid-Atlantic' accents, certainly upset a few local people by mispronouncing place names, but they were unlikely to broadcast with a text-book at their elbow. Aimed at meeting the statutory requirements, but still "reluctant to lay down rules", was a note about how to establish education in programming, to allow the Authority to show how this aspect of the statutory expectations was being met.[235]

There was a separate Radio Technical Consultative Committee (RTCC), emphasising again the role of the Authority as the broadcaster and transmission provider. However, technical matters reached RCC early in 1976, with the implications of the international conference on medium wave usage persuading the IBA – which in turn sought to persuade the companies – that VHF would assume steadily greater importance as interference on medium wave intensified. The IBA

231 RCC Minutes 7(74) of a meeting on 6 November 1974 p 3.
232 Letter from Cecilia Garnett, AIRC, to Tony Stoller 17 February 1976.
233 Ibid.
234 Radio Paper 10(75) April 1975.
235 Radio Paper 11(75) April 1975.

was starting to have ambitions for ILR beyond mere survival by this stage. It sought to establish a tape archive for the independent stations, based on keeping 'special' recordings, plus a sample one-day-a-year of 'normal' output, and even tried to interest the ILR companies in quadraphonic sound broadcasts. For the first years, most of this future-gazing was consultative, and the companies felt able to decline some of the IBA's ideas, but the IBA's directly interventionist tendencies were to strengthen.

It had looked for quite a time as though even mere commercial survival might be beyond ILR, although that sense was distorted by the particular difficulties of the two London stations and their high profile. Broadcast advertising generally was hard to come by for television as well as radio. Even in television, the IBA in February 1975 had to reduce the cost of networked programmes to the smaller ITV companies. The year 1974 had been tough all round, although from the summer onwards there were some signs of improvement for ILR. By ILR's first anniversary in October, the *Financial Times* was concluding that "it could have been worse", while noting that for the advertising market as a whole there was "every likelihood that things will get worse in the next 12 months". [236]

Radio had not yet won over the major advertising agencies, in part because it took time to set up a full audience research system. ILR had attracted £3.3m in advertising revenue in its first year, enough to give Clyde and Piccadilly an operating profit after six months, but others were struggling. The best hope of catching the attention of advertisers and their agencies, pending full audience figures, came from the case studies. Fine Fare tested on Clyde, and then went national, Birds Eye continued to support the medium. The sales agency heads tried to sound positive. Eddie Blackwell at Air Services argued that "throughout the world commercial radio has always done better than the competition when the advertising industry was in recession".[237] Terry Bate at Broadcast Marketing Sales was outrageously over-claiming, on the basis that "in Canada radio accounts for 14 per cent of advertising expenditure, and in the US it is 10 per cent".[238] The stations had begun to make audience claims – 2 million for Capital, 1 million for Piccadilly – and, although these carried an insufficient sense of authenticity, they made the system start to seem more credible.

The *Financial Times'* gloomy anniversary forecast proved wide of the mark. After some grim early months, broadcast advertising had revived by the end of 1975, with the IBA reporting ITV revenues up by some 20 per cent.[239] The Authority was forecasting that ILR revenues might even exceed that growth rate for the coming year, to reach £10m compared with the take in 1975 of £7.5m. By ILR's second birthday, it was tempting to take at face value the *Economist's* claim that "outside London, commercial radio is a hit".[240] Even though the last three ILR companies were still struggling to raise their start-up money, Clyde had made a profit of £105,000 in its first eighteen months, and Piccadilly had lost only £30,000 in its first year compared with a forecast loss of nearly three times that amount. When Swansea Sound had offered extra £1 shares at £1.60, to raise a further

236 Anthony Thorncroft in the *Financial Times* 10 October 1974.

237 Ibid.

238 *Financial Times* 3 August 1974.

239 *IBA Annual Report* 1975-76 p 7.

240 *Economist* 26 July 1975.

£25,000, it had sold them within days. Clyde was claiming a weekly audience reach of 64 per cent, Hallam 48 per cent and BRMB 39 per cent. Radio advertising was still only at around 1 per cent of total advertising, but it looked to be "on a high growth track as more stations go on air and as the conservative advertising industry learns how to use the medium".

Outside London then, there were some modestly positive indicators by the start of 1976. After the intense gloom of the previous two years, it provided a heady combination of relief and optimism, which could be seen both in the programming output and in the improved political standing for the system. The flagship London stations continued to flounder, but at long last, in March 1975, the permanent medium wave transmitters started up at Saffron Green, in North London. LBC and Capital could start to promote their main services on 261 and 194 metres respectively. The temporary 'radio clothes-line' transmissions ran in tandem with the permanent services for a further two months.

At that time, Capital also re-thought its music policy again, offering more pop music and rather less high-minded talk. That was to bring down on to its head a slew of criticism, from external pressure groups – and, to a degree, from within the IBA too – but it was necessary if the audiences and the appeal to advertisers were to be boosted. The audience response was swift. In May, Capital was able to claim a weekly adult audience of just over 3 million.[241] It was probably from that point onwards that the company's fortunes started looking up. So swift was the change that, by the spring of 1976, Capital was starting those add-on ventures which were designed to reinforce Whitney's vision that "if you aren't listening to Capital, you aren't part of London life".[242] *Help a London Child*, the Wren Orchestra, Capital's *Jobfinder* and its *Helpline* services all began during that summer.

For LBC the problems were more deeply rooted, and since the weaknesses at Gough Square had a direct bearing on the Independent Radio News service for all the other ILR stations, the IBA could not escape close involvement. Labour relations were permanently bad. All the efforts of Selkirk's Bill Hutton kept striking the rocks of NUJ intransigence, and that also poisoned local relationships between the out-of-London ILR stations, and their own NUJ chapels. Ralph Bernard, later to be firmly on the management side, recalls that as 'father of the chapel' (FOC, effectively 'shop steward') at Radio Hallam, he came to London for national meetings. He was joined there by the FOC for Radio City, John Perkins, later to be managing editor of IRN. "It was all very militant ... LBC was bad, Trent was bad, Beacon was bad for labour relations."[243] Incremental pay scales in the national union agreements, plus 'cost of living' rises in highly inflationary times, made employing journalists prohibitively expensive for ILR.

The IBA had announced in March 1975 that it was to waive LBC's rental payments for the first nine months of the year, amounting to £168,000. It was presented as an "exceptional measure ... designed to aid the company in a situation where LBC has been attracting a growing and appreciative audience, and during a period when the company's trend of advertisement revenue shows a significant increase – but with costs continuing very greatly to exceed the sales income".[244]

241 *Who is Listening?* Tony Stoller in *Independent Broadcasting* August 1975.
242 Tim Blackmore in an interview with the author, 23 March 2006.
243 Ralph Bernard in an interview with the author, 24 January 2008.
244 IBA News Release 21 March 1975.

This came with a promised deal with the unions for cutbacks, and a retrenching of programming to concentrate on daytime output of rolling news, pulling back from 24 hour broadcasting, and with classical music and then phone-ins filling the evening hours. To get to that point, managing editor Marshall Stewart had to resign and then withdraw his resignation.[245]

Selkirk's Bill Hutton returned to his parent company in Canada at the end of June, although he remained on the LBC board. The challenge passed to Patrick Gallagher, LBC's marketing director and the only internal candidate. The May JICRAR research provided some comfort, showing a weekly audience of 1.6 million, up from 1.2 million in October 1974. The programme output at least was starting to attract positive attention. In January 1976, alert to the party political winds, Marshall Stewart arranged the first live radio phone-in for Conservative leader, Margaret Thatcher, who managed a gentle set of calls and the never-gentle Brian Hayes with ease. However, labour troubles were never far away for the nation as a whole, nor for LBC, and the station actually went off air for 28 hours in August 1976.

The individuality of the ILR music stations' output generally took a while to settle down during these first few years, although the concept of 'flow programming' rather than distinct programmes characterised the daytime output from the start. Also settled from the start was the contrast between daytime output – when the services were trying to be popular and populist – and the evenings, when they set out to meet their more formal obligations. However, ILR could not take the extreme version of this approach which BBC Radio One followed, since it was required to be a full service operation most of the time. Apart from in London, where the news and entertainment franchises were split, each ILR station had to provide something close to the full range of BBC services, tailored for its own locality, and with only commercial resources. The stations were rather quicker to appreciate the key importance of the on-air presenters than to understand the particular black arts of music selection and scheduling.

ILR programme controllers were reasonably at ease with populist approaches to the provision of news and local information, which was actually rather well done. BBC Local Radio had shown that there was a wealth of local content, but it was ILR which made it widely accessible to a mass popular audience, with a younger bias. This steady diet of news and information throughout daytime programming made the most significant speech contribution. Certainly, the stations made genuine efforts in documentary and discussion programmes, the latter much augmented by the free-ranging phone-ins largely pioneered by ILR, but the energy and resources needed to sustain these over a longer period were more needed in developing good, mixed daytime programming.

ILR's drama efforts were led by Alan Bleasdale's *Scully* on Radio City, some years before he found fame with *Boys from the Blackstuff.* These hour-long programmes featured Franny Scully, "an underachieving schoolboy from Liverpool who was obsessed by Liverpool Football Club".[246] Kathy Barham's history of Radio City recalls that "on 2 January 1978, the character Scully graduated to BBC1, when 'Scully's New Year's Eve' was broadcast as part of the *Play for Today* series".

245 *Guardian* 21 March 1975.

246 Kathy Barham, *Radio City; the Heart of Liverpool*, Lulu.com, 2006.

Programme director Gillian Reynolds paid Bleasdale £75 to write, present and permit one repeat of Scully each week. Philip Shakeshaft describes also how Roger Harvey at Metro adapted a number of pieces for radio, working with Peter Wheeler and Geoffrey Freshwater of the RSC, and with Edward Wilson, who later went on to found the National Youth Theatre. Metro Radio productions continued until the later eighties, "when it was dissolved along with all other [structured] speech content in a re-branding exercise".[247]

The general approach of the stations also lent itself splendidly to charitable and social initiatives. Capital Radio's launch in Easter 1976 of *Help a London Child* represented the first occasion when the American-style charity marathon was deployed in the UK. With this innovation, ILR paved the way for the huge 'telethons' of later years. In what was then a revolutionary format, although it now seems commonplace, listeners were invited to pledge money by telephone for a particular record to be played; for the services of a DJ to wash their car; or take part in an auction. £12,000 was pledged, although in a reversal of what was to become the norm, less was actually received. Metro ran a similar event over the August bank holiday weekend that year, and a new British broadcasting staple was quickly established. The IBA moved to put in place rules for how these might be done, needing oversight from its Charitable Appeals Advisory Committee, but these were mostly practical and helpful protections against abuse or allegations of abuse.

In keeping with the mood of those years, all the ILR stations paid more than lip service to social action through radio. Phone-in programmes dealing with personal matters often needed off-air follow-up. Programmes such as Alan Nin's *Open Line* on BRMB established listeners' clubs with professional facilitation, or linked in with existing services. Another type of social action might be undertaken where a station perceived a need and addressed it directly. Radio City, for example, ran a regular *Job Spot* in its breakfast show, dealing with employment for teenagers. ILR stations were also active in linking those who needed help with those who offered it, such as local social agencies, through signposting both on and off air. Capital's *Helpline*, enjoying resources unavailable to smaller stations, was an outstanding example, but there were many other effective – if more modest – efforts across the ILR system. The IBA and the industry were turning their minds to making the case to the Annan Committee and the government for the continuation and expansion of ILR. Initiatives such as the on-air appeals, social action projects and the general impact of the stations in their local areas, all helped the cause, but it would be wrong to conclude that they were undertaken solely or even mainly as a cosmetic exercise. This was the type of radio that most of those who were attracted to work in ILR stations wanted to provide. It gave them a real buzz, and made the privations worthwhile.

The prudent companies began with no more than the level of programming ambition which they could reasonably sustain on start-up revenues. None of them was entirely cynical about seeking to discard their public service obligations – nor would the IBA, with its prior approval of all schedules, have permitted that – but in the harsh economic circumstances it made sense to acquire wealth before practising overmuch virtue, in a phrase much copied from the Annan Report.[248]

247 Philip Shakeshaft, *Independent Local Radio Drama, 1973-1990*, unpublished dissertation, 2008.
248 *Report of the Committee on the Future of Broadcasting* , March 1977, Cmnd 6753, p 158.

This was what the *Guardian* astutely called "the central dilemma of commercial radio – whether it is possible to reconcile the element of public service ... with the relentless pressure of the market place".[249] Keith Skues at Radio Hallam was clear that it all depended on whether the balanced content approach delivered audiences in sufficient numbers. "When we have got the listening figures we will be able to afford to experiment. If [balanced programming] fails then I will scrap it and go all out for a top 40 format." It is significant that – whether out of idealism or because of the regulatory regime – it was to be the public service path that was attempted first, and by and large this worked well for the ILR stations, once they began to manage their music policies competently.

Music remained the enigma for many of those involved with the start of ILR. Few of the hopeful groups had gone into any detail in their applications about their music policies, although popular music was to be the staple element of ILR's output. The Authority felt it had to work constantly to counter suggestions that ILR was merely a 'radio juke-box', so when it turned its attention to music it was normally to reassure itself that independent radio was far removed from the one-track sounds of commercial and pirate radio. The one small paragraph on the subject in the IBA's first annual report stressed that "most of the stations cover a range that includes many types of pop, light instrumental, classical, jazz, blues, country and western, folk, soul, reggae and – a distinctive feature in the North – brass bands".[250] It praised relays by Radio Clyde of the Cleveland Quartet and the Scottish Proms, and *lieder* recitals on Radio City.

Certainly, significant live music promotions were a valuable addition to the output, and the promotional portfolio. Apart from the classical concerts which the IBA always liked to put first on any list – such as broadcasts by BRMB of the City of Birmingham Symphony Orchestra, and by Radio City of the Royal Liverpool Philharmonic – the 1975 lists included Linda Lewis, George Melly, Elton John and George Morrison on Capital; folk, jazz and brass band concerts on Piccadilly; traditional Welsh music on Swansea Sound; Gallagher and Lyle, Ralph McTell, Seals and Croft, and Tangerine Dream on Clyde; and Magna Carta, Sassafras and Soft Machine on Radio Trent.[251] Such music promotion was an effective way of 're-cycling' the 3 per cent of revenue which had to be spent on the employment of musicians, and also gave stations the opportunity to be patrons of music in a way that might occasionally challenge the BBC. Radio 210, made, broadcast, and offered to the European Broadcasting Union network the first live concert recording of the Eurythmics. Given our deal with the local Hexagon Theatre in Reading, it was a very profitable night out, as well as making one small ILR station very briefly a European music pioneer. For Capital Radio, by way of contrast, its increasing prominence cut little ice with Herbert von Karajan, conducting the Berlin Philharmonic at London's Royal Festival Hall, promoted as part of the station's Great Orchestras of the World series. At the interval, he demanded half his fee in cash before he would continue, and John Whitney had to take it backstage in a briefcase.[252]

However, throughout the history of radio (with the possible exception of AM

249 *Guardian* 22 January 1975.
250 *IBA Annual Report 1974-5* p 38.
251 *IBA Annual Report 1975-6* p 35.
252 Quoted in Sean Street, *A Concise History of British Radio*, Kelly Publications, 2002 p 123.

popular speech radio in the US), popular music has been absolutely central to the success of popular radio. First of all, it has been quite literally the medium within which the service existed: the music is the medium is the message. Second, it is in the language of popular music that radio presenters engage with their listeners, that marketeers sell radio, and that the audience thinks about radio. Third, music is the true profession of popular radio; working in it, you have to be immersed in popular music, indeed synonymous with it. And fourth, popular music has enabled popular radio to brand itself, offering its listeners peer-group identity which has been central to the marketing of radio almost from the beginning.

Given its crucial importance, it was wholly counter-intuitive for popular music to have been the denied virtue of ILR – to be excused away, or hidden behind the prestige projects of symphony concerts – yet that was what the political realities of the seventies were thought to demand. For ILR, there was an overwhelming need for each station to establish a consistent station sound, and the one major lesson to be learned from the offshore pirates was that this must come in significant part from the daytime music policy; yet the stations were encouraged to forge their identities through their more specialist music output, with pop music just the background accompaniment. Even when the radio people in the Authority slowly came to understand that "music is the programming base … the way the music is broadcast by each station contributes to the individual sound and distinctive style",[253] and tried to help, it was with a very partial understanding. The IBA's questions, following its conference-style 'Consultation' with the companies on 24 and 25 March 1976, were rather earnest and didactic: How can music output best be given local emphasis? How and to what extent should rock be separated from folk, and jazz from classical music, or how far should they be mixed together?

As some of the programmers realised from the outset of ILR, but were rarely encouraged to express then – and as the contemporary techniques of computer-playlisting also often fail to recognise – music on the radio is something else. In 1973, as ILR was starting, Richard Carpenter wrote *Yesterday Once More* for Karen to sing. That was about personal identity, not a political debating-point or a marketing construct. Good popular radio offers the music which helps listeners to locate themselves in time: the feelings, the memories, the sounds of their lives. Music on the radio provides a roadmap for listeners' lives; it supplies a vocabulary of the day-to-day, and even helps reconcile listeners to their times and circumstances. Once the ILR stations could offer that, along with their local output and specialist programmes too, they had found the secret of popular success. But it is clear now, that all *three* elements were necessary conditions of that success. Once commercial radio regressed to providing only the music, it could all too easily be overtaken by the arrival of MP3 technology, leaving nowhere for the iPod generation to locate their yesterdays.

253 *IBA Annual Report 1975-56* p 34.

Chapter 6

Is there anybody there?

Audience research

The loneliness of the late night radio presenter is one of the enduring images of radio. You sit at your control desk in the studio, just you and – depending on the era – your turntables or CD players or hard disc music playout system, talking into the microphone, with perhaps no one else even in the building. However self-sustaining your ego may be, there are moments when you ask yourself, "Is there anybody there?". You can invite callers or dedications, and hope that the lights go up on the telephone control unit. What you really want, though, is to be re-assured that you are not just talking to yourself. More to the point, those managing the programme output need to know what is attracting audiences and what is not, and to build and change their programme schedules to optimise those audiences.

For the radio businesses and their commercial management, audience research is even more crucial. You might, just might, be able to programme a station on gut feeling only. In order to get advertising revenue, however – at least after any initial post-launch goodwill has been spent – you have to demonstrate a sizeable audience, or the delivery of difficult-to-reach demographic groups. That must be proven by credible research, and identified in proper demographic detail to meet the *soi-disant* scientific approach of those buying media advertising. The higher up the food chain of media-buying is the person you aspire to address, the more sophisticated the research you will need. Once into the offices of major national advertising agencies, you had better have a sales pitch impressively garlanded with research data, tools and analysis.

It had taken the BBC into its second decade to contemplate audience research, and then only when driven to it by the successes of pre-war commercial radio.[254] The BBC had no audience research function until 1936, when Robert Silvey arrived from the statistical department of the advertising agency, the London Press Exchange. Reith's determined reliance on the public service ethic, untainted by the opinions of the common man, had held sway.[255] Even by the seventies, and despite the impact of the offshore pirate stations, much of BBC Radio's research effort went into Appreciation Indexes (AIs) – for which they maintained panels of self-selecting BBC radio fans – rather than into counting numbers of listeners, which was thought a very inexact science.

254 See chapter 1.
255 Sean Street, *Crossing the Ether,* p 24.

For the independent stations, that science had to be mastered. Early on, several stations used individual 'dipstick' surveys, relying on face-to-face questioning of a random sample, and then their own diary-based research. Such *ad hoc* surveys were not enough for the advertising industry, which was used to television measurement based on meters installed in people's homes. Even newspapers had a convincing audit of their sales, carried out by the Audit Bureau of Circulations, before the later 'massaging' of the figures through extensive free and discounted copies undermined that. The press audit was backed up by a separate National Readership Survey, allowing newspapers to claim a certain number of readers per copy sold. The ILR companies knew from early on that they needed a system which was at least as robust as those of their competitors for national advertising, and that it had also to be a joint, industry-wide scheme. The early appointment by AIRC of respected media research consultant Tony Twyman[256] was evidence of their determination to get this right.

AIRC brought together representatives of the advertisers through the Incorporated Society of British Advertisers (ISBA), the advertising agencies through the Institute of Practitioners (IPA) and the companies themselves, to form, in 1974 the Joint Industry Committee for Radio Audience Research (JICRAR). The ISBA and the IPA were supportive, and without that support this important plank of the ILR platform could not have been put into place so effectively. JICRAR was the equivalent of television's JICTAR, although it is of note that the 'A' in the television body stands for 'advertising' not 'audience'. From the start, the radio companies wanted programming information as well as commercial data, in return for their very substantial investment. The cost to radio of conducting proper research was very high, and proportionately far more than for television or the press. Even in the eighties, ILR was paying six times more for its audience measurement research as a proportion of its revenues than was ITV.[257]

The early radio companies saw no reason to extend the research community to include the BBC. Deanna Hallett, who was to lead radio research first from within National Opinion Polls Ltd (NOP), and then from within the sales agency Air Services (both owned by Associated Newspapers), recalls that "we didn't actively try to co-operate with the BBC because we were surveying one bit of geography, and they were surveying another".[258] The early ILR stations were effectively unconcerned with the BBC as a competitor, either for revenue or audiences at that time, even though the readiness of the BBC to rubbish any figures produced by ILR was an irritation. The BBC "slipped out" audience research figures in November 1973 which indicated minimal audiences for both Capital and LBC,[259] even though by its own admission, it was only three years later that "the revisions to the suite of computer programmes needed to incorporate ILR listening" were completed.[260] Matters threatened for a while to get out of hand. When, in September 1974, the BBC released audience data damaging to ILR, Jimmy Gordon wrote to the *Times*,[261] challenging the BBC to substantiate their

256 See chapter 4.
257 Deanna Hallett in an interview with the author, 5 September 2007.
258 Ibid.
259 *Financial Times* 24 November 1973.
260 *BBC Report and Handbook 1977.*
261 *The Times* 5 November 1974.

claims. Brian Emmett, head of BBC Audience Research, replied by wondering, "How is a 'listener' to be defined?". [262] AIRC briefly considered legal action, but wisely withdrew.

In contrast with the rather dirty war being fought by some of the foot-soldiers, the people designing JICRAR, led by Twyman and Hallett, went about their work with a high seriousness. They were determined not only to meet the commercial needs of their industries, but also to create a robust, respectable and innovative research programme. Two preconditions had to be settled before JICRAR could get going. The first was the methodology to be used, and it was a North American study which was to be the defining factor here. In 1965, the All Radio Methodology Study (ARMS) had researched very extensively the different ways of recording radio listening, comparing each with actual listening checked by telephone calls, and had shown widely differing results for the various techniques. It concluded that self-completion diaries, personally placed with a random probability sample of people in the station's area, produced results closest to the median, and it was this technique which was adopted by JICRAR. People, chosen at random, were asked to record their listening, quarter-hour by quarter-hour in pocket-sized, diaries. The industry standard required that there should be a JICRAR-controlled study in each ILR area at least once a year, in the spring, thereby producing directly comparable trend data, in addition to the 'absolute' listening information.

The second critical issue was the geographic area to be surveyed. There was the formal franchise area, defined by the IBA in accordance with a VHF signal strength of 1mV/m, and a daytime medium wave signal strength of 3 mV/m; yet everyone acknowledged that radio signals do not just stop at a technical boundary, and that there was a great deal of listening going on beyond that so-called 'VHF area'. It is one of the features of radio audiences that if they want to listen to a radio station, they will do so even when the quality of signal is well below that which engineering purity considers acceptable. It was, however, a formal contractual requirement that the companies needed the approval of the IBA to be able to claim the wider and larger audiences they knew they were getting.[263]

After discussion at the October 1974 RCC,[264] the Authority came up with the notion of a 'total survey area' (TSA). Provided stations could demonstrate at least a 10 per cent weekly reach, they were able to incorporate these localities outside their VHF contract areas, for research and marketing purposes. The Authority was concerned to avoid "the growing danger of confusion with regard, for example, to the possible overlap of coverage, coupled with some rather ambitious claims by one or two companies".[265] The outcome was a TSA which, "while it may not enjoy constant signal reception to the best standard, nonetheless has enough listeners to be considered within a major audience research study".[266] I remember that this formulation, my own, had been necessary to placate the more formally-inclined engineers and officials within the IBA, whose television experience inclined them to be less flexible. Even so, the IBA remained a stern taskmaster. Deanna Hallett recalls "poring over maps with Colin Day of Capital Radio in the NOP offices

262 *The Times* 6 November 1974.

263 ILR standard contract Part IV para 13(8).

264 *Radio Paper 23(74)* October 1974.

265 *Radio Paper 3(75)* January 1975.

266 *Who is Listening* Tony Stoller in *Independent Broadcasting* August 1975.

saying, 'well let's try another little go'. We did it in three goes, pushing the TSA out and pushing it out, proving that we had at least a 10 per cent reach in each band, until we determined Capital's current total survey area."[267]

By the spring of 1975, the full JICRAR specification was agreed and ready to roll, although the first surveys under the auspices of JICRAR had been conducted in October 1974. That was none too soon. The sniping between the BBC and ILR had largely devalued earlier dipstick studies – and there had been some self-inflicted damage, too. Rumours abounded that Capital had some earlier dipstick research which it had declined to publish, until the audience had started to build properly. The management were said to have been urged to "go away and come back with something better".[268] LBC had released an NOP diary dipstick from November, claiming for itself a million listeners[269], but then Capital published its own NOP study in January 1974 showing a weekly audience for itself of 1 million, and for LBC of 448,000. This was potentially very damaging to ILR, and it was with relief that AIRC announced in May that it had agreed with ISBA and IPA to adopt a common and verifiable standard. The October surveys showed weekly adult audiences for Capital Radio of 2,040,000, and for LBC of 1,198,000, impressive by the standards of commercial radio internationally, and especially so in the debilitating context of ILR's troubled start.[270]

By August 1975, the IBA was prepared to allow its radio staff to make it clear that "Radio is about audiences ... not necessarily the largest audiences, for that depends on the type of radio service being broadcast ... even so, if a radio service is to be convincing in its claims that it is doing a worthwhile job, it needs to show that people are listening. This is radio's acid test." When I wrote that, it was celebrating the vindication of ILR provided by the May 1975 JICRAR surveys of LBC, Capital, Clyde, BRMB, Metro, Hallam, City and Forth. Together with a JICRAR study of Piccadilly in November 1974, and a dipstick survey of Swansea Sound, there was good evidence that ILR had already attracted a weekly audience of some 10 million adults, listening on average for a substantial 10-12 hours each week.

In Deanna Hallett's words, the research showed "amazingly high weekly reaches".[271] Clyde was shown to be listened to by two-thirds of the adults in its area, Swansea by an even higher percentage (although by a less reliable survey).

267 Deanna Hallett interview op. cit.

268 Email from Deanna Hallett to Tony Stoller, 7 January 2009.

269 *Adweek* 17 May 1974.

270 There are three different key measures of radio audiences: 'reach', 'hours listened', and 'share'. *Reach* represents the number of different people listening to a radio station for any length of time above five minutes in any quarter hour at any time during the designated measurement period. It is usually expressed as a percentage for ease of comparison, and for JICRAR and subsequently RAJAR research the period is one week – thus stations refer to their 'weekly reach' as being X per cent of the adult population in the areas surveyed. *Hours listened* is simply what it says, the number of hours listened to a station in the research period, normally a week, and in the JICRAR methodology obtained by aggregating quarter hours marked on the diary. *Share* is total hours expressed as a percentage of all radio hours. Typically, ILR stations were keen to demonstrate through their reach figures that advertising can be heard by large numbers of local people, although as commercial radio markets have become more competitive, share has become the most important tool when selling to national advertisers. The BBC has tended to talk more about share of listening, although that seems to militate against their self-stated objective of providing something for all licence fee-payers, which is better shown by the reach figure.

271 Deanna Hallet interview op. cit.

Hallam, Forth and City were the second most popular stations in their areas, and Piccadilly and BRMB both had large major-city audiences, with only Metro seemingly lagging behind. Capital had increased its audience since the previous autumn to exceed 3 million, and with its permanent transmitter now in operation its future was close to being assured. LBC's 1.6 million listeners belied its continuing commercial and labour relations woes. Even though these stations were now clearly seen as something other than the natural successors to the offshore pirates – or perhaps in part because of that – they had begun to make a lasting impact on audiences.

This was a huge achievement, especially when set in the context of the BBC's 50 year monopoly of domestic radio. Perhaps it should not have come as a surprise, given the appeal of non-domestic commercial radio in earlier years. Yet the audience levels for pre-war commercial radio had been largely air-brushed out of the popular memory, and the offshore pirates were ephemera whose claims were easily discounted. ILR, for all the troubles which attended its launch, had passed the acid test. It had shown that it could attract mass audiences, notwithstanding the limits on its popularity, supposedly imposed by the public service obligations of the statutory and regulatory regime, and the limits on its scope to provide popular music.

Two years on, subsequent JICRAR studies confirmed these first results.[272] Swansea and Clyde both achieved a weekly reach of 63 per cent, with small newcomer Orwell claiming 71 per cent from an early dipstick study. Audience research was not at all easy to conduct house-to-house in Northern Ireland in the late seventies, so there was some uncertainty at the time over Downtown's notable 68 per cent dipstick reach, but this was proved to be wholly credible by later research. Metro had recovered to register a 44 per cent reach. Across ILR areas as a whole, TSAs being now the standard reporting area, 43 per cent of all adults were listening to their local ILR station each week. In London too, despite the strength of the BBC's national services, Capital and LBC between them also reached 43 per cent of adults. This was remarkable audience penetration, confirming that these radio stations were major arteries for the information and entertainment which sustained local life.

The independent companies now also began to talk in terms of 'share' of audience, asserting that this showed increased listener loyalty, even in areas where a very high initial 'reach' was unlikely to be improved upon very much. In Glasgow, the share figure was an extraordinary 50 per cent, so that half of all radio listening in the city and around was to Radio Clyde. In London, the two ILRs held a 26 per cent share, a figure virtually matched in all the major cities where ILR stations broadcast. The share comparisons in the JICRAR research showed ILR at 25.2 per cent and BBC local radio at just 5.8 per cent, in areas where both operated (based on the ILR TSA only), suggesting that BBC local radio, except perhaps in its further reaches of Cumbria and the South West, was struggling to make real impact. And there were no BBC local radio stations in Scotland, Wales or Northern Ireland at that time, where ILR was achieving its highest reach scores. For all its cult status among a few fans, BBC Radio London had never won much of an audience, so

272 *As Millions Hear* Tony Stoller in *Independent Broadcasting* April 1977, with apologies for the 'Ealing Studios' style title … .

that Capital and LBC had taken their quarter of all radio listening from the BBC in its network heartland.

Once gained, the ILR stations largely held on to these sorts of listening levels for the next 10 years, until the arrival of competition from 'incremental' stations within their individual areas began to cannibalise audiences, although for a number of years the extra competition seemed to be adding inexorably to the total share won by the private sector services. The first nineteen ILR stations gained a combined weekly reach of 52 per cent in their areas in the spring of 1980. When new stations came on air, after the removal of the freeze to nineteen stations, they typically replicated the achievements of their predecessors in quickly building large mass audiences. By the time of the JICRAR survey in the spring of 1982, with 34 stations then on air, the ILR weekly adult audience had grown to over 17 million. That represented a share of around a third of radio listening across the UK in ILR areas, which by then covered three quarters of the UK population.[273] By 1987, as the expansion of ILR gathered pace, and as new stations became established, the total UK audience was as high as 18 million adults each week. The final survey under the old JICRAR specification, in the second quarter of 1992, showed a weekly adult audience for ILR of 22.5 million.[274] By then, JICRAR was surveying a universe of 43 million adults, showing how wide was ILR coverage of the UK, and the network had settled down for a while to a share of between 36 and 38 per cent of radio listening.

The make-up of the audiences was broader than a merely popular music service might have attracted. In the early stages, ILR was listened to within local areas by an audience which quite closely mirrored the demographics of each local area, except among those above 55 years old. In 1975, the IBA could boast that no station was drawing more than one third of its audience from the 15-24 years age group.[275] By 1977, that pattern was confirmed, together with a slight bias towards the lower socio-economic groups, but still the Authority could vaunt over 3 million listeners to ILR aged over 55.[276] At this point, with the strength of ILR audiences now established, and the JICRAR research methodology also thought respectable, the IBA was able to say publicly of the fairly regular disagreements between the BBC and ILR on audience figures that "public squabbling is unhelpful ... the situation shows that a system of centralised audience measurement would be appropriate". There is a hint of 'nanny knows best' in this comment (mine, as it happens), as in quite a bit of the IBA's regulation of ILR, but it was correct and forward looking. The IBA had argued for a common audience research system in its evidence to the Annan Committee in September 1974, but it was not to come about for radio until the arrival of RAJAR in 1992.

The audience research conducted by the companies was primarily concerned with counting the number of listeners, and identifying the demographic make-up of audiences. Attitudinal research was a different matter. Companies applying for ILR franchises were expected to undertake some research into what people might want to hear, but had little cash to spare for such research once they began broadcasting. However, the IBA had the funds, the resources and the inclination

273 *JICRAR* network survey spring 1982.
274 *JICRAR* network survey April–June 1982.
275 *Who is listening?* op. cit.
276 *As millions hear* op. cit.

to fill that gap. Audience research occupied a substantial department within the IBA at Brompton Road, and studies of public attitudes to television were staple fare. While radio was rather below the radar for the head of research, Ian Haldane, his deputy – the characterful Mallory Wober – showed interest, and this was taken up in particular by David Vick when he moved across to join the Radio Division staff in March 1977. Insofar as anyone in that small team had a specific task, Vick's was audience research, and he set about it with vigour.

Throughout the rest of the seventies, the IBA undertook two different types of attitudinal research. Each of the first tranche of ILR stations was treated to a fairly standard study into 'audience attitudes and patterns of listening', a series first conducted mainly between July 1975 and November 1979. A further group of studies, beginning with Cardiff in 1982, covered newer services, ending with Signal Radio in February 1988. By then, the imminent new dispensation for regulation was clearly going to preclude this type of co-operative work. These studies were in the main "gratefully received"[277] by the companies, at least up to the point where the relationship changed, and they filled a gap in the knowledge ILR had about its audiences.[278]

Of more general interest, and with relevance both to the way in which the IBA set about its regulatory tasks and to how the ILR companies tried to operate their franchises, was a series of attitudinal studies across the network as a whole, or into individual projects. Many of these were conducted to inform the annual get-togethers of ILR programme controllers under the auspices of the IBA. These looked at, for example, "the need for speech",[279] perceptions of ILR news output,[280] and attitudes to music[281]. The results were rarely dramatic, routinely reporting the value which listeners placed upon local news and information, and their openness to a range of content within programming. A more wide-ranging study in September 1980 showed that ILR listeners were distinguished from other radio listeners by the importance they gave to local news. 87 per cent cited this as a major reason for listening, compared with 69 per cent for all radio listeners. ILR's outstanding characteristic was perceived in this study as "best at reflecting the interests of people in this local area".[282] That unique selling point sustained independent radio through all the ups and downs of the seventies and the eighties. This research could have told them the risk of discarding genuine localness, which in the late nineties would start to destroy that unique point of appeal.

The IBA was building up a body of knowledge, specifically tuned to its own concept of *independent* radio, which was to a degree self-validating, but – as the example of local relevance shows – had high applicability then and later. The attitudinal research provided an academically sound basis on which to promote ILR as genuine public service broadcasting, and as a medium breaking new ground. It also distinguished ILR from ITV. Research the following year into how people

277 David Vick in an interview with the author, 24 June 2008 .

278 *Audience attitudes and patterns of listening* A series of 39 studies of the same title, or similar (*Audiences reactions to radio XXX)* conducted and published by the IBA between July 1975 and February 1988

279 *The Need for Speech*, A study of listeners' attitudes, IBA May 1977.

280 *Good News?* Perceptions and importance of ILR news, IBA March 1978.

281 *An in depth qualitative study of listeners attitudes to music on radio*, The Schlackman Research Organisation April 1979.

282 *The public impression of ILR* IBA December 1980.

used radio produced the key conclusion that "the sense of personal involvement and loyalty that many people seem to feel for their favourite radio station may become increasingly evident as a strength unique to radio among the broadcast media. Already the choice of television channel to watch is determined by decisions made and revised from hour to hour, and programme by programme ... With the fragmentation of the evening television audience, the distinctive role played by radio should become more prominent."[283]

Research into the new Basingstoke/Andover service, which was added to Radio 210 in January 1987, had particular significance. Unlike all other additions up to that time, this new service was broadcast on VHF/FM only. It was possible, therefore, to examine how listeners reacted to a single waveband offering. Many suspected that the listening patterns were changing. Basingstoke provided the opportunity in July 1987 to research how real that change was. The results showed a level of FM-only listening which was "phenomenal".[284] The immediate effect was to persuade the IBA's Radio Division to press ahead with new FM-only licensing for areas such as Oxford, Yeovil and Dumfries, where a matching AM service was impractical. It also opened the road further towards splitting frequencies for existing contractors.[285]

Such commissioned research continued to be a valued feature of the IBA's stewardship of ILR. [286] It unspectacularly served to inform regulatory decisions, and to boost the sector's knowledge of itself. It is therefore all the more surprising that the Shadow Radio Authority (SRA) specifically set its face against doing any such research. True, the regime it was asked to implement was intended to be 'lighter touch' than its predecessor, and it could no longer draw upon hidden cross-financing from television to fund a research infrastructure. Yet the SRA went so far as to request an amendment to the Broadcasting Bill to try and prevent a duty to conduct research. This amazed former IBA chairman George Thomson. "I am surprised to hear what the minister has said, and to hear that the provision is at the request of the new Radio Authority ... How on earth can a government ... take away the duty to carry out research into the effects of programmes on listeners? The effect of broadcasting is a fundamental issue, particularly as regards television, but it is equally true of radio. I am astonished that the government have *(sic)* rolled over and given in to the Radio Authority."[287] The amendment failed, and what became section 96 of the new Broadcasting Act required the Radio Authority to undertake audience research to "ascertain the state of public opinion" about the services it licensed.

The prime mover in the attempt had been David Vick, who was in an even more prominent position within the new organisation regarding research. In the IBA's Radio Division, attitudinal research had been his especial care but he now

283 *The use of radio – a research perspective* David Vick in *Independent Broadcasting* October 1982 reporting on IBA qualitative research findings and an analysis of JICRAR commissioned by the IBA from ASKE Research.

284 David Vick interview, op. cit.

285 See chapter 12.

286 The level of the IBA's commitment to research in the eighties is illustrated by its funding of a Radio Academy survey listing and providing summaries of BBC, IBA, other UK and international research into radio between 1975 and 1985. Published in September 1986, Josephine Langham's *Radio Research: the Comprehensive Guide* lists 703 pieces of work.

287 *Hansard* House of Lords 24 July 1990 col 1427.

took the view for the Radio Authority that any such requirement was "inconsistent with the general thrust of the legislation, and at odds with the Thatcherite zeitgeist (of which I was a firm advocate) which was to devolve responsibility for such matters on to 'the market': i.e. the companies themselves".[288] The opposite view was that the regulator should research each licence area before advertising it, to identify among other things which format would be best. However, against the threat that "if the service provided by the successful licence applicant subsequently failed, it would be the Authority that attracted much of the blame, for having proposed the wrong type of service in the first place" the SRA members accepted Vick's advice. "Ascertaining the state of public opinion" was to be achieved by careful analysis of the JICRAR (subsequently RAJAR) research, and examining the studies which applicants themselves would be conducting and funding. This freed the Authority from arguments about whether its own research should have primacy over that done by the companies. However, the lack of more general research weakened the regulator, both in terms of its actual and of its perceived levels of knowledge.

Meanwhile, the radio industry was gearing itself up for a major change in the way it counted and reported on its audience. The bad-mouthing between the BBC and AIRC on research figures had largely disappeared by the start of the nineties, replaced by an awareness on the part of the Corporation that it needed authentic audience data to defend its position in the upcoming Charter review. BBC Radio's Jenny Abramsky also saw such data as essential in her strenuous efforts within the BBC to protect and progress radio. AIRC was about to be extended by Independent National Radio (INR), and needed a wider survey base. If that could be achieved in a joint venture with the BBC, the credibility of the audience data among advertisers – the essential 'currency' of commercial radio – would be further enhanced. Both sides had every interest, therefore, in agreeing to move towards a new, industry research standard.

The new Radio Industry Joint Audience Research (RAJAR) specification came into operation for the final quarter of 1992. This used an improved version of the seven-day radio listening diary pioneered by JICRAR. That not only provided a degree of continuity and familiarity for ILR and its customers, but implicitly recognised its practical superiority over the aided-recall technique deployed up to this point for BBC audience measurement. It continued unchanged until 1999, when a revised specification was agreed. RAJAR was designed expressly to permit "the publication of results for the BBC and Commercial Radio national and regional services (with adult populations of 4 million or more) on a quarterly basis. Results for most local radio services, both Commercial and BBC, were published for each Quarter 2 and Quarter 4. Audiences for the smallest stations (with an adult population of under 300,000) were measured once a year, in Quarter 2."[289] In practice, commercial radio now had an enhanced currency to deal with advertising agencies, and the larger stations had some detailed data on which to base programming decisions as well. The smaller stations gained only a broad guide to what programmes were working or failing, but they still enjoyed the commercial currency at an almost affordable cost.

288 David Vick in a note to the author, 8 July 2008.
289 www.rajar.co.uk

That does not mean that RAJAR was cheap. Commercial radio paid the lion's share of the RAJAR costs; the BBC was always ready to plead poverty, with the implied threat that it would withdrawn the prestige it brought by being part of RAJAR. Deanna Hallet reflects that, "the relationships with the BBC were good, although there was a sense with hindsight that AIRC might have got a better deal. I don't think we were tough enough with the BBC financially."[290] All in all, however, it was a good deal all round, and was one of the key elements in radio's commercial golden years, which were to arrive in the mid-nineties.

RAJAR was top rate. David Vick, never one to offer easy praise to AIRC/CRCA, regards RAJAR as being "very, very high quality survey research". Its reliability was strikingly good. "Whenever RAJAR got anything wrong there was always a big kerfuffle, but actually the error rate was miniscule." [291] As well as providing the basis on which advertisers could buy radio, it at last brought the BBC into a common system. For a while, from the radio companies' point of view, it had the crowning advantage of demonstrating the audience growth of commercial radio in surveys also endorsed by the BBC.

ILR's audiences rose to their peak in the late nineties, and then began their sharp, seemingly inexorable decline. One less attractive immediate outcome of the new RAJAR specification was that it reported a lower nationwide weekly audience for ILR than JICRAR – 21.1 million as against 22.5 million, comparing the second quarters of 1992 and 1993. Neither was right or wrong. Both surveys used self-completion diaries, but different research methodologies produce different results. The crucial thing was to use a consistent methodology going forward, and not to chop and change, and above all not to devalue the 'currency' by criticising the research in any public way. On most of this both the BBC and the commercial companies proved reasonably well-behaved, although there were occasions when disappointed companies would blazon in the trade press that the researchers must have got it wrong. Over the years, the research contractors – RSGB, then Ipsos-RSL, and now Ipsos-MORI and RSMB – were careful and capable, with Roger Gane in particular an outstanding lead researcher for RSGB and Ipsos-MORI.

RAJAR also provided reliable data about in car listening. The 2000 research, for instance, showed that, in the course of a week, 45 per cent of the adult population (21.3 million people) listened to radio in their cars,[292] with in-car listening accounting for nearly 15 per cent of all listening hours. At that point, the audiences for BBC and Commercial Radio were roughly the same size, but in the car it was the commercial services that dominated, with 53 per cent of all such listening. For the age group 15-44, that share rose to fully 60 per cent, a critical and potentially crippling issue for DAB which even in 2008 had not yet achieved significant car radio penetration.[293]

A small change in methodology from 1999 had the effect of showing increased radio audiences generally. Commercial radio's reported adult weekly audience rose from around 28 million in 1998 to its all-time high of 31.8 million in 1999.[294] Although handy in setting advertising rates, this was an illusory peak in terms of

290 Deanna Hallett interview op. cit.
291 David Vick in an interview with the author, 24 June 2008.
292 *RAJAR*, quarter 1 2000.
293 See chapter 20.
294 *RAJAR topline results* quarter 2 1998 and 1999 RAJAR/RSL/Hallet Arendt.

the achievement of the network. The true high point had been in the spring of 1998, when commercial radio attained a share of all radio listening in the UK of 51.1 per cent. It was to be downhill from there. By 2000 that had fallen away to 47.2 per cent. The radio companies blamed their fall on the repositioning of Radio Two, which had indeed been done with flair and money, and untrammelled by anything approaching ILR's promise of performance restrictions.

The Radio Authority, for all that it commissioned no research itself, watched the RAJAR figures with assiduous attention. Its diagnosis was that the fall was also caused by under-investment in programming quality, and that it was important for the industry that the companies reversed the direction by renewed attention to their product. When the Radio Authority directors met the AIRC Board in the trade association's Shaftesbury Avenue offices on 17 February 2000, I recall that we found ourselves in what was virtually a shouting match. It was the first such meeting for the new AIRC chairman, John Eatwell. He effectively told his regulators to mind their own business, as the companies knew far better how to run their radio stations than did the bureaucrats. However, since that date the melancholy decline has continued almost unabated. By March 2008 commercial radio's share of UK radio listening was as low as 41.1 per cent.

It was Kelvin MacKenzie who first challenged RAJAR. He had acquired the ailing Talk Radio station in November 1998 and re-launched it as TalkSport in January 2000. TalkSport faced particular problems of confusion with other speech stations, notably the high profile BBC Five Live. Private research studies showed that when TalkSport was the only station carrying cricket coverage of the England test matches in South Africa, people recorded in their diaries that they had listened to it on the BBC. MacKenzie could not afford to believe the audience figures which RAJAR provided, and experimented with a wristwatch meter.

Deanna Hallett, one of the founders of RAJAR and one of the leading lights in UK radio research, is disinclined to apportion blame. "The meters had been around for a long time … and as an industry we have been very slow to make changes or evaluate new things generally. I think Kelvin was right to push for a meter."[295] Her concern, and that of other specialists, was that his high profile might unsettle the industry as a whole, and that he had picked the wrong meter. "If you were going to go with a meter, of all the options we knew about and evaluated, the wrist watch was the worst. It was vital that a device should be able to differentiate platform as well as the station."

The idea of meters had been a regular feature – both pro and con – at the annual European market research conference since the early nineties. The leading proponent of the wrist watch version was Matthias Steinmann of Switzerland. He was chairman of the Swiss audience research board, and Switzerland was the only European country so far to have adopted this method of measuring radio audiences, although Arbitron use a different meter system in some parts of the USA. The great problem for meters has been their failure to record early morning listening. In all the trials it appeared that people just were not remembering to switch on their meters at a key time for mass radio listening.

Beaten down by MacKenzie's persistence, and worried that his flair for publicity was damaging its credibility, the RAJAR board commissioned its own

295 Deanna Hallett interview op. cit.

experiment with portable people meters, in collaboration with the Broadcasters' Audience Research Board (BARB), the television research body. Starting in 2007, research company TNS was commissioned to run a two-year pilot project, which failed even to run its full course. By the spring of 2008 the RAJAR board had already concluded that people meters did not work. It abandoned the project "due to significant concerns regarding issues of panellists' behaviour and the feasibility of meter measurement to deliver a credible UK currency for the entire industry, at an affordable cost".[296] RAJAR set out instead to test an online, interactive diary to form part of a new research strategy.

This was late in coming, but essential to restore credibility, which was threatened for the first time since 1975, or at least since the introduction of RAJAR in 1992. Commercial radio had known for many years, certainly since the launch of Digital One, that it must find a means of measuring 'platforms', but introduced such measurement only belatedly. Almost from ILR's inception, by embracing the pukka research of JICRAR, the radio industry had demonstrated a seriousness of intent which stood it in good stead. It showed that it was prepared to fund quality research, even though this was at a disproportionately higher cost than for ITV; later, with RAJAR, that it was willing to bear the burden of the deal with the BBC, and still support high quality research. This was a characteristic of *independent* radio which *commercial* radio would jeopardise at its peril.

296 RAJAR News Release 28 April 2008.

Chapter 7

Now we are nineteen

Autumn 1976 – 1979

The late seventies continued to be fun in ILR, but dysfunctional for the nation as a whole, which was approaching the end of an era. The post-war political consensus was worn out, exhausted by labour troubles as much as anything else, and exposed by the end of the first long post-war boom brought about by the oil shocks. The youth surge of the sixties and early seventies had stalled, but its successor was not yet in sight. Although the Sex Pistols displaced Cliff Richard from the top of the album charts in November 1977, metaphorically the old music continued playing until the 'winter of discontent' of 1978/9 brought everything to the point of change.

The half-expressed public appetite for things to be different, and dissatisfaction with the established institutions, provided ILR with its great opportunity, and it began to feel that it was getting somewhere after a rough start. There was a growing confidence that the system would avoid abrupt political termination. Just as had happened for ITV, large audiences seemed to include plenty of those in the demographic groups who had been most likely to vote Labour, and the party political split over independent broadcasting was fudged by electoral realities. Most of the stations came off the 'commercially critical' list after the first few years, and plenty began to make modest profits. Those working in the early stations felt a sense of belonging to an 'exclusive club'. There was the buzz of being part of something which had been uncertain, but which now looked to be going all right; where the day-to-day work provided a real adrenaline buzz; and where encounters with listeners and advertisers – especially local advertisers – offered consistent encouragement.

In those years, ILR newsrooms, when not distracted by scratchy industrial relations, had plenty of copy, although an IRA ceasefire provided a couple of years of relative peace. In the tabloid newspaper world, the *Sun* overtook the *Daily Mirror* in circulation in 1978, and the arrival of the true 'red top' tabloids finally put an end to discretion as a virtue in popular journalism. Sport was starting to provide the running soap operas which are now routine, but were then still a novelty. Within six weeks in 1977, England cricket sacked its captain, Tony Grieg, and Manchester United its manager, Tommy Docherty, for rather different breaches of convention: Greig for setting up a rival cricket circus with Kerry Packer; Docherty for setting up home with the wife of the club's physiotherapist.

The old certainties were gone, and for a new young medium these were heady days. ILR was able to interact with the news in previously unimaginable ways.

When Elvis Presley died in August 1977, Roger Moffat told his Radio Hallam listeners that the body was to be acquired by the station, stuffed, and put into a glass case in the foyer. The young Jon Snow, accredited by LBC/IRN to go to Uganda to interview Idi Amin – and reportedly sleeping on the floor of another journalist's hotel room to save money – claimed to have faced and declined the chance to shoot the dictator in his sleep.[297]

The ILR stations were interacting with their local communities in a new way for popular radio, helped by increasingly extended broadcasting hours – although those still needed the approval of the IBA, which was by no means automatic. However, the expectation of something close to round-the-clock broadcasting was crucial at times of local emergency. Plymouth Sound drew wide praise during the blizzards of early 1978, for tracking down three bus-loads of children who were stranded between Bristol and Plymouth, and putting the minds of anxious parents at rest. The *Daily Mail's* Martin Jackson, his own car stuck in a snowdrift, wrote that the station "demonstrated magnificently how local radio can rise to a crisis, making way for emergency messages and opening the airwaves to parents asking about children stranded at school, and husbands telling wives they couldn't get home from work".[298]

Once a station had faced a local crisis together with its community, it took a quantum step upwards in the closeness of its relationship with the locality. Every station needed the equivalent of a *Snowline*, reporting on closed schools and stranded transport, to make that link beyond its regular listeners with the local society as a whole. The fierce winter of 1978/9 provided plenty such opportunities. BBC local radio had met these challenges before, but ILR had a mass audience, reaching especially the younger and middle-to-lower class demographics, whose taste for popular music radio had previously meant that they were largely excluded from the ready provision of local information. It was, therefore, to ILR that they tuned, when the need arose. This was a unique exception to the rule that broadcast audiences switch to the BBC when there is an emergency.

Music programming was hampered by copyright restrictions, especially the 'needletime' limit to 50 per cent of output, up to a maximum of nine hours per day,[299] and by a continuing reluctance from some record companies to offer ILR stations anything more than basic material. The major labels would much have preferred things to go on as they were, dealing only with Johnny Beerling and a very few BBC producers at Radio One. Beerling, controller of Radio One, remembers that "we were being visited by various record company pluggers who would bring down the current 'record stars' to promote their latest records ... There was always someone with a guitar, and after dinner it would turn into a singalong with silly games."[300]

The ILR stations each selected their own music, sometimes through a committee of the presenters, sometimes by the *dictat* of the programme controller. Even once the playlist was agreed, by no means all the music played came from it. On many stations, for many of the DJs, what they played was what they liked, plus – often reluctantly – whatever percentage of playlist tracks they were required to

297 *Independent* 15 February 2000.
298 Quoted in *Television and Radio 1979* IBA p 138.
299 See chapter 13.
300 Johnny Beerling, *Radio 1 The inside scene*, Trafford Publishing 12–8, p 103.

include. That made ILR music quite imaginative and unpredictable. It also militated against the establishment of a single music 'sound' for ILR at that time, which commercial radio in the US and Australia deployed when seeking highest audiences in the target demographics. If there had been – as for a while seemed likely – a single national commercial pop channel, it would have been largely a clone of Radio One. ILR, though, was able to try and be a range of different things, and in its early years it managed to be just that, with almost as many different approaches to playlists as there were stations.

From the start of 1977, there was better news in London. Capital was finding its feet, even though that meant losing some of the earlier stars, notably Roger Scott and Kenny Everett. It had a settled breakfast show across the seven days, with Graham Dene on weekdays and Kerry Juby at weekends. Special programmes gained a new profile. Capital began a monthly series of live debates, *Headline*, and promoted pop and rock concerts at the Drury Lane Theatre as well as a major orchestral concert series. Late in the year, it launched Maggie Norden's outstanding series of O level revision programmes in the Sunday afternoon children's slot, *Hullabaloo*. This was the perfect illustration of what independent radio was meant to be, using the wide popularity of the station, especially among young people, to inform and educate. "Pop go your O levels" ran the headline, as literature came alive on air.[301] Norden and co-presenter Jill Freeman found relevance for youngsters studying *Romeo and Juliet* in the current events in Belfast, for Graham Greene's *Brighton Rock* among London's alienated youth, and for *Richard II*, scooping up a handful of earth as he returns from Ireland, using the same actions as the Czechs when faced with invading Russian tanks.

LBC saw its "forward bookings running at twice the rate of the same time last year".[302] It unveiled a new afternoon schedule with Barbara Kelly, Sarah Dickinson and Claire Rayner. This was an unintentional counter-balance to an angry protest by Women in Media, which accused Capital of "pushing its women broadcasters into the background", which would have been quite an achievement given that it included the likes of Anna Raeburn, Gillian Reynolds and Maggie Norden.[303] The two London stations had lost their initial driving forces on the programming side. Michael Bukht left Capital in July 1976, and Marshall Stewart finally took leave of Gough Square in May 1977, but the stations were able to attract those who were also to become major radio figures: Ron Onions took over from Marshall Stewart in May as editorial director of LBC and IRN, while Capital recruited Tim Blackmore from the BBC in June, to provide the programme-making substance to back up Aidan Day. The Wren Orchestra and the *Collection* programme showed what Capital could do in terms of classical music promotion and scheduling, once it had some profits to work with. Its April 1977 AGM reported a profit of £767,000. It made such apparent luxuries a good deal easier to attain, although DJ Tony Myatt would insist on prefacing trails for the *Collection* with "and now here's something terribly posh".[304] Even a three-day news blackout of LBC and the IRN service, as a result of local NUJ action, failed to dent the upbeat mood. This was business as usual for ILR in London, almost for the first time.

301 *Evening News* 22 December 1977.
302 *Broadcast* 7 March 1977.
303 *Evening Standard* 2 March 1977.
304 *Broadcast* 11 July 1977.

Most of the out-of-London stations had got into their stride quickly, in terms of programming and their ready acceptance among listeners. The remarkably swift building of mass audiences is testimony to that.[305] The most successful were those who quickly found their own style. Piccadilly Radio in Manchester was sneered at for offering "pop music, more pop music and football",[306] but still managed to win the Society of Authors' award for the best radio documentary of the year – *If The Bombs Dropped Now* – competing with BBC and ILR. The very formal radio critic John Woodforde noted that this was "a huge achievement for a small commercial station".[307] Piccadilly was not that small, and had skilled BBC-trained programme leadership from Colin Walters, but it made its point. It had also not lost its own style. On the first day of a summer heat wave, it offered a prize to be claimed by the first bikini-clad girl to kiss DJ Roger Day. By the allocated time, there were queues of bikini-clad girls outside the station, while a number of decoys in jeans and t shirts distracted attention from the sober-suited Roger Day, who strolled around unnoticed. All very seventies, all part of the fun that was ILR at the time.

Radio Clyde never slackened from its good start, offering remarkably varied programming with wide popular appeal. Radio critic Anne Karpf, a perceptive observer and a stern critic too of commercial broadcasting, observed that "to those who think local radio is made up of inane jingles and testy phone-ins, interspersed with worthy but dull local gen, Radio Clyde comes as a bit of a shock ... it has become the 'jewel' of the Independent Broadcasting Authority".[308] Clyde's *Towards 2000* series of programmes, which included Prince Philip speaking about how life might be at the end of the Millennium, drew requests for transcripts from 110 networks and newspapers in 40 countries. Equally to the point, its popular music programming was second to none. In England's Thames Valley, Radio 210 – untroubled by News International's decision to sell its 45 per cent shareholding in the off-the-wall small station – received the first royal visit from the Queen in March 1978. [309] She was shown the son of one of her racing pigeons, which had been presented by the Royal Lofts to the station when it started its own pigeon-racing team.[310] That was the style of ILR, blending deference with a level of cheekiness in a way which precisely caught the mood of the times.

Ironically, the one exception to that was its reception among the musically-aware young. The deliberately subversive nature of punk posed a particular problem to UK popular radio, not just in a socio-political sense, but musically as well. Actually, the whole British broadcasting establishment found punk difficult to handle – which was, of course, the point. When Malcolm McLaren was exploiting the shock value of the new Sex Pistols in 1976, ITV's Bill Grundy famously fell into the baited trap, egging on John Lydon ('Johnny Rotten') to a

305 See chapter 6.
306 *Campaign* 29 April 1977.
307 *Sunday Telegraph* 10 December 197?.
308 *Sunday Times* 13 November 1977.
309 *Radio and Record News* 1 April 1974.
310 This Radio 210 story is probably true, even though the press report is dated April Fool's Day. The Queen continues the royal family's long association with racing pigeons, which began in 1886 when King Leopold II of Belgium made a gift of racing pigeons to the British royal family. In recognition of her interest in the sport, the Queen is patron of a number of racing societies, including the Royal Pigeon Racing Association.

four-letter-word tirade. The BBC's approach to punk music on the radio was hardly any more skilled. ILR tried largely to ignore it.

Punk drove forward two changes in radio. One was mainly mechanistic. As punk records began to feature extensively in the charts, so those became less valuable as a source of data for playlisting. This was to tie in with the general sidelining of the national charts, which was to be a key change during the eighties. In the mid-seventies, though, it just made the on-air mix ever less predictable. The other was a serious challenge to ILR's programming brief. If younger listeners were demanding the Damned, the Clash, Buzzcocks and the Undertones, how could ILR stations – which were intended to be all things to all listeners – then cope with adult distaste for this music? Generally, ILR stations built mass audiences with a reasonable diversity of pop music, but they could not match the aspirations of the progressive wing of the British music scene as a whole. This is probably the point where ILR lost the support of those aficionados who had hoped it might be the natural heir to the offshore pirates. The realisation that radio – even in its new, expanded guise – could not keep up with the more outré offerings of the music industry, illustrates that ILR itself had started to become part of the broadcasting establishment.

Up and down the four home nations, ILR stations were attracting improbably large audiences, and stirring a mixing pot of brash entertainment, aspirational programmes, quirky stunts and the occasional disaster, into a heady and attractive mix. Furthermore, profits did not only mean dividends for shareholders, they also meant that secondary rental became payable. That was starting to fund a range of programmes, concerts, events and enhancements which added lustre to the stations, and the sharing of programmes of merit between stations. ILR was fun radio, but it was also beginning to do things which made it serious when it needed to be. Secondary rental projects became an even more prominent aspect of ILR in the early eighties.[311]

Much of the regulatory activity centred upon the system of 'rolling contracts', set out in the 1972 Act. This meant that each company was awarded a franchise for three years, and at the end of the first year its performance was formally reviewed. If that was deemed satisfactory by the IBA, a year was added to the contract. The process was repeated year after year, and the agenda of the Authority saw regular papers about each station's performance, and the potential rolling of its contract, which provide a regulatory almanac of the progress of ILR. The first few reviews went smoothly. The IBA announced in April 1975 that it had extended the contracts of Clyde, BRMB and Piccadilly and then, in April 1976, those of all the first ten stations. However, for both LBC and Capital, for whom the IBA had determined that their contracts would commence only when medium wave broadcasts began on the permanent frequency, that is to say in March 1975, in neither instance was the extension straightforward.

LBC's troubles were well-known to the Authority members, but so also was its crucial position within the ILR system. The paper which preceded the decision to roll encapsulates the entire rationale for the IBA's dealing with LBC/IRN throughout these years, and reveals John Thompson's understanding of how closely their fortunes were bound together. "LBC's significance derives from being

311 See chapter 9.

more than just historically the first ILR station. It represents an experiment in providing a news station for London, together with a news service for ILR, which is unique in Europe. As such, LBC has enjoyed (and needed) the encouragement, assistance and candour of the Authority, while the company has sought to establish itself both as a service of quality and a potentially viable operation. There are now some favourable signs, but members will be aware that it will be a considerable time before LBC/IRN can be regarded as soundly established, at least from a commercial viewpoint."[312] On that basis, the LBC contract was duly 'rolled'.

The IBA came under fierce attack from the political left over Capital. The Local Radio Workshop, which was to campaign against Capital in particular over a number of years, argued that there was an "almost total lack of local input in Capital programming"[313] which comprised "a diet of pop records and advertisements played ad nauseum *(sic)*. No access, no participation, no involvement. The rush for dividends, euphemistically termed 'viability', has seen to it that all local and creative material is squeezed out by the nice, safe, low cost pap that Capital and their advertisers prefer." In the confrontational language which characterised the labour relations of the period, and which is a reminder of the constant background of industrial strife which the NUJ inflicted on ILR, the NUJ's Denis McShane attacked the IBA for having "just laughed in our face" in allowing Capital to close its newsroom, "because they wanted to increase their profits at the expense of their duty to fulfil their franchise obligation".[314] This drew the rebuke from Radio City's Terry Smith, himself a member of the union, that the NUJ's approach was "vicious and nonsensical".[315] McShane's language shows the nature of the political environment within which the IBA had to work, and goes some way towards explaining its obsession with pressing the public interest delivery of ILR.

I remember, though, that many officials at the IBA felt concern at the long leash being allowed to Capital, and there were several drafts of the crucial paper considering the rolling of the contract. The resulting report highlights the Authority's dilemma. It notes very fully the criticism Capital faced for abandoning some of its more ambitious programming plans and for being overwhelmingly a popular music station, and makes a wide range of criticisms of the quality and seriousness of the programming. Yet it also acknowledges that "in practical terms, bearing in mind the experience gained in the last two years, the Authority may probably wish to have a service which is clearly and predominately 'entertainment'. This entertainment programming would be largely recorded music, designed to be popular with the largest number of listeners ... ".[316] The Authority agreed to roll Capital's contract; it had no real choice.

The first sign of troubles outside London came with the IBA's decision in the summer of 1976 not to extend the contract of Radio Trent, at the end of its first year on air. Managing director Denis Maitland, unafraid to be thought out of step, had found himself in dispute with his programme controller, Bob Snyder; with two unions, NUJ and ABAS; and with AIRC itself. This last was over his refusal to join the trade association, since he wanted to avoid the then ubiquitous national

312 *IBA Paper* 29 January 1976.
313 Local Radio Workshop, January 1977.
314 *Broadcast* 3 May 1976.
315 *Broadcast* 10 May 1976.
316 *IBA Paper* 33(76) 29 January 1976.

union agreements. Falling foul of the IBA was a step too far, however. The Authority, at its 10 June meeting, declined to roll Trent's contract and required the company's chairman, Norman Ashton-Hill, to attend a meeting at Brompton Road on 25 June. The implication of that meeting, and Lady Plowden's follow-up letter, was unmistakable. "You must please now take all practicable steps to achieve an effective resolution to the recent internal problems at the stations ... you may feel free to explain to your management – and to your staff – that until the station can be heard and seen to be concentrating on supplying the best possible local radio service within the resources available, and with a clear longer-term strategy, the Authority will not be able to consider rolling the contract."[317] Not until a year later was the Authority prepared to extend Trent's contract. Beacon faced similar challenges. The Authority took the then unprecedented step of meeting the entire board of Beacon Radio in July 1979.

The contract roll system gave the IBA real power, and lent emphasis to the 'master and servant' relationship which existed between the Authority and the companies. All the ILR MDs felt that their contracts were continually at risk, a factor restraining them from challenging the Authority in any significant way until 1984, when they were sure they were backed by their chairmen.[318] AIRC felt that "under the rolling contract system, the Authority's teeth are bared all the time".[319] Whether the IBA would ever have contemplated terminating a contract in the seventies is open for question. There could be no certainty at that time that government approval would be given for re-advertisement, and a regulator-imposed failure could have done real damage to ILR's credibility.

When, however, in the wake of Annan Report and the lifting of the freeze on new franchises, the IBA proposed in October 1977 adding a ten year 'break' into the rolling contract, the companies were affronted. The then Chairman of AIRC, Piccadilly's Philip Birch, wrote to Brian Young pointing out the disruptive effect of a contract break on a company's operations "for a period of up to two years prior to the break-point".[320] Aspirant groups would suborn existing staff, and boards of directors would become even more short-term in their planning. In a foretaste of the regular antagonism that existing radio companies would show at any time towards any potential new radio services, BBC or commercial, AIRC argued that since the original nineteen stations had borne the "heavy costs of the news service", and been prevented from becoming a national advertising medium while Annan deliberated, they should be spared a contract break. However, the matter should be left open for any new companies only "until such time as the government have made their views ... a little more clear".

The RCC meeting that October exposed some differences in view between Brian Young, who said that "his personal opinion was that break-points would help to meet the 'public interest' arguments, and John Thompson who was "very conscious of the practical issues involved".[321] AIRC's subsequent submission to the IBA argued that "it is more than self interest which motivates us to say that whatever doubtful political benefits there may be in introducing 'break-points',

317 Letter from Bridget Plowden, IBA, to Norman Ashton Hill, Radio Trent 28 June 1976.
318 See chapter 10.
319 AIRC submission to the IBA 11 November 1977.
320 Letter from Philip Birch, AIRC, to Sir Brian Young 18 October 1977.
321 RCC Minutes 19(77) of a meeting on 19 October 1977.

they will be much, much more offset by the harm [done] to ILR". Despite the companies' urgings, break-points were duly introduced by the new Broadcasting Act in 1980.

The other key regulatory issue, which was to resonate down the years, was over what was dubbed 'meaningful speech'. This notion was often thought to have been sparked by the arrival from the BBC into the programming job in the IBA's Radio Division of Michael Starks. He had joined the IBA in the summer of 1976, replacing Bob Kennedy, and had brought a rather more formal approach into the previously rather free-wheeling Radio Division. However, the Authority members had already earlier in the summer discussed the general pattern of ILR programming output, and had "expressed the hope that the radio companies would aim, during the coming year, to develop the 'meaningful speech' element in their programming"[322] – so the original phrase probably comes from Bob Kennedy or John Thompson.

There were two strands here. First, there was an entire department at the IBA devoted to compiling statistics about the output of the ITV companies. This reflected the quota approach to television regulation, contained in the television contracts. Programme statistics were a feature of the IBA's approach to regulation. It hoped to find something similar for ILR, and had therefore conducted a quasi-statistical exercise, working through the ILR schedules, to try and compile such data. Second, once the categories of output had been identified, and a measurement system established, the Authority was in a position to direct the stations – at least *post hoc* – in respect of their programming output. This was to surface, in various different guises, as a repeated point of tension between the regulator and the stations right up until the present day.

Facing this approach for the first time at the Radio Consultative Committee meeting on 21 July, the companies were uncomfortable, but their sense of subservience to the Authority remained. They queried the methodology used to gather the figures, and the categories identified. Within the definition of 'meaningful speech', the Authority staff had included (in this order): current/social affairs, community information, features/arts/interviews, listener/consumer service, traffic, weather/public service, competitions/quizzes, religion, education, programme information and serials/drama. With some miscellaneous items, and adding in Welsh-language programming in Swansea, the analysis of ten stations gave an average of 19 per cent of total output. Music represented 46 per cent of output, news 10 per cent, sport 3 per cent, advertisements (only) 6 per cent, and 'other' 16 per cent.[323] The last category was by definition 'meaningless speech', and the companies were set the target for the year ahead of "achieving more local focus and more contemporary professional crispness" within this element of output. In other words, the IBA, sensitive to the jibe that ILR was mere 'pop and prattle', sought to minimise the dominance of the 'pop' and was determined to reduce the 'prattle'.

A year later, the IBA was acknowledging that "one of the most significant achievements of ILR is its impressively high audience appeal" and was pressing the companies to "broaden the range of their speech and music output as their financial position permits", and to "strengthen where necessary the 'meaningful speech',

322 RCC Minutes 14(76) of a meeting on 21 July 1976.
323 Radio Paper 31(76) 12 July 1976.

information content and local relevance of their programming".[324] A year further on, the April 1978 monitoring showed music up to 53 per cent, advertisements at (a much healthier) 10 per cent, and meaningful speech at 18 per cent.[325] What was now dubbed 'light speech' was at a politically defensible 9 per cent across the week. The Authority was – not without justification – feeling confirmed in its view that what seem to be remarkably strict requirements on independent radio stations produced large audiences and a highly credible quality of output. The July 1978 RCC meeting was to be the high point for the salience of 'meaningful speech', and marked the departure back to the BBC of Michael Starks. As his successor as head of radio programming, I was much more preoccupied with the opportunities and obligations of an expanding ILR network. The pressure this put on the IBA Radio Division's modest staff resources obliged us to leave more of the policing and monitoring to the companies themselves.

Beyond the annual monitoring exercise, the chief mechanism for keeping ILR companies to their obligations was prior approval of programme schedules. Stations had to submit proposed schedules quarterly, with additional proposals for Christmas, Easter and any other times of significant change from normal output. The detail required was formidable, including regular feature items within each programme; the amount of music normally played within programming; regular competitions and promotion spots; time and duration of news bulletins; non-British content; live music slots; and any networked, syndicated or purchased material. Along with the fully exercised right of the IBA to approve any changes in broadcasting hours, this was highly interventionist regulation, and involved a lot of paperwork. IBA Radio Division staff actually checked all the incoming schedules, and not infrequently went back to stations to require changes. In practice, though, the relationship was more sympathetic and informal than the apparent rigidity of the system implies.

Included in an RCC Paper in 1978, in what was for those days a rare flash of lightness, was a supposed "caricature" of a scheduling discussion.[326]

> ILR programme controller: 'Can I have approval to give the Top 40 show on Sundays to Bill Smith, so that Fred Jones can stand in for Ian Brown doing jazz on Wednesdays?'
>
> "IBA radio officer: 'We don't mind too much about that, but isn't it time you included some local education?'

That was not so far from reality, as I recall it, revealing the depth of involvement of the regulator in the programming process.

At last the financial position was improving, certainly for most of the majors. For much of 1977 and the first half of 1978, the UK economy enjoyed a respite from the worst of its economic and industrial troubles. Advertising money started to flow again, with ITV in 1977/8 almost doubling its take over two years.[327] By the end of March 1977, ILR annualized advertising revenue had reached £14.5m, up 67 per cent on the previous year and enabling fourteen of the companies to trade either in profit or at break-even. Adjusting for the stations not on air in the earlier period, the increase was 46 per cent, so that – despite the express train of

324 Radio Paper 27(77) 15 July 1977.
325 Radio Paper 37(78) 12 July 1978.
326 Ibid.
327 *IBA Reports and Accounts* 1976/7 and 1977/8.

inflation at some 15 per cent – there was real growth. Even LBC increased its advertising revenues sufficiently to resume paying transmitter rental in October 1976.

For the year to March 1978, ILR revenue of £22m was up a further 50 per cent. Clyde and Capital paid dividends to their shareholders, five of the major companies repaid some or all of their loan stock or redeemable shares, and eight companies had recouped their start-up costs and any past losses. Radio's share of total display advertising spend had reached around 2 per cent by the middle of 1978, where it was to remain stuck more or less until the launch of INR in the early nineties.

The Annan Report into the Future of Broadcasting was published on 23 March 1977.[328] This report was intended to herald the future for the social liberal approach to broadcasting in the UK, but in the event it was to be the last hurrah for the old dispensation. As such, it included the distilled essence of the liberal consensus which had run Britain since the Second World War, and which was about to be swept away by what we now regard as the Thatcherite revolution, and a belief in the primacy of the markets to deliver public benefits. The Annan Report was expected to provide a series of signposts to a liberal future. In the event, although it permitted movement on local radio, and introduced a very non-market new television service which became Channel Four, its informing philosophy was to become outdated within a few years.

That was wholly unforeseen. After a false start before the 1970 general election, the committee was set up in April 1974 by home secretary Roy Jenkins "to consider the future of the broadcasting services in the United Kingdom ... and to propose what constitutional, organisational and financial arrangements and what conditions should apply to the conduct of all those services".[329] Jenkins seems to have chosen mainly members with whom he would have been happy to dine. Noel Annan, provost of King's College, Cambridge, peer, don, wartime intelligence officer, biographer, Trustee of the British Museum – and purveyor of the "higher gossip"[330] – was representative of the old liberal establishment. His committee likewise comprised just those people who might naturally have served on one of the previous periodic enquiries into broadcasting, which began with Sykes in 1923. They included further academics, Professors Sims and Himmelweit, a proper representation from the home nations, good political balance, one ethnic minority member in Dipak Nandy and at least one person – broadcaster and author Marghanita Laski – who had appeared regularly on BBC radio shows.

The report contained 174 recommendations, ranging on the commercial side from requiring the Authority to co-ordinate training, to excluding advertisements from children's television programmes. It hugely upset the IBA's *amour propre* by proposing that it should be re-named the 'Regional Television Authority', as part of an envisaged three-tier system of regulation – national, regional and local – within which there would be a new Local Broadcasting Authority. Annan envisaged a continuation and reinforcing of the arrangements whereby public authorities should intervene where necessary to provide the viewing and listening public

328 *Report of the Committee on the Future of Broadcasting (Annan Report)*, March 1977, Cmnd. 6753.
329 *Hansard* House of Commons 10 April 1974, col 212.
330 *New Statesman* John Gray's review of Annan's book *Our Age* 1990.

with what they ought to see and hear. It saw its overall aim as helping towards a system "which gives broadcasters the greatest freedom and independence to produce *what is best for the audience* [my italics]". Government and social policy was conceived in the seventies as a series of rules and structures designed to direct constrained enterprise.

The recommendation which had the most impact was for the establishment of a second advertising-funded television channel, but one which would operate as a publisher rather than a broadcaster, and would only be a little bit commercial. This led to Channel Four, and the first steps towards a multi-channel future for television; but with extensive safeguards and an elaborate structure, intended to avoid the risk of US-style commercial television competition with ITV. The television companies were more than pleased, as it seemed likely that they would retain virtual monopolies over television advertising. The BBC's DG saw it as merely "a centre for every type of experimental broadcasting",[331] and eventually felt relieved. The IBA bridled at the notion that Channel Four was to be the responsibility of a new Open Broadcasting Authority, accusing the report of "inconsistencies, where the conclusions or recommendations appear not to follow from the argument, or where evidence appears to be no more than hearsay". The IBA provided a long annex to its own response, listing "certain errors of fact and statements which appear to us to contain particular misjudgments".[332] Notably, none of those referred to radio.

The Annan Committee was clearly rather taken with local radio, but not with its structures: "local radio is in a mess".[333] The report urged the expansion of local radio, so that most people in the UK should receive at least one service. It recommended that the BBC should no longer be responsible for local radio with the existing BBC local radio stations being replaced by a system of non-profit-distributing trusts. Those services and ILR should become the responsibility of a new Local Broadcasting Authority, together with the licensing of hospital, university and cable services currently done by the Home Office directly. In a substantial chapter on local broadcasting[334] the Committee urged "breaking the present mould of financing broadcasting" and noted that there needed to be a new public body to do that because "local broadcasting is a different animal from network broadcasting, and [that phrase again] needs a different sort of keeper".[335] However, the committee was unable to reach complete consensus. It had been divided ten against six on whether to split the BBC into two, to handle national television and radio. Then trade unionist Tom Jackson and BBC panellist Marghanita Laski entered formal notes of dissent over the proposal for a Local Broadcasting Authority, the latter on the grounds that "commercial local radio stations have offered ... jolly circuses" whereas in local radio "the need is for bread before circuses".

The reactions of each of the main interested parties were forthright. The BBC

331 Quoted in *A Licence to be Different; the story of Channel Four* Maggie Brown BFI p 14.

332 *The Annan Report, The Authority's Comments* published in *Independent Broadcasting* July 1977.

333 *Annan Report op cit* p 205.

334 Ibid. pp 205-228.

335 The late Phillip Whitehead MP, an influential member of the Annan Committee, told me many years later that he thought there had been a real opportunity in the late seventies to bring all radio under a single, separate control, and thus to anticipate and perhaps improve upon the creation of the Radio Authority in 1990. He and his lobby, however, were distracted by the prospect of a new type of television channel, and the chance slipped by.

mounted a huge campaign, encouraging their listeners to write in to 'save' BBC local radio. That was directed by Solent station manager Maurice Ennals, who had the not inconsiderable advantage of being the brother of David Ennals, who was then secretary of state for social services. It is said to have generated thousands of letters. The ILR companies made sure that they were not neglected, urging a rapid expansion of ILR, and that although BBC should keep local radio it should be limited to its existing twenty stations. They also took the chance of a sideswipe at the BBC's lobby, with Radio City claiming that "a careful examination of the reports and letters supporting BBC local radio, published in newspapers throughout the country, shows that the vast majority have come either from BBC staff or from members of [BBC] local advisory committees".[336]

The IBA's response was more nuanced, not least because John Thompson and others within the radio arena were not averse to hedging their bets on the outcome. For Thompson, the need was simply "that ... the institutional niceties are given closest attention", hardly a ringing endorsement of any effort to keep radio within the IBA.[337] In a striking contrast with Robert Fraser's equivocation in 1970[338] however, Brian Young's official Authority line was a determination to hang onto its radio responsibilities, pressing for an urgent expansion in the number of ILR services and arguing that "the IBA has the knowledge and resources to do this ... the IBA could carry out the duties the committee envisages for the regulatory authority in a more knowledgeable and cost-effective manner than any other body".[339] The IBA pointed out the range of services already included within ILR, and its readiness to experiment further with associated companies of various sizes, and small-scale neighbourhood radio. Far from having its radio tasks hived off to a new quango, and to prevent the horrors of being relegated to mere 'regional' status, the Authority argued that "it seems to us inconsistent that a committee which decided not to separate BBC Radio from BBC Television should have recommended splitting the IBA's much smaller radio and television operations".

From the publication of the Annan Report in the spring of 1977, the political manoeuvering and lobbying continued apace, although it grew increasingly frustrated as no decisions seemed to be forthcoming. The two-page Independent Broadcasting Authority Act in July 1978 extended the Authority's life until the end of 1981, and allowed it to roll forward the ILR contracts which were set to expire with the nominal IBA end date in July 1979. ILR had been the first new commercial broadcasting venture since the mid-fifties, but it was already clear that it was now going to be followed by others. For nine weeks coinciding with the publication of the Annan Report, the IBA permitted Yorkshire Television and Tyne Tees Television to broadcast an experimental sequence of breakfast television programmes for an hour between 8.30 and 9.30 am. The radio companies were understandably concerned to be challenged at their peak listening hour of the day, so soon after they had begun to establish themselves. Research conducted jointly by the IBA and Trident (the owners of YTV and TTTV) suggested pretty low ratings for the new television service. Still, 13 per cent of listeners to Metro, Tees, Hallam and Pennine said that their radio listening had reduced as a consequence of watching the new

336 *Financial Times* 18 July 1977.
337 *Listener* 7 April 1977.
338 See chapter 2.
339 *Independent Broadcasting op c it* p 15.

TV service.[340] In the event, it was to be February 1983 before TV-AM was launched,[341]but the experiment was a reminder that the IBA had been first of all a television authority, and radio still sat uneasily within it at times.

ILR, however, did not need to wait to resume its expansion. The Callaghan government's white paper in 1978 envisaged a way forward, for both BBC local radio and ILR, which was sufficiently close to the pre-existing arrangements that both felt able to press the government to be allowed to move towards establishing new stations. The Local Radio Working Party (LRWP), set up by the Home Office in the wake of the white paper, was to be the motor for this expansion, and the point at which the IBA and BBC Radio engaged with each other both competitively and co-operatively. This meant, essentially, teams led by John Thompson and BBC controller of local radio, Michael Barton, chaired by civil service mandarin Shirley Littler. Thompson and Barton got on well, and worked informally towards a solution. Barton felt that "at times it was almost as if we belonged to the same organisation. Indeed, I sometimes felt I had more in common with John Thompson than with many of my colleagues at the BBC."[342] In October 1978 the LRWP produced its first report, which included recommendations to the home secretary – in effect, authorisation – for nine more ILR franchises, as well as further BBC local stations. The new ILR areas approved were Cardiff, Coventry, Gloucester & Cheltenham, Peterborough, Bournemouth, Exeter/Torbay, Dundee/Perth, Aberdeen/Inverness and Chelmsford/Southend.[343]

The allocation of new stations between the IBA and the BBC was largely a matter of horse trading, with the successive secretaries of state key figures: Merlyn Rees up to 1979, and Willie Whitelaw afterwards. It was clear that both ILR and the BBC wished to be in the major conurbations, subject to the arcane processes of frequency planning. Annan's idea that localities might have either a BBC or an independent service was never seriously contemplated as a long term opton. For the IBA, there were a number of gaps in the ILR map which needed filling swiftly, notably Leeds, Bristol and Cardiff. The BBC continued to show almost no interest in local radio outside England – apart from Inverness, Derry and the Channel Islands – so the IBA could select unhindered in Scotland and Wales, although the sectarian situation in Northern Ireland largely ruled out further expansion into Derry until the time of the Radio Authority, chiefly because of the difficulty of site-finding and building transmitters.

Keen to make a quick start, and get a good number of the new services on air quickly, the IBA advertised Cardiff and Coventry before the end of 1978, with a closing date for applications of 12 February 1979, and the intention that both stations would begin broadcasting during 1980. Peterborough, Gloucester & Cheltenham and Dundee/Perth were advertised in February 1979; Bournemouth and Exeter/Torbay in March. Problems were encountered in finding frequencies for Chelmsford/Southend and the LRWP agreed that Leeds might be substituted.

340 IBA Audience Research Department/Trident TV reported in Radio paper 5(78) 17 January 1978.

341 See chapter 8.

342 Michael Barton in an interview with the author, 14 February 2008.

343 The nomenclature for franchise areas – and later for licence areas – is relevant. Where there are two localities linked with a backward slash /, that means separate transmitters with the potential for some degree of separate service. Where they are linked with an ampersand &, the intention is that a single service shall cover both localities, even if more than one transmitter is needed.

Six more franchises were advertised around the end of 1979: Aberdeen/Inverness, Leicester, Southend/Chelmsford, Bristol, Luton/Bedford and Ayr. By the end of 1979 the IBA could point to an imminent ILR network of 32 stations, all to be on air by the end of 1981.

It seemed to be a strong political position, which the IBA determined to reinforce with its approach to the first few of the new tranche of services. These owed much to the expectations generated by the Annan Report. North-East Scotland presented the IBA with an early opportunity to demonstrate flexibility. It was advertised – perhaps rather disingenuously – on the basis that it might be run as one station or two separate stations, and the split award of two franchises on 27 December 1979 duly gave ILR its smallest station to date, Moray Firth in Inverness. 'Twinned' franchise areas such as Exeter/Torbay represented "one way the Authority can bring ILR to less populous parts of the UK".[344] However, it was in one of its licensing decisions that the Authority was most dramatically led astray.

The IBA had received six applications for Cardiff, and five for Coventry. In Cardiff it selected a broadcasting 'co-operative' of the type dreamed of by the proponents of community radio before the arrival of ILR. Membership of a trust, or participation in local workshops, permitted direct participation by activists in the running of the station and its programming. The IBA, its judgment undermined by the need to head off the Annan Report's notion of a rival authority, and by its expectation that the old political dispensation was going to continue, saw these as "fresh techniques for strengthening links with the community". Actually, they ensured from the start a complete disaster. Cardiff Broadcasting Company (CBC) began broadcasting on 11 April 1980. Politically, it was a year too late, and was already an anachronism, along with the Annan Report and most of the assumptions of the liberal broadcasting caucus.

In the meantime, for the established ILR stations the world was looking much brighter. The fifth anniversary of LBC and Capital in October 1978 was the time for some satisfaction. Capital had established a 43 per cent reach of adults in London, which included 69 per cent of all those aged under 35. ILR in London and elsewhere was providing them with the sounds of their lives. Capital's social action projects were flying. This was central to Capital's philosophy, and that of most ILR stations. In Michael Bukht's words, "we were interested in the lives and welfare of our listeners. I have always regarded good radio as being an enormous spanner in the tool-box of social engineering."[345]

The fourth *Help a London Child* appeal in April had raised £70,000, the *Helpline* had taken 136,000 calls in its two years. 120,000 copies of the flat-share list had been given out, and more than 10,000 people had come to Euston Tower where Capital's studios were located to speak to Employment Agency staff who manned a desk in the foyer. The merchandising side was going well too. Capital had sold more than 200,000 T-shirts, and given away two million car stickers.[346] Profit for 1978 exceeded £3m, and even when £1.2m had gone to the IBA as secondary rental – more than half of which was to come back to the company for 'approved projects' – there was still enough to buy London's ailing Duke of York Theatre as an

344 *IBA Annual Report* 1979/80 p 46.
345 Michael Bukht in conversation with the author, 2 February, 2009.
346 *Music Week* 14 October 1978.

investment. Capital's weekday presenter line-up in 1978 was Graham Dene, Michael Aspel, Dave Cash, Roger Scott, Adrian Love, Nicky Horne and Tony Myatt. Remarkably, three of those – Cash, Scott and Myatt – had been part of the 1973 schedule, with the first two still in the same time slots. For the large majority of Londoners it was meeting John Whitney's original aspiration that "if you are not listening to Capital, you are not part of London life".[347] Bukht recalls a moment of pure pleasure, when a listener said that "Capital Radio fitted my life perfectly".[348] Truly, this was ILR providing all the sounds of your life.

LBC had found the going much tougher, and still did so. Nevertheless, it also was able to report a profit for 1978, of £326,000. This left an accumulated deficit of around £2m, but the company had survived and built a vociferously loyal audience. Returning to 24 hour broadcasting in February 1978 had cemented the relative recovery, and it was a surprise to the industry when MD Patrick Gallagher, who was widely credited with turning the station round, was ousted in January 1979, reportedly after "a boardroom coup inspired by LBC Chairman, Sir Geoffrey Cox".[349] Gallagher had been scarred by his four years of union troubles, estimating that labour problems had occupied close to 40 per cent of his time.[350] His replacement, broadcaster George Ffitch – like Cox an old ITN hand – was charged with taking the station forward editorially. LBC's board felt the worst was behind them, and wanted to grow their product.

Then came what the *Sun* named the 'winter of discontent'. For the UK as a whole, just as for ILR, 1977 and the first half of 1978 had seemed to be going a bit more smoothly than before. Industrial strife was never far away, as the government tried to impose limits on pay settlements amid high inflation, but there seemed to be hope of a more settled future. The trauma of the government going to the International Monetary Fund for a loan in 1976 had passed, and the stern economic measures which were demanded in return by the IMF seemed briefly to be having some effect. An IRA ceasefire held for most of that year, and although it was brutally shattered early in 1978 there were no further atrocities in England until the spring of 1979. However, this fragile optimism proved misplaced, as strikes flared up in opposition to the government's 5 per cent pay limit from the start of August 1978. In September, workers at Ford began a long strike, followed by bread workers in November which led to brief bread rationing as panic buying ensued. Lorry drivers struck in December, and together with a strike of tanker drivers who also blockaded oil terminals, they threatened widespread disruption. Train drivers and railway workers swiftly followed, and more than a million workers were laid off as industry staggered to a stop. The most grievous damage to the public mood was done by strikes by waste collectors, leading to mountains of rubbish piling up in the streets. Then gravediggers went on strike in Liverpool, the city council hired a warehouse in Speke to store unburied coffins, and the headline writers rejoiced.

The TUC leaders eventually reached a general settlement on 14 February 1979, rather against the wishes of their more militant union members. However, the damage was done. In May 1979 Callaghan was faced with the unknown quantity of Margaret Thatcher, in a general election which was well-nigh

347 Quoted by Tim Blackmore in an interview with the author, 17 October 2007.
348 Bukht, op. cit.
349 *Broadcast* 22 January 1979.
350 *Broadcast* 12 March 1979.

unwinnable by Labour. On 3 May the Conservatives duly gained a majority of 43 seats, with the largest swing in votes since 1945. The nation was set for a greater shift in its political economy than it realised. This was to be revolution, complete with its violent confrontations – such as with the miners at Orgreave colliery – and its gurus too, in Keith Joseph and Milton Friedman. Yet whereas the previous decades had been concerned with the revolutionary left, this was revolution from the right. Its themes have been pursued by governments now for more than 30 years. Annan's notions for institutional change – and much more – were discarded, an indication that the social model of broadcasting was nearing its end.

In 1978, Feargal Sharkey,[351] once a Derry teenager and lead singer of the Northern Ireland punk group the Undertones, wrote *Teenage Kicks*. Punk music is an apt metaphor for the years of political punk which followed. The new zeitgeist was far from the social market model of the previous thirty years. Punk had entered into the mainstream of UK life and politics, with destructive energy, iconoclasm, but scant regard for long-established comfort zones. It was an approach which lasted largely up to the financial crises of 2008. Sharkey and his song were given renewed fame with the death of John Peel in October 2004. Peel, pirate DJ turned Radio Four stalwart, named the song as his all-time favourite and had its opening line engraved on his tombstone. "Teenage dreams so hard to beat" is a neat epigram for what the first nineteen stations had achieved in making such a success of independent radio.

351 There was considerably more to Feargal Sharkey than his status as a punk pop icon might suggest. He was to become an effective, unconventional member of the board of the Radio Authority in December 1998. He subsequently chaired the UK government task force, the 'Live Music Forum' in 2004, evaluating the impact of the 2003 Licensing Act on live music performance, and was appointed chief executive of British Music Rights in 2007.

III. The independent radio experiment 1980 – 1989

Chapter 8

Victories and losses

1979 – 1985

As the new decade arrived, Margaret Thatcher's victory in May 1979 seemed to mark a new start in many sectors. Remarkably, it was to take more than ten years before the full impact of that pivotal election changed independent radio, requiring ILR to run through yet another ferocious recession, before the commercial benefits could start to flow. Between 1979 and 1985, ILR was still chiefly concerned with keeping its audiences and adding to its revenues, but with the additional theme now of renewed expansion of the system. The roll-out of new ILR franchises dominated these early years of the eighties, and it will consequently occupy much of this chapter. The other three major themes of those years – the use of secondary rental to enhance programming output, the climactic Heathrow Conference of 1984, and the failed community radio experiment the following year – are considered in the next three chapters.

The new government had wasted no time in reinforcing the decision of its predecessor to permit the IBA to establish more ILR stations. In addition to the nine new franchise areas announced by Merlyn Rees in 1978, Willie Whitelaw in November 1979 allowed a further fifteen franchises to be advertised: Ayr, Barnsley, Bristol, Bury St Edmunds, East Kent (in principle, subject to more work on finding frequencies), Guildford, Hereford/Worcester, Leeds, Leicester, Londonderry, Luton/Bedford, Newport (Gwent), Preston & Blackpool, Swindon/West Wiltshire, and Wrexham & Deeside. Since Aberdeen and Inverness were to be separate, the original nineteen ILR franchises, joined by Rees' ten and Whitelaw's fifteen, presented the prospect of a network of 44 independent stations. Even adopting the fairly restrictive coverage areas represented by the ILR franchise specifications, between them they would cover 75 per cent of the UK population.

The remaining 25 per cent lived mostly in communities of 200,000 or fewer, and the composition of the 1979 list – despite coming from the new Conservative Government – still reflected the Annan approach in many respects, most notably in the trials for smaller stations than had hitherto been thought viable, such as Ayr and Bury St Edmunds. Twinned franchises, and the potential for associating one small franchise with a larger neighbour, were specifically identified as possibilities, even though they would involve a form of subsidy which the new governing political philosophy should not admire. It illustrates the time lag in the UK between policy and its effect that many of the more experimental features of *independent* radio were brought into existence after the electoral change which heralded the eventual move towards *commercial* radio principles.

The IBA had completed the first new round of franchise applications with noticeable speed, announcing nine new contractors during 1979, even though an additional layer of process had been introduced for the post-Annan phase; public meetings ahead of every franchise award and re-award. This was not a statutory requirement, but was seen by the IBA as helping it and the programme companies "avoid mistakes and provide a better ILR service".[352] Seven new stations started broadcasting in 1980: CBC in Cardiff, Mercia Sound in Coventry, Hereward Radio in Peterborough, 2CR in Bournemouth, Radio Tay for Dundee/Perth, Severn Sound for Gloucester & Cheltenham and DevonAir for Exeter/Torbay. Two of the remaining three from the first batch of new stations, Essex Radio for what was now called Southend/Chelmsford and North Sound in Aberdeen, came on air in 1981. Moray Firth Radio in Inverness, the smallest of the new services, had to wait until February 1982, by which time Radio West in Bristol, Radio Aire in Leeds, Centre Radio in Leicester, and Chiltern Radio for Luton/Bedford, had all joined the airwaves. Later in 1982, stations opened for Bury St Edmunds (Saxon Radio), Hereford/Worcester (Radio Wyvern) and Swindon/West Wiltshire (Wiltshire Radio), making a total of 38.

In the meantime, the Home Office Local Radio Working Party was proceeding apace to identify new areas. Significantly, by December 1980 it had recommended, and the government had accepted, that there was no longer any need for an area to have to choose between an ILR or BBC local station. This was partly a response to the pressure from both the BBC and the IBA to complete their coverage of England and the UK respectively, but it also acknowledged that ILR was now an established and separate system. The LRWP's third report made that point explicitly. "Two quite distinct forms of local radio have grown up. We see no reason of principle why the two authorities should not develop different patterns of local radio services, provided that these meet public demand and their plans do not place unreasonable burdens on the frequency spectrum."[353]

This was a milestone in the official acceptance of independent radio as a medium in its own right, and it encouraged the IBA early in 1981 to put forward to the home secretary a new list of 25 locations for further ILR stations. After a period for public consultation, the first time that such a procedure had been followed in the selection of localities, Whitelaw approved the list on 14 July 1981: Aylesbury, Basingstoke & Andover, the Borders (Hawick) with Berwick, Brighton, Cambridge & Newmarket, Derby, Dorchester/Weymouth, Eastbourne/Hastings, Great Yarmouth & Norwich, Hertford & Harlow, Huddersfield/Halifax, Humberside, Maidstone & Medway, Milton Keynes, Northampton, North West Wales (Conway Bay), Oxford/Banbury, Redruth/Falmouth/Penzance/Truro, Reigate & Crawley, Shrewsbury & Telford, Southampton, Stoke, Stranraer/Dumfries/Galloway, Whitehaven & Workington/Carlisle, and Yeovil/Taunton. The IBA was at last aiming to establish ILR stations in some of the iconic BBC localities from which they had been excluded by the previous process of negotiations: Brighton, Oxford, Cambridge, and Norwich.

The Home Office was still determined to keep its hands on the detail. The IBA's proposal in 1982 to link Northampton and Milton Keynes, even though separate franchises had been approved in the list of the previous year, needed to be sanctioned by a statement to the House of Commons by the minister of state,

352 Michael Johnson, *Public consultation; what's the point?* in *Independent Broadcasting*, June 1982.
353 *Third report of the Local Radio Working Party* Home Office December 1980 para 8.2.

Timothy Raison, to the effect that "the IBA has been re-examining the likely viability of these two stations and has, in consequence, sought his agreement to a revision of these two proposals so that Northampton and Milton Keynes would be serviced by separate stations, but associated so that some programming could be common to both".[354] Similarly, in the following year, when there was public support for a franchise in Doncaster which could affect frequency availability for Barnsley, Whitelaw found himself making a formal statement to the House of Commons that he was asking the IBA "to look at the development of ILR in South Yorkshire as a whole".[355] Heavy touch regulation remained the order of the day for the government in its relationship with the IBA, just as for the IBA in its relationship with the ILR companies.

That is not to suggest that Whitelaw was unsympathetic; quite the reverse. He took a personal interest in local radio, and understood it considerably better than many political grandees appreciated the intricacies of broadcasting generally. As MP for Penrith and the Borders from 1955, he had been an active supporter of BBC Radio Cumbria, and regularly took the opportunity to appear on air to speak to his widespread constituency. He was equally sympathetic to independent radio, and was a real asset to those who wanted to grow local radio generally in the early eighties, as had been his Labour predecessor, Merlyn Rees, another who understood local radio. However, when Whitelaw was succeeded by Leon Brittan following the 1983 general election, he proved a much less traditional, more libertarian Tory. Brittan flirted with community radio,[356] and shared his prime minister's mistrust of both the BBC and the IBA.

The new ILR companies owed most of their style and approach to that which had been developed by the first nineteen, and tested in the fire of the seventies. The franchises were all awarded to broadly-based consortia with an emphasis on local shareholdings and directorships.[357] The first groups were notably representative of their localities. Centre Radio reflected the textile trade in Leicester, through a 10 per cent holding from Palma Textiles and smaller shareholdings from the National Union of Hosiery and Knitwear Workers, and Montfort (Knitting) Mills. Essex Radio featured the makings of its own TUC, with the Society of Allied and Graphic Trades (SOGAT), the Association of Professional, Executive, Clerical and Computer Staff (APEX), the Electrical, Electronic, Telecommunication and Plumbing Workers Union (EEPTU) and the National Society of Operative Printers, Graphical and Media Personnel (NATSOPA), alongside Keddies, the local Southend department store. There could be little mistaking the Aberdonian roots of North Sound, with the Northern Cooperative Society and the Hugh Fraser Trust alongside banking and offshore oil interests. Max Bygraves introduced a hint of stardust, by owning 4.8 per cent of Lord Stokes' Two Counties Radio in Bournemouth, while there were echoes of a mercantile past in Imperial Tobacco as the largest shareholder in Bristol's Radio West.

354 Minister of state (Timothy Raison) written reply to parliamentary question 19 July 1982 *Hansard* Commons cols 22/3.

355 Home secretary (Willie Whitelaw)written reply to parliamentary question 28 February 1983 *Hansard* Commons col 4.

356 See chapter 11.

357 Information on shareholdings, directors etc comes from successive *IBA Annual Reports* for 1979-80, 1980-81, 1981-82, 1982-83, 1983-84 and 1984-85..

The new stations drew heavily on their predecessors in recruiting their management. John Bradford left Radio Tees to launch Mercia Sound in Coventry, and Allen Mackenzie escaped from Beacon to run Radio Tay for Dundee/Perth. Cecilia Garnett left AIRC to run Hereward Radio in Peterborough, Colin Mason went from programme controller at Swansea Sound to be MD of Chiltern Radio. Eddie Blackwell left the national sales agency Air Services to lead Essex Radio in his beloved Southend, and Chris Yates moved from Radio 210 in Reading to start up Radio West. That pattern was to be repeated as more stations came on air, and the names of those who were to be at the heart of ILR's future development began to appear with increasing prominence. Hereward recruited Ralph Bernard from Hallam to be the head of news and information. Paul Brown became head of programmes at Radio Victory, when John Russell moved there as MD after its troubled start. Similarly, Richard Findlay had moved from Clyde to Forth, then become its MD when Christopher Lucas headed back south. Each move opened up opportunities for others. I myself succeeded Cecilia Garnett at AIRC in 1979, and then Chris Yates at Radio 210 in 1981.

The most dramatic move of all was the appointment of John Whitney as director general of the IBA in 1983, following the retirement of Sir Brian Young. Lord Thomson, former cabinet minister in Callaghan's government, EC Commissioner and eventually of the Liberal Democrats, had succeeded Lady Plowden as chairman of the IBA in 1981. *The Times* noted that Thomson "turned the IBA chairmanship into more of an executive position".[358] Colin Shaw, then director of television at the IBA, and John Thompson may have felt that the succession to Brian Young as DG probably lay between them. I was told shortly afterwards that they had met the evening before the selection board, and in their very gentlemanly way had agreed that each would be prepared to serve under the other. Neither of them – nor to be fair anyone else – had foreseen the outcome. Whitney's diary tells the story of his first meeting with the IBA, a "bizarre occasion". He had to be smuggled in to the IBA's Brompton Road offices via the car park. "I was given instructions that I had to enter by a side door. I rang the buzzer, and the door was immediately opened by Lord Thomson's PA, Carol van Herwaarden, who was waiting for me on the other side of the door. We took the lift to the 8th floor where Lord Thomson met me ... and the world turned upside down."[359] Whitney's appointment wholly confounded expectations. He was a commercial man, an impresario rather than an administrator, and a senior radio executive rather than a television mandarin (although he had enjoyed great success in making television dramas, notably *Upstairs Downstairs*).

To begin with, the ILR companies enjoyed the more relaxed regime. Young had been kindly but carefully formal, meeting the radio companies on an informal basis only when strictly necessary. He was avowedly 'old school'. Thomson had recruited Whitney precisely in order to find someone from a newer one. However, Whitney inevitably was soon drawn into the innate formality of the IBA and its officials. When the relationship between the Authority and ILR came under severe stress, over the period of the Heathrow Conference, he was unable to help the companies and everyone felt uncomfortable and rather sad as a result. Whitney's

358 *The Times*, in its obituary of Lord Thomson of Monifieth, 5 October 2008.
359 John Whitney in an interview with the author, 11 October 2007.

own trials as DG were only beginning. *Death on the Rock*, and the mother of all handbaggings from Margaret Thatcher, still lay ahead.[360]

In 1979 the key players in ILR seemed well established. Philip Birch had retired to Florida, but others of the 'founding fathers' remained, notably John Whitney, Jimmy Gordon, David Pinnell and Terry Smith. Prominent within ILR were Richard Findlay and John Bradford, who were leading figures in the Heathrow Conference. The company MDs were joined in AIRC debates by Mike Vanderkar, Eddie Blackwell and Dick Seabright, who were leading the three national sales agencies: respectively Broadcast Marketing Sales (BMS), Air Services and Radio Sales & Marketing (RSM). Theirs was not always a harmonious relationship. Something of a counter-balance could be found in attempts to bring stations together in a regional sales offer, to go up against ITV regions, but the sales agency rivalries bedevilled these when stations with differing representation wanted to link up. The most successful and enduring example at the start of the eighties was Scottish Independent Radio Sales (SIRS), which was to be a Scottish sales agency representing all the Scottish stations. AIRC had a couple of goes at launching a radio marketing operation for the whole of ILR, but these foundered.

The smaller station MDs were characters in their own right, but with a clear sense of scale. For Charles Braham, the local newspaperman who had stepped from a non-executive board role to become MD of Swansea Sound in a pre-launch crisis, ILR needed "a bit of entrepreneur, a bit of Barnum and Bailey, and a bit of old-fashioned hand-on-the-purse".[361] Plymouth Sound's Bob Hussell echoed that approach. "It's a peanut business, and your attitude has to fit the scale. You can't start thinking of grand ideas." Yet all saw it as more than just a business, and therein lay their real satisfaction. At the other end of the scale, while he was at Capital John Whitney was adept at adding a dramatic flourish to the essential message. "We have realised that being part of the community is actually good for business. It's the most marvellous opportunity to serve two masters – God and Mammon."

There was genuine camaraderie between the senior executives in different companies: the managing directors, attempting to keep their companies alive while still relishing the fun of being on the shop floor; the programme controllers, each with a draft schedule for a national network in their knapsack; the sales directors, later mainly bright young women, contributing directly to the style and heartbeat of all aspects of the stations; and the chief engineers, sober hands-on techies who had often built the studios they now kept going. Almost everyone could be a broadcaster. Bill MacDonald, MD of Hallam, started the trend of the managing director being the classical music presenter. The ILR companies were virtually all cottage industries, and they relied greatly on informal networks of support, as well the more formal overview enjoyed by the Canadians and such as Associated Newspapers.

Recruiting on-air talent was very challenging. Accomplished broadcasters and news editors could be lured to the larger stations, but only into management roles and not always with success. Some were thought to be a real catch for but then seemed to lack the necessary dynamism to programme an independent station. A number of junior presenters from BBC local radio enjoyed the chance to work

360 See chapter 12.
361 *Management Today* September 1979.

with more freedom, and for longer hours, in the independent sector. Some broadcasters had been local club DJs, radio enthusiasts or from hospital or student radio. UBN – the United Biscuits Network, in-factory radio stations for biscuit companies – was a significant source of major talent. This unlikely source yielded presenters such as Roger Scott, Adrian Love, Nicky Horne, Graham Dene, Roger Day and Dale Winton before the network closed in 1979. Some DJs who had fallen out of the national media might find a home in a local station for a while. Bob Harris, once of *The Old Grey Whistle Test*, and Howard Pearce from Radio Luxembourg, were both presenters on Radio 210.

Still, all companies – and the smaller stations in particular – had to grow their own talent, and there was an obvious need for training within ILR. The establishment of the National Broadcasting School in November 1980, started by Capital but funded from secondary rental,[362] was a major factor in keeping the quality of ILR broadcasting to at least an adequate level, and often much higher than that. In these years of 'meaningful speech', each presenter had to drive the broadcasting desk without help from a technical operator or producer; handle multiple external feeds from IRN, AA Roadwatch, British Airways, external contributors and agencies; play music on turntables, advertisements from cartridge machines – and *still* engage the audience individually. It required real skill and fluency, and was a talent under-rated by all except those who needed to employ them.

Journalists usually came from local newspaper newsrooms or from BBC local radio, which meant at least that they had a grounding in the basic rules governing defamation and contempt of court. The IBA produced regular guidance notes, usually through the RCC papers, to help the companies. ILR was very exposed by the junior nature of their news staff, but with the help of those guidelines they mostly avoided the worst pitfalls. I presented a major paper setting out guidelines on privacy, gathering of information and crime reporting, to the RCC in January 1979.[363] The routine work of the quite substantial (by modern standards) local news teams, and some investigative programming, were welcome to the IBA, but it was also conscious of the pitfalls. The guidelines explained the rules on using interviews, recording telephone interviews and the use of hidden microphones, as well as more generally fairness and impartiality. These 17 pages were intended to help stations avoid contempt of court, to know how to handle crime reporting, and how to deal with the occasional highly sensitive kidnapping or security issue. This was the ninth in a series of such papers, stretching back to January 1974, and where newsrooms saw them – which happened mostly but not everywhere – station staff found them a useful substitute for the panoply of senior management and legal advice on which the BBC and newspapers relied. It was a not inconsiderable achievement to have so many small, new ILR companies on air, all broadcasting regular news which they had originated themselves, and so few disasters.

Finding – and keeping – advertising sales teams was a major challenge for all but the largest companies. Recruiting was mainly either from local newspapers, or of salesmen and women from other industries. The pay was not high, bonuses depended on the erratic local sales, and a successful recruit looked for quick

362 See chapter 9.
363 *IBA Paper* 58(79) 11 January 1979.

promotion or found that they themselves were a desirable commodity within the local labour market. Then at the bottom of the food chain, but steadily climbing, were the unpaid volunteer assistants. To get a job in radio, the best thing a youngster could do was to hang around their local station, or one of its outside broadcasts, in the hope of being asked to carry something or get the tea; exploitation, but a way in to an ILR career.

The seventies closed on a good commercial note for ILR. For the second year running, all traded profitably in the financial year to the end of September 1979, and by the end of March 1980 all except LBC had paid off their pre-operational costs and accumulated losses from the early years. Ten companies had done well enough in 1979 to pay secondary rental, totalling £2.71m. Advertising revenue for 1979 was up 37 per cent on the previous year, and for the year to September 1980 rose to £44.3m. Piccadilly was reporting annual advertising revenue of £3m, and Clyde £2.5m, in what was starting to be described as a "bonanza".[364] Even though Capital's break from BMS in June 1979 had meant that there were four separate agencies trying to get into national agencies and advertisers to sell ILR, it seemed for a while that the market could bear that. All the ILR companies had benefited from strike action at ITV in August and September 1979. The entire commercial television network (except Channel TV) was blacked out from 10 August for 10 weeks, not resuming until 24 October. Estimates put the revenue increase for ILR in that time at between 20 to 30 per cent, with some stations unable to find enough time from their statutory maximum of nine minutes advertising per hour to satisfy demand. With the first of the new ILR stations ready to roll off the IBA's 'production line' in the spring of 1980, and the start of the Performing Right Tribunal hearings on 19 November, the year was ending on an unusually optimistic note.

The financial pattern continued into the start of 1980. Again all the original nineteen traded profitably, as did Mercia and Hereward, and by the end of March 1981 LBC had covered its accumulated losses. By then, though, there were signs of a slowing down in advertising growth, and the 1980 Broadcasting Act had introduced yet another tax on ILR profits; a levy set at 40 per cent of profits after secondary rental but before tax, over a threshold of £250,000. This further burden was one of the factors which tipped the ILR companies into overt dissatisfaction with their regulatory lot. From October 1979 the IBA had also re-cast the basis on which it assessed primary rental, to the benefit of the smaller and medium-sized companies but the disadvantage of larger companies; and had done it again, with the same effect, from January 1982. The top rate of secondary rental also increased in 1980 from 50 to 55 per cent. This was all in order to fund the building of transmitters and other costs associated with the new ILR stations, given the prohibition on cross-funding within the IBA from television to radio, but it was provoking discontent. Controls also remained strict on trading in company shares, irritating the companies, with the IBA continuing to insist on local owners (with the awkward exception of overseas 'white knights' such as Standard Broadcasting). The Authority felt obliged to scrutinise such earth-shaking deals as "a 3.9 per cent holding in Hereward Radio ... transferred from Grill Floors Ltd to a variety of new

364 *Sunday Telegraph* 1 October 1979.

and existing local shareholders".[365] and Mr W S Fyfe's sale of his 6.5 per cent holding in West Sound to McGhie's Dairies Ltd.[366]

Meanwhile, just as ILR had been brutally battered, soon after its launch, by the down-turn which followed the Arab oil embargo in late 1973, so the new-found ILR prosperity and expansion of the early eighties was hit by another damaging recession. Revenue for the first nineteen radio companies fell 14 per cent in 1981, compared with the strike-enhanced figures of the previous year. Comparable national advertising was down fully 23 per cent, and it was only the revenue from the new stations which kept the network fall overall to just 5 per cent. Profits fell or disappeared for the established companies, reducing the secondary rental take to £1.3m. All but three of the new stations found themselves operating at a loss. There was some recovery in advertising revenue in the first half of 1982, but it was not sustained. There were widespread staff cuts. Only four companies paid secondary rental in 1982, a total of just £800,000, yet three of them – Capital, Piccadilly and City – also paid levy to the Exchequer of £500,000.

Nationally, the optimism with which 1980 began soon disappeared amid a sea of troubles. The structural weakness of the UK economy had not gone away just because there had been a change of government. Unemployment, which had been falling at 1.2 million when the Conservatives took office, rose to 2 million by November 1980. Manufacturing output fell 16 per cent that year, inflation rose again to 20 per cent and a deep recession took hold. It barely seemed to respond to the orthodox monetarism with which Margaret Thatcher's Treasury tried to meet the challenge, and many, including in her own cabinet, thought that worsened an already bad situation. Despite the prime minister's assertion at the 1980 Conservative party conference that "the lady's not for turning", the government backed off from a confrontation with the National Union of Mineworkers (NUM) in February 1981 over proposed pit closures. Still, a defiantly non-Keynsian Budget a month later made it clear that the proposed economic medicine would be unpalatable.

Violence stalked politics. Thatcher's close friend and adviser Airey Neave was killed by a car bomb in the House of Commons car park in 1979. Scion of the establishment, Lord Mountbatten, was assassinated in Ireland in October. IRA men interned in prison – disappointed in their expectation of negotiations – began a hunger strike in which their leader, Bobby Sands, and nine others died in 1981.[367] There were riots in Brixton in London and Toxteth in Liverpool, and yet more strikes. It was starting to look as if the Thatcher revolution had been tried briefly and was to be discarded. But then Argentina took a hand, and transformed the fortunes of the prime minister and, as a by-product, the course of independent radio.

On 2 April 1982, the Galtieri junta in Argentina invaded the Falkland Islands,

365 *IBA Annual Report* 1980-81 p 49.

366 *IBA Annual Report* 1981-82 p 52.

367 The government stood firm against Sands, but had been much less obdurate the year before in a circumstance more closely linked with broadcasting. Some proponents of a separate Welsh language television channel instead of Channel Four, had resorted to direct action, including raids on transmitters in 1980. When the 68-year-old president of Plaid Cymru, Gwynfor Evans, declared he would begin a hunger strike the government capitulated, with home secretary Willie Whitelaw uttering the heartfelt phrase "We can't let poor Gwynfor die". Thus, as Maggie Brown relates in *A Licence to be Different, the story of Channel 4*, Wales got S4C and ILR another competitor.

a British territory in the South Atlantic. Against orthodox military strategy, and with distinctly lukewarm backing from the US, Margaret Thatcher's new 'dry' cabinet resisted what seemed another inevitable climb-down. A British naval task force, with its supply lines extended halfway round the world, re-took the Falklands on 14 June. Thatcher's popularity rose at once, and the so-called 'Falklands factor' sustained her in office for the rest of the decade, most immediately to a landslide general election victory on 9 June 1983, against an opposition hopelessly divided between a militant-weakened Labour Party and an opportunistic Social Democrat Party. It also provided the government with the public support – and the political backbone – to accomplish its three defining aims: the defeat of the NUM, and with it militant unionism elsewhere; EC budget re-negotiation; and – through privatisation – the first dismantling of the economic structure of the Attlee years. The prospect of escaping the seemingly endless and debilitating labour relations troubles heartened those running the ILR companies; the concept that the state (in the form of the IBA) might be made less intrusive, emboldened them. This was the point at which – with hindsight – it is possible to identify the start of the move across the fulcrum, as *independent* radio began to aspire once more to be *commercial* radio.

Yet for ILR, the immediate gain from the Falklands War was in its public service response. Just as the local stations had gained whenever there was a local crisis, harsh snow or local floods, so this national emergency confirmed ILR's status as a credible news medium. Each local station was able to offer reliable news of the war – through a very competent IRN – heightened by local relevance. It was personal, local, authoritative and immediate. For IRN, Antonia Higgs in Buenos Aires and Kim Sabido on board the *Canberra*, and then from the Falklands, provided vivid reports, available to each ILR station as their 'own'.

The most poignant story comes from Radio Victory in Portsmouth. In the home of the Royal Navy, any failure would have been devastating, success disproportionate. Invoking the shade of Nelson, which befitted the station's own name, MD John Russell sent a memo to his staff telling them that 'England expects'. The station broadcast round the clock for the first time, wheeling in all its freelancers to do live overnight shifts. Penny Guy, whose husband Derek was serving on *HMS Sheffield*, was reading the early evening news on 4 May when the MoD announced that the ship had been attacked. She read that story without any apparent tremor, although it was to be a long time before she knew that her husband was not among the 20 dead or 24 injured. When *Sheffield* sank, it was the first British warship to be lost for 37 years. Chris Carnegy recalls that "legend has it that Derek emerged from the Sheffield with only the clothes he stood up in and a cassette of Barbra Streisand's *Greatest Hits*".[368] The story is pure ILR, although Radio Victory was to survive *HMS Sheffield* by only just over four years.

Buoyed by its Falklands prestige, ILR was given another pointer towards a freer commercial existence by the first public flotation of an ILR company. On 27 May 1982, Radio City (Sound of Merseyside Limited) was listed on the Unlisted Securities Market. The flotation was particularly successful, being hugely oversubscribed. Nevertheless, there was no real sign of any sustained upturn. In the harshness of the continuing recession, advertising would increase for a short while

368 Chris Carnegy in an interview with the author, 18 December 2007.

and then fall back again. CBC, Radio Aire and Centre Radio had all needed deferment of their IBA rental payments, as they struggled to cope with the economic winter compounded by their own management failures; and, in the case of CBC, its hopelessly over-wrought 'democracy'. CBC was re-structured and eventually taken over, Radio Aire similarly 'rescued' by Red Rose. The IBA was starting to bow to what seemed inevitable, and allowing stronger companies to own weaker ones, although still characterised as 'rescues'.

There was no rescue for Centre Radio, however. The station had been bedevilled with problems since before its launch. The IBA awarded the franchise in May 1980 to a group chaired by the BBC radio personality Lady (Isobel) Barnett, a former Lady Mayoress of Leicester. Convicted of shoplifting amid a glare of publicity in October, she committed suicide. The company lost its next chairman, Kenneth Bowder, less tragically – in May 1982 he walked out in the middle of a board meeting after "fundamental and irreconcilable differences of opinion".[369] MD Ken Warburton quit, followed by most of the senior directors over the following months, leaving Standard Broadcasting's Bob Kennedy holding a very sickly baby indeed. Centre Radio went off air on 8 October 1983 after just two years. This was the first ILR company that the IBA had actually allowed to go out of business, although many had been in a similar plight in the past. Its epitaph was written by journalist Terence Kelly, a wise observer of ILR: "Leicester and its surrounding area can quite patently support an ILR station as the figures ... prove. The board was unable, or perhaps unwilling, to raise more capital to keep Centre going ... . [I am] left with the conclusion that some boards know nothing about radio and have no business being in ILR."[370]

There was an historic irony in the first ILR failure happening in the city which had seen the first local radio station in the UK, BBC Radio Leicester, back in 1967. The IBA re-advertised the franchise. A new station, Leicester Sound, came on air on 7 September 1984, with Radio Trent in Nottingham as its major shareholder, sharing programming for 12 hours a day, accounting, commercial production and administration. Such almost covert takeovers began to be routine as the eighties progressed.

There were at last some modestly cheerful signs for ILR as a whole, by September 1983. The first 32 companies were showing an increased advertising take of 11 per cent, which rose to a network increase of 18 per cent when more new stations were taken into account. Of those first 32 companies, 24 reported trading profits for the year ending September 1983, and 17 of those paid dividends to their shareholders. Seven companies paid secondary rental, totalling just under £1m. LBC, however, was finding the going particularly tough, with continuing union and staffing problems, a sharp decline in listenership during 1983 and severe competition from other media. The IBA members were warned in September that "LBC/IRN's problems are acute; and will call for careful, calm assessment over the next few months".[371]

However, across ILR, advertising revenue growth remained unreliable. It had looked promising in the first half of 1984, but slipped back again in the second half,

369 *Leicester Mercury* 18 May 1982.
370 *Radio News* 21 October 1983.
371 *IBA Paper* 147(83) 28 September 1983.

giving an overall like-for-like increase of just 5 per cent, boosted to 10 per cent by ten new stations. In the first months of 1985, revenue actually fell for the established stations by 5 per cent. For the financial year up to the end of September 1984 careful cost control had allowed some modest increases in profit here and there. Of the 36 established companies, 27 had reported trading profits, and 17 of these paid dividends. Seven of the newest 14 companies were trading profitably, with Red Rose and Chiltern being among the 10 most profitable ILR companies. Nevertheless, there was still no real sense of significant progress, and as the companies and their shareholders cast around for reasons, the sense of being thwarted by the IBA loomed larger. The IBA had once more 're-aligned' primary rentals, with effect from July 1984, and again a disproportionate increase fell on the largest companies. That was offset by a reduction in the top rate of secondary rental from 55 to 45 per cent, but the indignation of the ILR companies boiled over. That took the form of a direct challenge to the IBA, following the Heathrow Conference in June 1984, in circumstances discussed in chapter 10.

As described there, the IBA took the unprecedented step in response of reducing primary rental by 10 per cent from April 1985. That may have been a useful tactical response to the antagonistic stand-off with the companies, but its effect was to choke off the supply of capital to the IBA's engineers to build new transmitters. That virtually brought to an end the second phase of ILR expansion. The IBA's achievement had been substantial. Since the seventies embargo had been lifted, it had identified, cleared frequencies for, built transmitters, advertised and awarded 32 new franchises in the 7 years between 1979 and the end of 1985. This had been done at high speed, through another challenging recession and in difficult political circumstances. The efforts of the Radio Division team – and their continuous pressing of the IBA finance and engineering staff to go to the very limit of what was possible – lay at the heart of the licensing achievement.

In 1977 the IBA had proposed the introduction of 'break-points' into ILR contracts after 10 years, when "there would be a presumption in favour of opening the franchise to competition",[372] and that was included in the 1980 act.[373] As AIRC had foreseen, this introduced additional uncertainty for the companies, and a constant programme of work for the IBA, from 1982 onwards. The Authority was obliged to review fundamentally every contract after 10 years, opening each up to new competition. Generally, the re-advertisement and re-award of the franchises to the incumbents proceeded for a while with little controversy. The first franchise to be re-advertised was that of LBC/IRN on 31 March 1982, at the company's request a year earlier than the legislation required. The company was re-appointed for a further eight years at the end of September, against competition from London Sound. The great prize was clearly Capital's London (General & Entertainment) franchise, and two companies challenged the incumbent when applications closed on 21 June 1983: A2Z and Metropolitan Radio. Given the extent to which the IBA had identified the fortunes of ILR with Capital, there was little surprise when they were re-appointed.

Only two contracts appeared vulnerable, Beacon and Victory. Beacon faced no competition, leaving Victory potentially very exposed. Victory had not begun

372 *IBA Minutes* 413(77) quoted in *IBA Paper* 14(78) 19 January 1978.
373 Broadcasting Act 1991, s 19 (2).

well, back in 1975, and the Authority had delayed 'rolling' its contract at the end of 1976, troubled by programming quality, local news and the capital base of the company. Selkirk Communications came to the rescue financially, its investment taking it to a 16 per cent holding, and the contract was eventually extended in February 1977. Programme controller Dave Symonds had departed in August 1976, and his successor Eugene Fraser left the following March. The station did not have much luck. When it tried to launch a property free-sheet paper, it upset the IBA's delicate relationship with the Newspaper Society. Victory found itself vigorously opposed by Robert Stiby who, although a major figure on the Capital board, was also the new president of the Society. He argued that "if the IBA says it is a free world then we should be free to own radio stations, but we would prefer that local radio stations are prohibited from launching local papers".[374]

By that stage, the IBA was re-advertising some of the franchises for a wider geographic area, as a way of spreading ILR coverage into localities which were not "felt to be financially capable of supporting a 'free-standing' ILR station". The costs of the extra transmitters were to be paid for on the 'forward funding' principle, which involved applicant companies paying the IBA directly for many of the costs involved in setting up transmission facilities. Nottingham, Belfast, Reading and Wolverhampton were offered the potential to increase their coverage, to take in Derby, some of the west of Northern Ireland, Basingstoke & Andover and Shrewsbury & Telford respectively. It was still not a complete solution to the post-Heathrow constraints. The Authority was told early in 1985 that "apart from the willingness [or otherwise] of contractors to find funds, the pace of forward funding is largely conditioned by the resources of the engineering division".[375]

Victory had struggled to build the type of audience levels enjoyed by other stations away from the London conurbation, and the 1980 JICRAR showed a weekly reach of just 35 per cent, one of the lowest of any station apart from LBC. John Russell had taken over as managing director on 3 September 1980, with Paul Brown joining him from BRMB. At that stage the IBA's assessment was that "matters at Victory are very much betwixt and between ... the company now has a good chance, through recent management changes, to put itself on a much firmer basis".[376] The Falklands performance and general improvements suggested the station was gradually doing better, but the IBA's decision to offer an extended Portsmouth/Southampton franchise at re-advertisement seemed ominous.

As the IBA team assembled to interview the three applicant groups for the widened franchise on 2 October 1985, they had in front of them impeccably objective briefing papers, but with a potent sting in the tail. "Members will appreciate that this Portsmouth and Southampton ILR franchise is the most briskly contested radio contract for some while; indeed, probably not since the re-advertisement of the two London franchises (in 1982 and 1983) has an existing contractor seemingly faced such serious competition from a fresh applicant. The Ocean Sound group in particular, with several years of preparation (initially expecting to bid for a Southampton-only franchise) and some helpful guidance from the successful ILR company in neighbouring Bournemouth, appears to have mounted

374 *UK Press Gazette* 9 May 1983.
375 *IBA Paper* 18(85) 1 February 1985.
376 *IBA Paper* 284(80) October 1980.

a credible application for this new, combined area. The existing Portsmouth contractor, Radio Victory, although it has built a rather stronger reputation during the last few years, had a bumpy beginning, both commercially and in programming. These two main applicants have evident strengths and weaknesses, which members will wish to weigh carefully before reaching a decision."[377]

The Authority reviewed the three applications for the extended franchise, and on 17 October 1985 awarded it to Ocean Sound. The loss of the franchise by Radio Victory drew from the locality some of the support which had been lacking beforehand. A petition with nearly 50,000 signatures was presented to the IBA. The *Portsmouth News*, studiously failing to mention that it was with Selkirk the largest shareholder in Victory, claimed that "the IBA ... was always going to pull the plug on the Fratton Road based station".[378] The station's chairman, Michael Poland, was to claim that "we had a warning from an official of the IBA, which was minuted. It said 'the IBA to show it has some teeth left will probably make a sacrificial lamb of Radio Victory'."[379] Actually, the franchise award to Ocean Sound seems well-justified by the detail and argument in the IBA papers, but the decision left anger in Portsmouth, and discomfort within ILR generally.

Victory failed to sustain broadcasting up to its nominal franchise end date of 13 October 1986, and closed down on 28 June. Its farewell compilation programme struck one of relatively few dignified notes from the company in the whole episode. The final track played was Neil Sedaka's *Our Last Song Together*. After a valedictory commercial from the Television Parlour retail chain, and the sound of the Portsmouth Guildhall clock striking twelve noon, Chris Carnegy signed off: "Serving southern Hampshire, West Sussex and the Isle of Wight for ten years and 257 days, this has been Radio Victory".[380] Even now, it is hard to listen to without choking up at least a little.

These years also saw the arrival of new broadcasting competition. Some of it did not amount to much. In 1985 the Home Office set up and then abandoned in 1986 an experiment in 'community radio', which threatened to outflank both the IBA – which was probably its intention – and less intentionally the ILR companies too. In the event, it failed without a single new service taking to the airwaves.[381] The Cable Authority, from December 1984, did its best to promote 'wired Britain', but since free-to-air transmissions work notably well in the small islands of Britain, there was little appetite yet among television audiences to pay for the privilege of getting something they already enjoyed free.

Channel Four arrived on 2 November 1982 in a form surprisingly close to that envisaged by the Annan Committee, and unlike any of the ITV franchises. Established by the 1980 Broadcasting Act to be a publisher of television programmes rather than a programme company, its board was to be appointed by the IBA itself. In her seminal history of the channel, *A Licence to be Different*, Maggie Brown records that the IBA's Lady Plowden asked Capital's chair, Richard Attenborough, to become chair of Channel Four. He declined, although he remained its "godfather" [382] as deputy chair. ITV was to be obliged to fund Channel Four

377 *IBA Paper* 126(85) 25 September 1985.
378 *Portsmouth Evening News* 2 June 1986.
379 *Portsmouth Evening News* 26 June 1986.
380 *Farewell to Victoryland* Chris Carnegy.
381 See chapter 12.

and sell its advertising, thus reducing commercial competition within television when compared with the effect of an ITV2, and – as an unintentional by-product – easing what would have been a difficult competitive situation for radio. The main effect on ILR would be the greater availability of television advertising minutage, which meant that there was less chance of advertising which could not find an affordable television outlet being 'squeezed' onto radio. Still, an additional commercial television channel had seemed inevitable for quite a while, and the unusual structure was to radio's advantage. When it was launched, ILR prepared to weather any early storms.

The ILR companies had been more nervous about the prospect of direct competition from breakfast television, from the very start of independent radio. Before Capital had even begun broadcasting, Richard Attenborough was writing to John Thompson, relieved that the new company was not going to have to face that from the start. "I was really thrilled to gather from you that our fears, in terms of a concurrent opening date as far as Breakfast TV is concerned, can be eliminated. Obviously one fully understands that this particular innovation is inevitable; it was merely the timing that was causing us concern."[383] This was a consistent theme in the correspondence surrounding Trident Television's experiment with a breakfast-time programme in May 1977, with the added complication that this clashed with one of the early national JICRAR research sweeps. The companies objected to being presented with "something of a fait accompli", and tried in vain to get a caption broadcast at the 9.30 close-down each morning, cross-promoting to ILR, during the experiment.[384]

By the end of 1979 it was clear that, buoyed by its new lease of life and having seen off the main threat of the Annan recommendations, the IBA would be advertising a breakfast television franchise swiftly. LBC felt especially threatened. It feared that "the advent of BTV [breakfast time television] is likely to have a seriously adverse effect on our revenue."[385] It contemplated bidding for the service itself, but after detailed consideration had concluded that "the uncertainty of revenue levels [for BTV] in the early years ... was beyond our financial resources". It therefore pitched to the IBA that they should ensure that the winning TV applicant offered LBC a small shareholding – they posited five per cent – in exchange for editorial and programming input. The IBA's own assessment was that "there would be a general threat to ILR companies, but it is probable that it could only develop to a serious level in the case of LBC because it is metropolitan, and because it supplies a service similar to that proposed for the new franchise".[386] Privately, the IBA considered that "this might be countered to some extent by LBC having a stake in the franchise-holder, or supplying some services – for example local news if there were an opt-out in due course for London".

Already faced with the prospect of the start of Channel Four at the end of 1982, AIRC argued that breakfast television would seriously affect the financial viability of the new stations, and urged that it should be delayed until ILR coverage had

382 Maggie Brown, *A Licence to be Different; The Story of Channel 4*, BFI London 2007, p 29.

383 Letter from Richard Attenborough, Capital Radio, to John Thompson 21 June 1973.

384 Letter from John Bradford, Radio Tees, to John Thompson 24 February 1977.

385 Letter from Sir Geoffrey Cox, Chairman LBC to Lady Plowden IBA 7 May 1980.

386 *IBA Paper* 261(80) 28 August 1980 Roy Downham (IBA Director of Finance) assesses the financial implications of breakfast-time television.

reached 85–90 per cent of the UK, which would not be until at least October 1984. The Authority was strongly urged by its head of radio finance, Peter Faure, in May 1981 to delay the launch. "One doesn't want to be over-protective towards ILR, but I can see no reason for altering the view that the later BTV starts, the better for ILR."[387] Initially the Authority relented, and when it awarded the franchise to TV AM in 1981, that was on the basis of a launch in autumn 1983. However, pressure from Peter Jay on behalf of the self-styled 'famous five', persuaded the IBA in June 1981 to bring the launch date forward to May 1983. This dismayed AIRC, which felt that "when the Authority has made a decision stating that it had regard to the effect on Independent Local Radio … it should abide by that decision when the circumstances are not materially different from those that pertained when the decision was made".[388]

Dismay turned to anger when an advance of a further three months was agreed in May 1982. A delegation from the ILR companies, led by the current AIRC Chairman Terry Smith, met the full Authority on 4 November 1982 where they raised issues over the many apparent concessions to TV AM, and their own straitened circumstances. They gained nothing on this point, and TV AM launched on 1 February 1983, two weeks after the BBC's *Breakfast Time*. Despite the shambles of the launch, and all TV AM's subsequent financial and reputational troubles, the IBA's repeated concessions to the television company reinforced the growing view within AIRC that they were poorly served by the IBA. New television opportunities were becoming dominant, which was to their direct disadvantage.

Similar issues arose early in 1984, when AIRC got wind that ITV was pressing for a move into 'coffee time television' on weekday mornings, shifting its schools programmes to Channel Four. It looked to the ILR companies as if the IBA was going to let them down again, to go off with its television friends. When in July 1985 the issue of diversification into magazines and periodicals went sour, Richard Findlay wrote to Whitney, as much in sorrow as in anger: "When the Authority decided to proceed with Channel Four, we warned that a sudden glut of television advertising minutage would have an adverse effect on our revenues. When the Authority decided to proceed with breakfast television, we warned that the BBC would not sit still, and the combined effect of the two breakfast television services would have an adverse effect on our audiences. Both predictions have come to pass. … One of the reasons why radio was placed with the IBA was that in a *regulated* system of independent broadcasting, the Authority was able to maintain a balance between the needs and aspirations of each part of the system … Why does the IBA do something like this, particularly at a time when we are trying so hard to establish harmony between us?"[389] That changed nothing. Despite AIRC estimating a loss of 300,000 listeners each week and revenue shortfall of £3.6m a year,[390] ILR lost the 'coffee time television' battle too, with schools' programming shifting to Channel Four from September 1987.

By the mid-eighties, for all the expansion in the number of services, mainstream popular radio seemed short of energy, waiting for change. Queen's drum-

387 Memo from Peter Faure to Roy Downham, May 1981.
388 Letter from Bill Coppen-Gardner, Director AIRC to Sir Brian Young, 30 April 1982.
389 Letter from Richard Findlay to John Whitney, 19 July 1985.
390 Letter from Richard Findlay, Chairman AIRC, to John Butcher IBA 16 September 1985.

mer, Roger Taylor, locked himself in a room in Los Angeles and wrote down his frustration with contemporary music radio, in a song which became *Radio GaGa*. He saw radio as having declined into a secondary medium – bland, background, ordinary – and a poor successor to the supposed golden age of 1950s Radio Luxembourg and the 1960s pirate ships. That was not how ILR was in 1984, but the seeds of just that change had been sown. The frustrations of the radio companies were reaching the point where they would demand fundamental shifts, and run the risk of making Taylor's words come true.

Chapter 9

Doing well by doing good

Secondary rental and programme sharing

Nothing more clearly differentiates *independent* from *commercial* radio than what Gillian Reynolds has called "John Thompson's masterly invention",[391] the device of 'secondary rental'. This meant, in essence, that once a company made a profit on turnover above a percentage threshold, it would pay a proportion of that profit back to the IBA, over and above the fixed 'primary' rental payable for its franchise. The IBA would then use this fund for the benefit of the ILR system. This mechanism applied from the start of ILR up until 1988.

The wide sense of nervousness in the seventies that the new radio services might turn out to be American-style radio meant that there had to be a structure in place in the legislation and its regulation to prevent 'excess profits'. Given the narrow support base which ILR then enjoyed, these concerns had the potential to prevent it happening. Personal enrichment was not a comfortable topic in the early seventies, when super-tax for the highest earner could total 19 shillings in the pound. It was to be 21 years after George Harrison had berated the *Taxman*[392] before the appearance of Gordon Gekko, and the 'greed is good' notions of the nineties. The provision in each ILR licence for secondary rental, as a way of removing some profit above a fairly low level, served the political purpose. Thompson's achievement lay in arranging that these moneys could be redeployed to enhance independent radio as a whole, rather than disappearing into the maw of the Treasury, as happened with the ITV levy. Secondary rental served quite quickly to add those special elements of programming, training and technical facilities which lifted ILR out of the ordinary, and for a while made it a genuine rival even to some of the more highfalutin characteristics of the BBC.

Those running the ILR stations thought less well of secondary rental. AIRC's position was that "secondary rental bites too hard and too soon",[393] while Clyde's Jimmy Gordon complained in 1977 that "out of a pre-tax profit of £500,000, for example, we end up with just over a third. At the top end, half the profit goes to the IBA in secondary rental, and then 52 per cent of the rest goes in corporation tax. The amount of money a company loses control over is ridiculous." In the high taxation regime of the time, and given the suspicion about 'a licence to print money', retaining one third was actually not too bad. It was, perhaps, Gordon's point about "control" that was really the matter at issue. Much of the money would

391 Gillian Reynolds in an interview with the author, 25 November 2007.

392 *Revolver,* The Beatles, 1966.

393 *Campaign* 9 December 1977.

go back to the company concerned for authorised projects, but it was the IBA's decision on where it should be spent; the company could only make proposals. The proponents of independent radio as public service saw it as a major opportunity to move that service to a higher level, in keeping with the social aspirations of the pre-Thatcher decades. It was also a valuable argument against a tendency of the BBC to claim sole guardianship of the public service broadcasting ethic in radio.

The mechanism built into the contracts from the start of ILR provided for a graduated additional 'tax' on company profits. The Authority was to use some of that itself to fund its radio work. It would deploy the rest for approved projects, which would benefit the system as a whole, and the communities served by the stations paying the additional charge. The rates set were 25 per cent on profits in excess of five per cent of net advertising revenue, after pre-operational costs and accumulated losses had been paid off; and then 50 per cent on profits over 10 per cent. In the first couple of years, the contractual provision was activated only for Radio Clyde, which was liable for £807 in 1975 and £25,000 in 1976. However, on the basis of their profits for the 12 months to September 1977, Capital, BRMB and Piccadilly, in addition to Clyde, all paid significant amounts, as did seven companies the following year, so that secondary rental quickly became very real indeed. Thompson had persuaded the Authority members in December 1976 that 40 per cent of the secondary rental due from a company would be used by the IBA itself "for the benefit of the ILR system", leaving 60 per cent "to be spent by the contractor on projects agreed with the Authority".[394]

The IBA's first spending decisions concerned the secondary rental payable by Radio Clyde. Thompson proposed to the IBA members returning £12,000 to Clyde for agreed items, with the IBA spending the balance. The IBA's list highlighted the obligation it felt toward supporting live music, given the history of the political bargaining over ILR. In case they risked forgetting that, Musicians' Union assistant secretary Stan Hibbert made public that they were pressing the IBA very hard for "a substantial proportion" of the secondary rental money to be spent on employing musicians. "No other competing interest", he argued, "has made such a substantial contribution to the success of ILR as music".[395]

This inital expenditure of secondary rental funded 11 programmes of classical music, to be recorded at the International Festival of Youth Orchestras in Scotland, and two chamber music series. £2,500 was to go towards the administrative costs of the Clyde '77 Arts Festival, with a further £2,500-3,000 to be expenditure on local musicians' employment. For the first two projects, the IBA had secured from AIRC an interest for these programmes to also be taken by other ILR stations. The medium-term projects were a series of 50-minute lectures by prominent public figures (including one by the Duke of Edinburgh which drew international interest), a booklet supporting a Clyde feature programme on orienteering, and an "educational experiment of a character yet to be precisely determined, possibly explaining aspects of popular music".[396] The balance of £8,635 was retained by the IBA for "capital expenditure associated with the sound broadcasting of parlia-

394 *IBA Minutes* 397(76) of a meeting on 2 December 1976.
395 *Campaign* 9 December 1977.
396 *IBA Paper* 156(77) 3 June 1977.

ment". There was also a notion for a possible "neighbourhood radio experiment", but this came to naught.

The approach contained in this first agreement for secondary rental expenditure set the pattern for the future. A good proportion of the money raised would be returned to the company paying it, but only for approved projects which bore the unmistakable hallmark of public service broadcasting. The money was thus kept within the ILR system, but on a very *dirigiste* basis. To set against the pain of reduced profits, there was prestige for the companies from the resultant programmes and projects thus funded. The IBA would announce the totals collected but not which projects were thereby funded, leaving it to the stations themselves to decide whether to acknowledge that their latest series of concerts, or the support form a new community project, had come from the "fairy gold".[397]

With Capital, Clyde, Piccadilly and BRMB all making substantial payments, secondary rental availability soon rose dramatically. It totalled £1m in 1978, a figure almost doubled in 1979. These were riches indeed, in the still precarious nature of ILR, but the Authority maintained its 40/60 formula. The IBA's expenditure focused on two main purposes: to enhance IRN, and to support transmission facilities. For the first, a total of £157,000 was earmarked to cover parliamentary broadcasting costs, to pay IRN the fees which Beacon was unable to meet, and to fund a return line to enable stations to feed good audio quality material into the national news service. This last was a real gain, and confirmation that IRN and the network could operate as a joint asset. To meet its second purpose, the IBA used almost a quarter of a million pounds to pay towards installing standby generators at the MF transmitters. The experience of power cuts arising from industrial action, and the threat of Irish or home-grown terrorism, had made this a priority for the Authority and – perhaps more surprisingly – for many of the companies too.

The sums returned to the companies in 1978 continued the approach of the previous year. Capital had weighed in with ideas for concerts and specialist music recordings, training, drama, and music composition awards; a set of proposals whittled down by the IBA to £260,000 from initially almost double that amount. Two major initiatives, the Capital Radio *Helpline* and its support for Operation Drake, were major recipients, as was involvement with the Young Vic for dramatised features. The *Helpline* was a 24-hour referral service for Londoners, linking on-air advice programmes with specialist agencies off-air. Staffed by teams of part-volunteer part-paid phone-answerers, and overseen by a distinguished but largely irrelevant advisory committee, this was a substantial undertaking. Operation Drake was a two-year expedition organised jointly by the Scientific Exploration Society and the Prince's Trust, involving sending youngsters on combined scientific exploration and community projects around the world. This was ILR showing its social conscience, validating its public service credentials, and getting some useful programming material and local affirmation in the process. Clyde received funds of £138,000 for music recording of easy listening material for overnight broadcasting, for concerts and recording generally, training, booklets to support programming initiative research and community service projects. BRMB's

397 Gillian Reynolds in *Broadcast* 21 January 1979.

modest £6,000 was to go on children's programming, Piccadilly's £60,000 to concerts, documentaries, community projects and a book on Piccadilly Radio.[398]

By the following year, based on profits from 1978/9, the IBA had over £2.7m to allocate, up from £1.9m the previous year, to ensure "the politically healthy and also socially conscious development of ILR as a self-financing system".[399] The IBA continued to honour what it saw as its commitment to support live music, involving projected expenditure of £1.2m out of the £2.7m, including establishing the *Great British Music Library*, originally recorded, non-copyright albums of music, with 200 copies of each for participating stations, to ease the needletime pressures. The list of projects and programmes to be funded by the companies included major classical concerts in London, at the Snape Maltings and in Aberdeen, Sheffield and Belfast; social action projects across the network; drama, and documentaries. This was a cornucopia of public service, and represented the apotheosis of the idea of independent radio, where the popularity, audiences, advertising and profits generated by radio enterprise would be re-invested in a brand new kind of public service broadcasting. Secondary rental, and the skill with which it was deployed both by the IBA and the companies at the end of the seventies, seemed to prove that ILR was working.

The 1980 Broadcasting Act provided a statutory basis[400] for the grants which the IBA was making to companies out of secondary rental, notably the funds going to those stations which were not themselves sufficiently profitable to pay it. Home secretary Leon Brittan confirmed that "there are no financial limitations on the amount that the IBA can pay a programme contractor"[401] and that "there is no statutory limitation on the IBA requiring it to issue contracts only to companies likely to make a profit".[402] There was pressure on the IBA to help establish 'community' stations, but it decided against direct funding of small stations, stressing that "groups applying for ILR contracts should assume that their services are to be provided on a self-financing basis unless special arrangements are said by the Authority to apply. Present contractors should assume that their operations must continue to be self-financing."[403] In a rare failure of foresight, John Thompson felt that "in the years ahead, working within the resources available at a given time, the Authority of the day could review its opposition in relation to this new section of the 1980 act". That time was not to come. The first years of the eighties were the high water mark for secondary rental, for the notion of social engineering through cross-subsidies, and in many ways for Thompson's concept of independent radio as a whole.

By the end of the seventies, the IBA had begun prodding the ILR companies about their training needs, partly in response to pressure from the NUJ. Secondary rental enabled some direct subventions by the IBA itself, and the Authority provided bursaries for post-graduate students at the National Centre for Orchestral Studies, and funded a radio lectureship at the School of Journalism

398 *IBA Paper* 167(78) 7 June 1978.

399 *IBA Paper* 357 (79) 12 December 1979.

400 Broadcasting Act 1980 section 27.

401 Home Secretary Leon Brittan reply to question from Philip Whitehead MP Commons *Hansard*, House of Commons col 449 – Standing Committee of the Broadcasting Bill 17 April 1980.

402 Ibid. col 452.

403 *IBA Paper* 285(81).

Studies at University College, Cardiff.[404] Then Capital Radio, not only as a result of being nudged by the IBA but also reflecting the aspirations of its senior management, opened the National Broadcasting School (NBS) in London in November 1980. It was to be established by Michael Bukht, who had completed his time as programme director at Capital and whose huge creative energies were ready for a new project. Typically, he was ready with an ambitious vision. "It will be run by experienced broadcasters from all the disciplines to help anyone who studies there to improve their own performance and so the service to the public ... I hope it will make a real contribution to understanding and entertainment for the doers and the listeners".[405] The NBS was to cover broadcast presentation, production, creative courses, technical, journalistic, business and management, for existing ILR staff and for university leavers and community groups. It operated from "quite impressive studios and facilities in Greek Street, London W1".[406]

The set-up of the NBS was financed very largely out of secondary rental, although this was tricky timing in terms of its availability. 1979/80 had been an exceptional year, with ILR revenues buoyed by the ITV strike. The rush to build new transmitters to enable the rapid expansion of ILR was to take £862,000 of the 1980/1 funds, increased to £1.3m when related expenditure was also met. Support for parliamentary broadcasting and other current IBA projects required a further £300,000, leaving only £1.1m available. The NBS was projected to be self-financing from fees charged, by 1983 or 1984. Meanwhile, the IBA through secondary rental was initially prepared to absorb £55,000 of the annual loss, with contributions of £80,000 coming from Capital itself and the other companies. Capital organised a meeting of the other ILR companies on 2 December 1981, getting 16 of the 33 invited MDs to attend, but little financial commitment. It consequently asked the IBA to underwrite it against loss for that year. With only £34,000 coming from other ILR companies, and £35,000 from Capital itself, this ended up as a subvention of £108,000 for the year ending September 1981.[407] The IBA's enthusiasm for the project was leaving it in a bind, just as secondary rental was becoming less available.

The NBS continued to do some terrific training, but the finances would not come right. It was still technically a subsidiary of Capital Radio, and the Capital board decided to cut off any financial support from 1 October 1982, from when "their directors concerned were no longer authorised to sign cheques for the NBS".[408] Peter Baldwin, then the IBA's deputy director of radio, worked hard to rescue the NBS by establishing it as a stand-alone company, but he was faced with legal advice that it should not become a subsidiary of the IBA. The negotiations with Capital reached an impasse. It was broken only when two of the executive directors of Capital – Brian Morgan and Keith Giemre – who were guarantors of the embryo company, NBS Ltd., agreed to activate the company with Bukht as a third member, if the IBA would agree to appoint a member and governor "as a token of commitment". That was also on the formal understanding that the IBA would make an annual grant of £120,000. Baldwin was eventually able to agree to

404 *IBA News Release* 29 September 1981.
405 *Record Business* 21 April 1980.
406 *IBA Paper* 253(81) 1 October 1981.
407 *IBA Paper* 268(82) 10 November 1982.
408 Ibid.

this, and was given a personal indemnity from the full Authority, meeting in late November that he was "acting properly on behalf of the Authority". A similar indemnity was agreed for any future IBA nominees to the board of governors of the NBS. Peter Baldwin himself became the chairman of the new company, as one of the three governors appointed by the IBA under the eventual structure.

Matters got no easier. By June 1984, the annual grant was no longer enough to keep the NBS solvent. IBA director general John Whitney – who had arrived at the end of the previous year from Capital Radio, which had introduced the NBS and now sustained it – agreed emergency action, providing a further £30,000 as an advance on the next year's funding. That also had to be agreed by the full Authority early in July. The IBA acknowledged that "while the IBA plays such a direct and dominant part in the School's affairs at board level, it is not, in practice, possible for us to distance ourselves from financial responsibility for any trading deficits".[409] In negotiation with the ITV companies later in the year, the Independent Television Contractors Association (ITCA) offered an annual subvention of £105,000 plus a non-returnable 'loan' of £100,000, provided there was a television curriculum, and subject to continuing support of the NBS from the ILR companies. There was real doubt over the last. Brian West for AIRC had already indicated that the "AIRC Council has very considerable doubt as to whether the level of expenditure [on the NBS] can be justified ... the limited financial resources available could perhaps be better utilized in other directions, e.g. by forging close links with colleges of higher education, and with companies being *provided with funds* [my italics] to expand their training programmes at Company level."[410]

The NBS of the eighties was a prestige project, typical of that middle period of Capital Radio, and supported by the IBA's assumption of future high levels of secondary rental. The relatively ambitious way the NBS was originally set up worked against its viability as a stand-alone operation when it was forced to rely upon its own resources, and that in turn put it out of tune with the new market approach for broadcasting. Capital Radio in the early eighties found a way to keep it going, to suit its philosophy of what a major radio station ought to be offering the network; Capital Radio in the nineties could never square that expense with what it perceived to be its obligation to shareholders. Once that driving philosophy had dispersed it was vulnerable; as Michael Bukht recalls, "when secondary rental failed, everyone bailed out immediately, and we died within two months".[411]

Support for live music persisted as one of the central uses for which the IBA would permit stations to spend 'their own' secondary rental, and for which it encouraged proposals from others. Foremost of all was Capital's founding of and support for the Wren Orchestra. A new classical orchestra in London went rather against the market trend, but it suited Capital's needs as well as its own. Capital provided an initial £50,000 support, and then in 1978 bought 50 per cent of the orchestra. It provided the sort of publicity which the more established orchestras could never afford, notably among Capital's young audience, and linked live recordings in with a weekly classical music programme, *Collection*. The chairman of the orchestra, Howard Snell, was in no doubt that the orchestra had captured a

409 *IBA Paper* 103(84) 28 June 1984.

410 Letter from Brian West, AIRC to Peter Baldwin, deputy director of radio IBA.

411 Michael Bukht in conversation with the author, 2 February 2009.

new audience "outside the static pool of people interested in concerts, theatre and ballet. You can tell by their response. They clap between movements. They look different. They apparently aim to enjoy themselves."[412] The Wren gave Capital an outstanding promotional tool and genuine respectability, as well as events to which it could invite the 'great and the good' who were rather too old to be impressed by pop music equivalents. The IBA saw Capital meeting its more ambitious contractual obligations, easing the pressure from the Musicians' Union, and making the continuing case that ILR was good thing, and a public service. For a short time, secondary rental had truly come into its own.

That was not to last. The Wren eventually fell victim to rationalization by Capital's new MD, Nigel Walmsley, who also felt it had slipped away from its intended cutting edge.[413] The secondary rental take decreased as the 1981 recession started to bite. 1980/1 secondary rental fell to £1.3m, and then again to just £794,000 in 1981/2. By 1984/5 the secondary rental total had fallen to £168,000, with only five companies making payments; Capital's contribution was down to a meagre £22,000 and Clyde paid nothing at all (on reduced charging levels). The following year the total rose to £497,000 as Capital produced £241,000. In 1986/7 the take was up to £781,000, as Capital's yield rose to £425,000 and Metro Radio added £56,000.[414] However, in the aftermath of all the pressure from the companies over their levels of profitability, and the removal by the Treasury of the ILR levy in 1987, the IBA decided to suspend secondary rental by setting the charge at zero from the year ending September 1988 onwards. Grants to parliamentary broadcasting were to continue, but funded from primary rental income, as was support for training and some other minor projects. Secondary rental was consigned to history, as the IBA abandoned its radio 'gold standard'.

Programme sharing

A structured system of programme sharing within ILR was to follow a similar trajectory to that of secondary rental. Not all the companies welcomed the idea of programme sharing in the early days, especially when it was material originating from LBC/IRN for which they had to make a (modest) payment, and which was accompanied by a degree of 'encouragement' from the IBA to take it. LBC had previously had little trouble syndicating Edward Heath's short sailing features, or the Percy Thrower gardening slot. However, the first significant attempt in 1976 at network-wide syndication of more substance met resistance. LBC's flagship *Decision Makers* was a 30 minute current affairs programme of good quality. This was for sale, not free of charge, costing £500 for a package of programmes. Ten stations bought it, but eight declined to do so, even when Michael Starks wrote to them all, asking how those stations not taking it "intended to fulfil their national

412 *Daily Telegraph* 2 October 1978.

413 Another, longer-lasting classical ensemble, which was given a start by secondary rental, was the Brodsky Quartet. The IBA wanted to enter an up-and-coming string quartet into the EBU's prestigious String Quartet Competition which was to be staged in Cambridge in 1983. A number of ILR stations suggested local musicians. From tapes submitted, Peter Black from Capital Radio and I selected the Brodsky, proposed by Radio Tees, as our ILR entrant. To the amazement of the BBC, which had assumed that a UK-hosted competition was its to win, the Brodsky triumphed, and has gone on to be a significant pioneering quartet. Secondary rental briefly enabled ILR to find a place even at the top table of European public service broadcasters, much as ITV had done through the Prix Italia and Golden Rose awards.

414 *IBA Paper* 56 (88) 14 April 1988.

current affairs commitment". The stations felt "heavy pressure"; the IBA saw this as merely "encouragement".[415]

The IBA had a statutory duty "to secure a wide hearing for programmes of merit".[416] It carefully avoided defining what 'programmes of merit' might include, John Thompson saying that he preferred "not to interfere with natural developments"[417] – but it clearly expected arrangements for both networking and syndication[418] which would, to a degree, parallel those for ITV, and involve payment. The companies accepted the value of some syndication, but continued to be resistant to networking, the small and short-lived Liaison Group of the RCC arguing that the latter "would be neither live nor specifically local".[419] There is an historic irony in the respective positions – the regulator promoting networking and the companies opposing it – in the light of the battles that were to be fought from diametrically opposite positions in the nineties and afterwards.

By the beginning of January 1978, the increased availability of secondary rental was encouraging the IBA to speculate on the practicality of making 'programmes of merit' more widely available. The Authority identified types of programmes suitable for sharing, including "music programming of particular distinction, a programme or series which contains interviews of general interest with people it would be difficult for 19 stations all to approach ... a production-intensive, light entertainment programme of general appeal, occasional dramatic material of general appeal; a well-produced documentary, or series, on a subject of general interest or a topical feature of general appeal; sports coverage of an international sporting event of general interest".[420] The examples chosen illustrate where the IBA felt most comfortable with ILR's output: Clyde's *Towards* 2000 Duke of Edinburgh programme; LBC's *Decision Makers*; Kenny Everett in *Captain Kremmen* from Capital; and Tees' Elvis Presley obituary programme.

The breakthrough came as a result of widespread interest at a meeting of ILR programme controllers, organised by the IBA on 15 February 1978. By then, it was accepted by the programming professionals that there could be no significant market in the sale of syndicated output, since the administrative costs would far outweigh any affordable level of charges. However, stations were increasingly keen to use material produced by others, and companies making 'programmes of merit' were more than happy to have those enjoy a wider hearing. Hallam's Bill MacDonald wrote to the IBA in June, after consulting the other ILR MDs, pointing out the practical issues. "We cannot use the IRN line because: (a) it is used for news items and precludes anything running for one hour solid, (b) even sending it in the early hours of the morning would not help because many stations do not have engineers to take the tape off, and (c) the mono line and its frequency range

415 Both quoted in *Broadcast* 25 October 1976.

416 *Independent Broadcasting Authority Act* 1973 s. 2(2).

417 *RCC Minutes* 13(76) of a meeting on 21 April 1976.

418 The term 'programme sharing' may include three distinct types of co-operation. Broadly, 'syndication' is the supply of a programme made by one company – or an outside production organisation – which more than one station then broadcasts, at a time and format of its own choosing. 'Networking' is the simultaneous relay of common output between two or more stations. A 'sustaining service' is a single stream of output, to which a station can default when it is not originating its own output, typically overnight.

419 *RCC Liaison Group Minutes* 3(78) of a meeting on 10 January 1978.

420 Paper produced for the ILR Programme Controllers Meeting 6 February 1978.

would weaken the programme immensely, great care having been taken to achieve a good stereo balance. Dubbing off ninety reels is not practicable … we do not have fast dubbing machinery available anywhere in the network."[421]

The matter was discussed at a formal consultation, effectively a longer conference involving the Authority staff and all the ILR companies, on 19 September at the IBA's Brompton Road headquarters. That confirmed general support for some form of central unit, funded by secondary rental, to undertake the notification, copying and distribution on tape of programmes to be shared. The resulting structure involved the IBA's Radio Division preparing a regular list of material suitable for sharing, with LBC and Capital arranging the copying, the whole funded out of secondary rental. The IBA had agreed in May 1979 that funding could "help towards establishing and operating a system of programme sharing and to provide finance for a selected programme or series".[422] £30,000 was allocated for 1978/9 to encourage "some further prudent and effective exchange of programming".

Unsurprisingly, given the 'master and servant' relationship which still existed between the Authority and the companies, ILR stations were keen to offer their best programmes to the system. AIRC's then new chief executive, Brian West, had quite recently succeeded Bill Coppen-Gardner, who had moved to LBC. He observed that "it tended to be the bigger stations that offered what they considered to be their best programmes, to get brownie points from Brompton Road, and the smaller ones that gratefully asked for copies".[423] I can attest from my experience at Radio 210 that one of the side benefits of taking programmes from the scheme was that they came on a free reel of tape, a useful perk when budgets were tight. We used to take all those offered, and then broadcast just the ones which suited our output; but all the tape was cleaned and re-used later.

The programme sharing unit took on a more unified and substantial existence in January 1984, when it was taken over by AIRC, which wanted to consolidate the genuine prestige accruing to the system from programmes such as Alan Bleasdale's drama productions for Radio City, Piccadilly's ground-breaking documentary on child abuse *Conspiracy of Silence*, and Ralph Bernard's award-winning examination of alcoholism for Hallam, *Dying for a Drink*. The stations' priorities varied, but almost all valued the material. Gillian Reynolds wrote of programme sharing that "interviews with rock stars are popular, so are concerts. Radio City's Paul McCartney concert went down especially well, as did Hereward's two hour Christmas Special from George Martin … Capital can offer tapes of most of the big groups who play in London."[424] Short features on topics of social concern were snapped up readily, but the biggest general demand was in the winter when colds and coughs strained presenter rosters: 'My jazz presenter has gone down with 'flu. Have you got anything we can use to fill his slot?'

AIRC's role was "to communicate with stations, seek out material, prepare preview cassettes to supplement information contained in the [information] sheets themselves and ultimately supervise copying of the required programmes".[425] The

421 Quoted in *Radio Paper* 34(78) 12 July 1978.

422 *IBA Paper* 70(79) 4 April 1979.

423 Quoted in *The Hidden History of Commercial Radio* by Sean Street in *Aural History – Essays on Recorded Sound* British Library, National Sound Archive 2001 p 98.

424 Gillian Reynolds in the *Daily Telegraph* 20 February 1988.

whole operation was funded out of secondary rental by the IBA, which mostly resisted the temptation to interfere with which programmes were offered and taken. The programme sharing unit was run by Felicity Wells, one of a number of IBA and later Radio Authority staff who went on to play significant roles in the ILR system.[426] It offered 25 programmes in its first month, yielding 100 hours on air, and the volume grew steadily until 1987, when cost-cutting at AIRC led it to cut this service among others. For a short while, Capital Radio then stepped in to underwrite the system, and it operated from the first floor at Euston Road. However, it failed to survive the watershed of 1990.

Programme sharing material was also supplemented by the availability of international concerts, through the European Broadcasting Union (EBU) scheme. The IBA had managed to wangle ILR membership of the EBU, which was overwhelmingly a club for public service state-funded broadcasting organisations, by stressing the 'national' nature of IRN. The EBU provided a window on to a slightly wider world, in which these programmes – plus even more valuable access to football commentary rights – were the main resulting benefit. For the IBA, and later for the Radio Authority, the ability to participate in some of the technical development work undertaken by the EBU was a considerable attraction, and this was central to the early and significant role played by the Radio Authority in the introduction of digital audio broadcasting, notably in the years before the 1996 Broadcasting Act.[427]

One innovation from the eighties survived the changes and survives still; the *Network Chart Show*. Piccadilly's Colin Walters had argued in March 1984 for a rival to BBC Radio One's then all-conquering *Top of the Pops*. "If there were one authoritative ILR chart with a unique formula and a top-flight presenter ... it just might dent the composure of Radio One and it would certainly attract advertisers."[428] On Sunday 30 September 1984, 44 ILR stations simulcast the first network chart programme, putting David 'Kid' Jensen up against Alan 'Fluff' Freeman, head-to-head from five to seven pm. Only Southern Sound, which had technical problems, and Pennine – which was unwilling to displace its traditional rugby league programme – declined to take the show. (LBC was not given the option.) Jensen himself was an immediate hit, winning the 'UK personality of the year' gong at the 1985 Sony Radio Awards. The changed regulations of the Radio Authority's regime allowed such programmes to be sponsored, and the programme became a major revenue-earner. At the start, though, the show's commercial target was simply a decent share of Radio One's six million listeners and "... in its first six months on air it virtually doubled ILR's Sunday afternoon audience".[429] With the first significant network sponsor allowed in ILR, the *Nescafe Network Chart Show*

425 Ibid.

426 Many of the tapes and most of the information sheets from the programme sharing unit survive, and are available through a new digital archive established as part of the Radio Archive of The Centre for Broadcasting History Research at Bournemouth University Media School at the end of 2007. Felicity Wells died in her early forties of multiple sclerosis, in 1998, and the archive is named for her. The University's 2007 booklet describes "the exchange of programmes between ... the ILR network, a growing group of stations which ultimately became the foundation of the changed geography of 21st century commercial radio" as a "memorable – and lost – period in UK commercial history [and a] unique record of a key time in the history of British ... radio".

427 See chapter 20.

428 *Marketing Week* 2 March 1984.

429 *IBA Radio and Television Yearbook*, 1986.

was a clear sign that *commercial* radio was on its way. Programme sharing, the hallmark of *independent* radio, was a thing of the past. This one offspring, later to be given a new boost as the *Pepsi Chart Show*, hosted by Neil Fox, was alone to flourish and become a multi-media offering.[430]

430 See chapter 16.

Chapter 10

London Heathrow calling

The Heathrow Conference and its impact

The vintage years of Thatcherism were 1984 and 1985. After her government's overwhelming second general election victory in 1983, the Conservative parliamentary majority of 144 seats confirmed that the mood of the country had swung decisively away from the intervention required in social liberalism, towards a freer commercial market model. Privatisation got firmly into its stride, demolishing the post-war consensus that certain industries needed to be run as much for their social impact as for their commercial wellbeing. British Telecom was the first public utility to be sold back to those who thought they already owned it, when 50.2 per cent of its equity was traded in November 1984. The defeat of the miners in March 1985, when the NUM returned to work, broken collectively and effectively starved back individually, marked the end of mass trade union power in the UK. This situation was confirmed when first Eddie Shah in Warrington, and then Rupert Murdoch in Wapping in February 1987, deliberately took on and broke the power of the once unassailable newspaper print unions.

However, in the mid-eighties, independent radio was showing few real signs of emerging from recession. In many locations ILR stations were struggling, and AIRC was campaigning for rental reductions. The companies began by seeking a 25 per cent rental reduction, although by January 1983 "they had refocused their sights on a 10 per cent reduction".[431] The IBA considered the scope for any reduction to be "very small", not least as it was facing negative cash flow on its radio capital accounts for six of the next ten years. The May cost of living increase was moderated, but this provided little comfort. By the following year, the IBA was receiving frequent requests to defer rentals. At just one meeting in early May 1984, the Authority agreed to deferments for Marcher Sound, Radio West and Radio Aire (for the second time).[432]

Alongside the floundering CBC in Cardiff, Gwent Broadcasting in Newport failed in April 1985. The IBA allowed Red Rose, the Preston contractor, to acquire both CBC and Gwent, and re-launch them in October 1985 as Red Dragon Radio, serving Cardiff and Gwent with just a little separate output. Red Rose also provided a 'rescue' for Radio Aire in 1985, as a nervous Authority began to allow take-overs which had previously been anathema.

Radio West's wretched start in Bristol left it vulnerable too, and a merger was agreed between it and the Swindon company, Wiltshire Radio. To the IBA this

431 *IBA Paper* 6(83) January 1983.
432 *IBA Paper* 62(84) 26 April 1984.

was simply to be a "closer association" between the contractors, but Ralph Bernard recalls it differently.[433] He had only just been "pitchforked" in as MD of Wiltshire, and was wholly new to such corporate issues. In February 1985 he met 'Duke' Hussey, the new and hugely distinguished chair of Radio West, at the Castle and Ball Hotel in Marlborough. Both were keen to bring about a merger. Hussey (who was to become BBC Chairman in 1986) had just overseen the bi-centennial of Murdoch's *Times*, so the merger plans were codenamed 'Rupert'. Bernard and John Bradford, then MD of Radio West, met at a private house in Calne to plan the details, creating what then became GWR. For a short while, they shared the running of the new company; in a matter of months Bradford had gone, an outcome mirrored with David Mansfield when GWR and the Capital Group 'merged' in 2004.

In most other respects the regulatory regime for ILR was no lighter than it had been from the start, and they felt unfairly treated. Telecommunications and newspapers were still free to conduct their own businesses primarily as businesses, and there was a renewed challenge from extra-territorial radio stations, notably the 'top forty' offshore pirate station Laser 558. Multi-channel commercial television also began inching its way into being. Although cable television was not to have its expected impact then, it seemed potentially significant and was expected to have a much lighter regulatory burden under the new Cable Authority from 1985. More immediately, breakfast television had appeared in the UK for the first time in January 1983, when the BBC stole a march on the new commercial service, TV AM, which launched the following month. Channel Four had begun broadcasting in November 1982. By February 1985, with an audience share of 10 per cent, its maverick chief executive Jeremy Isaacs was telling David Frost on breakfast television that "we have got there ... I can look ITV in the eye. We can pay our way."[434] The main sufferer in ILR from breakfast competition was LBC, and its MD George Ffitch was not going to suffer in silence.

The leaders of the independent radio industry generally were starting to wonder where all this left them. Expansion in the number of stations, following the 1978 green paper and the 1980 Broadcasting Act, saw 38 stations on air by ILR's tenth anniversary in October 1983. Yet the rules under which they operated remained much as they had been ten years before. Each of the companies was, in principle, limited to owning just one station – except where 'rescues' were allowed – and trading in their shares was very restricted. ILR's commercial growth was constrained by limits on diversification, especially into free newspapers. There were tight rules on sponsorship of programmes, although these were freer than for ITV, and the IBA considered that allowing such 'co-funding' of radio programmes was generously liberal. The companies were particularly exercised by their high level of 'fixed' costs, of which primary and secondary rental were directly imposed by the IBA. From the stations' point of view, the Treasury levy, their music royalties, payments to musicians and fees to IRN also came, in effect, from the same regulatory regime. These were the years of increasing commercial freedom in the newly-sanctified 'free market', and ILR wanted some of that too. At least

433 Ralph Bernard in an interview with the author, 19 November 2007.

434 Jeremy Isaacs interviewed by David Frost on TV-am 2 February 1985, quoted in *A Licence to be Different, The Story of Channel Four* Maggie Brown British Film Institute 2007 p 107.

one normally prosaic MD was moved to confused hyperbole. "We should not be afraid of freedom: ILR has nothing to lose but its chains. Freedom is unlikely to kill any cherished pheasants or make the cows go dry. It might radically increase their yields."[435]

It was the chairmen of the original 19 companies who took the first initiative early in 1984, forming themselves into a potential lobbying group and meeting together separately from their MDs, to the latter's considerable chagrin. Richard Findlay, then MD of Radio Forth, was that year's chairman of AIRC, and on its behalf met several of the ILR company chairmen to agree a common way of moving forward. It was agreed by them, and endorsed by the AIRC Council on 30 May 1984, to convene a special meeting of AIRC, and then to take forward their case to the IBA and to the government. That meeting, which took place on 23 June at the Sheraton Skyline Hotel, became known as the 'Heathrow Conference'.

This was the occasion when *independent* radio started to shift into *commercial* radio. Findlay himself is clear on this point. "If there was a moment when the seeds of change were sown, it was probably round that Heathrow Conference, when the industry found its own voice, which it hadn't had up to that point. It found that politicians and Home Office officials were very keen to have a direct dialogue, and that the industry could have an influence and be listened to. That's when the commercial momentum began. We could change things for our commercial betterment, and begin making money."[436]

Each of the ILR companies was invited to send two delegates to the conference, intended to be their chairman and managing director. All but three companies attended, the only absentees being Swansea Sound, Downtown Radio and Marcher Sound.[437] Ideas for what might be discussed were canvassed widely among the companies in the early weeks of June, and they were all invited to submit resolutions for discussion against the background that "AIRC Council believes that the industry urgently needs to determine its policy stance on a number of key issues, and then mount a positive platform of its requirements from the IBA and the government".[438] Thomas Prag, MD of the smallest of the stations, Moray Firth in Inverness, had sent in some ideas to John Bradford – who was chairing the Public Affairs Committee of AIRC – but doubted whether the long journey to attend in person would be worthwhile. Bradford replied with prophetic understatement. "I hope you will be able to make the trip south for 23 June. I know it's a long way but it might be quite important."[439]

As usual, debates within the independent radio industry quickly acquired a high level of intensity. Unless there were strong voices for moderation, the industry was capable of getting itself worked up into a lather, and this was no small matter. Jimmy Gordon, who thinks he might have headed off some of the antagonism, left the conference early to attend his daughter's birthday celebrations.[440] The conference debated eight resolutions, which in trade union conference language had been 'composited' from the expressed areas of concern. That

435 Conference Paper by David Pinnell, MD BRMB 23 June 1984.
436 Richard Findlay in an interview with the author, 15 January 2008.
437 Memorandum from Brian West to all AIRC members, 20 June 1984.
438 Memorandum from Brian West to all AIRC members, 31 May 1984.
439 Letter from John Bradford to Thomas Prag, 4 June 1984.
440 Jimmy Gordon in an interview with the author, 25 April 2008.

Saturday morning, the meeting considered the first four topics: that the development of community radio and/or independent national radio should take full account of the existing ILR system; that responsibility for independent radio should be moved to a new, separate regulator, with the companies owning and operating their own transmitters; that programming regulation should be drastically reduced; and that sterner efforts be made to curb land-based pirate radio. Fortified by a buffet lunch at the 'Poolside Patio', the afternoon session discussed four more issues: that ILR companies should have greater commercial freedom; that sponsorship should be permitted for radio; that advertising controls should be eased; and that broadcasting from commercial premises should be permitted.

It was without precedent for the individualistic and fissiparous radio companies to agree on anything so major so quickly. The presence of their chairmen helped, and the absence of the polarizing effect of national sales agencies probably assisted, too. Agreement there was, and by the end of that day. The conference unanimously adopted six resolutions.

> "1.AIRC is concerned that UK radio developments now being contemplated are examined in the context of all independent radio, and requires the government, and the Independent Broadcasting Authority, to take full account of the possible effects of any changes or additions to independent radio on the existing ILR system.
>
> "2.AIRC requires that any funds drawn from Independent Local Radio by the Independent Broadcasting Authority must not be used for the provision of transmitters, or to meet any other costs, associated with the development of Independent National Radio.
>
> "3.AIRC resolves to commission EIU Informatics, as a matter of urgency, to carry out in-depth research into the various levels and consequences of radio de-regulation. EIU Informatics will also be asked to examine the IBA and its relevance to the current and future needs of ILR.
>
> "4.AIRC totally supports the most recent letter from the chairman of the Independent Broadcasting Authority to the home secretary on pirate radio. The association calls on the government to increase resources immediately, and use all the powers available to it, in order to halt at the earliest possible moment the blatant flouting of the Wireless Telegraphy Act 1949 and the Marine &c Offences Act 1967 by illegal broadcasters. AIRC resolves that, in the event Government does not take such actions, the members of AIRC will reconsider their own various statutory and royalty payments, currently costing the industry in excess of £13m a year.
>
> "5.Recognising the nature of the market place, independent radio companies in the UK should be able to trade with the same degree of freedom as other commercial enterprises, limited only by the explicit requirements of the Broadcasting Act, the Companies Acts and the laws of the land applicable to all businesses and private individuals. Accordingly, AIRC resolves to enter into immediate discussions with the Independent Broadcasting Authority, and others, to secure:
>
> i. A substantial reduction in the current annual IBA rentals bill to companies of £6m and subsequent further reductions in rentals;
>
> ii. Reduction to an absolute minimum of IBA interference in company activities, including programming, broadcasting hours, advertising, technical standards, capital structures, shareholdings and diversification;
>
> iii. A clear understanding that all forms of control will be reduced in accordance with the very much easier controls emerging for cable, and likely to be obtained for community radio and other new forms of UK broadcasting.

> "6.AIRC calls upon the Independent Broadcasting Authority to acknowledge the essential difference between radio and television marketing opportunities, and relax the advertising control system which at present prevents ILR companies from seizing specific advertising and sponsorship opportunities."[441]

Such a mixed set of demands, highly protectionist in approach and with bags of bluster, reflected the emotion of the moment, the range of individual aspirations, and the accumulated frustrations of the industry. It was less a shopping list, more a first draft of a manifesto, but there can be no doubt that it shook the towers of the IBA's Brompton Road to their foundations. Almost more significant than the resolutions themselves was what AIRC director Brian West did with them the following Monday, sending them to the prime minister, home secretary Leon Brittan, industry secretary Norman Tebbit and (fourth on the list), IBA chairman George Thomson. Not content with questioning the "relevance" of the IBA to ILR, and commissioning the Economist Intelligence Unit to report into that, the ILR barons were signalling that they were no longer satisfied dealing only with IBA officials, not even John Whitney, once one of their own. The wording of the AIRC press release made clear their confrontational intentions. "Independent Local Radio has thrown down the gauntlet to the government and the IBA." The government, and in particular Leon Brittan, would have been neither bothered nor surprised, as Findlay had previously met him to explain ILR's concerns, and had been heard sympathetically.[442] Still, whatever the outcome of their imminent discussions, the relationship between ILR and the IBA was fractured, and would not be the same again.

A flurry of activity followed, the two London MDs being prominent in the developing lobby. Nigel Walmsley, who had followed John Whitney at Capital Radio was, with his Post Office civil service and marketing background, a skilled operator in political circles. LBC's George Ffitch had been editor of the *Economist*, and his wide contacts (and LBC's London premises) were helpful to the AIRC efforts. It was Ffitch who had brought in the Economist Intelligence Unit, with a brief to "conduct research into the regulation of radio broadcasting in the UK and, in particular, the effects of varying degrees of regulation."[443] That report was released by AIRC in November, although as discussed shortly it had by then been overtaken by other developments. It identified five elements of regulation – content; economic/financial; frequency allocation; ownership and management; and technical standards – and three themes: regulation itself; the effect of regulation on markets; and the need for at least a minimum level of regulation of frequencies. The study concluded precisely what anything from the *Economist* stable might be expected to conclude. It had found no evidence that profit and public service "are not compatible, or that regulation per se is necessary for the fulfilment of the public interest ... the present financial constraints on the radio companies are more largely caused by the custom of the IBA and the origin of the industry than by the letter of regulation".

441 Memorandum from Brian West to all AIRC members, 25 June 1984.
442 Richard Findlay interview, op. cit.
443 *Radio Broadcasting in the UK* Economist Informatics undated, but released by AIRC 23 November 1984.

The EIU report aimed in its own words "not to define conclusively the best direction for UK radio broadcasting, but to provide a framework for considering de-regulation and a basis for an informed debate". It rehearsed four scenarios: continuing with current regulation; two tier de-regulation, where less profitable or unprofitable stations would be more lightly regulated than the majors; regulation with "a medium touch", in essence substantially easing content, economic and technical standards, but keeping heavy control on financial matters, frequency allocation, ownership and management; and "regulation with a light touch" – a phrase which became a mantra for the next 20 years at least – which involved keeping only frequency allocation controls at any level of intrusion. The BBC being outside its terms of reference, the EIU made no significant observations on its future. Perhaps AIRC did not get much substance for their £20,000, but the companies hoped it might be a powerful totem.

AIRC began to enjoy having access to the government, a freedom it had not previously appreciated. Richard Findlay recalls that "up until then we had been very much at arm's length from any direct contact with the government. It was always pointed out that we were not the broadcaster, we were contractors to the IBA. It was almost an unwritten rule that we should not deal formally with the Home Office. We assumed that the IBA would relay our thoughts and concerns, but at this point we thought we needed to have contact with the government ourselves. That's exactly what we did, first of all with Leon Brittan, and then with Douglas Hurd."[444] Brittan had already been helpful to Findlay, and the two met again on 25 July when the home secretary visited AIRC at its Paddington premises. By September, Findlay felt sufficiently emboldened to set out the companies' concerns in a five page letter. This included considerable detail about costs, pirate radio and commercial restriction; and possibly too much detail in its continued goading of the IBA. "Regulation restricts the commercial freedom and independence of the privately owned ILR companies … the IBA should restrict its regulatory activities to those areas provided for in the 1981 Act and not extend and embrace the whole range of a company's activities, which it currently does, almost down to the colour of paper in a station's toilets."[445]

As he showed dramatically, in pressing for the directly-licensed community radio experiment of the same period,[446]Leon Brittan was unconvinced by the IBA's stewardship of independent broadcasting. Short of new legislation, however, there was little the government could do beyond offering general sympathy, although Brittan had done his best to apply some pressure. "I have, in the light of [some of the things you said] ... had a discussion with Lord Thomson and John Whitney about ILR matters … my impression is that the IBA is sensitive to many of the matters which AIRC feels at the present time, and wishes to examine them in an open-minded way. This seems to me very much the right approach, and I have encouraged the IBA to carry matters forward in this way."[447] By the end of 1985, Brittan was anyway engaged in the full-blown constitutional crisis of the Westland Affair, where his resignation from the cabinet on 24 January 1986 over the leaking

444 Richard Findlay interview, op. cit.
445 Letter from Richard Findlay to Leon Brittan, 11 September 1984.
446 See chapter 11.
447 Letter from Leon Brittan to Richard Findlay, 19 September 1984.

of a letter damaging to the Prime Minister's rival, Michael Heseltine, probably saved Thatcher's own political skin.[448] His replacement, Douglas Hurd, was more used to the subtler processes of diplomacy, and unlikely to move with haste.

There was, though, a discernable change in wider views about independent radio, especially within the advertising agencies' trade association, the Institute of Practitioners in Advertising (IPA). The advertising industry wanted more radio services, and although that was not a view shared by the ever-protectionist AIRC, it was an alliance of convenience to seek general change. To pursue that, the IPA held its own conference at Leeds Castle in Kent on 12 February 1985. The leading figures attending from the advertising agencies were known for their interest in radio, including Chris Dickens of Young & Rubicam, Ken New of Abbott Mead Vickers/SMS, and John Perriss of Saatchi & Saatchi. They met with five ILR MDs, David Lucas of County Sound and Ian Rufus of Mercia joining Findlay, Gordon and Walmsley. IBA DG John Whitney attended the conference only for dinner, so the Authority was represented by John Thompson, Harry Theobalds and David Vick. Bill Innes from the Home Office was also present, and the conference was chaired by Christopher Chataway, now in his role as chairman of LBC for Australian investors Crown Communications.

The outcome was to recommend a "comprehensive government review of the future development of radio, covering both BBC and ILR services as well as any proposed new services".[449] Forward-looking proposals will have helped the group towards consensus, but they also show that the IPA at least expected a lot of further growth in independent radio. The review was intended to cover "the inhibiting effect of high copyright restriction and charge", the provision of community of interest stations to ensure diversity, finances for community radio, pirate radio, and "whether the existing regulatory system is the most appropriate for carrying through any expansion of radio, or whether we now need a Radio Authority". The keen interest of the advertising agencies was not driven only by the potential of independent radio. They also saw changes in ILR regulation as a Trojan horse to gain access to the greater prizes available in television, notably sponsorship and co-funding.

Nevertheless, whatever external pressures they could summon, AIRC's route for the present still led directly to the IBA. The companies sensed, or hoped they sensed, that George Thomson and John Whitney were sympathetic towards them, but felt that John Thompson offered contradictory signals. "On the one hand he was a blockage, on the other hand privately he was quite enthusiastic."[450] Within the IBA, Thompson was the companies' most realistic supporter, and his efforts to ensure that the IBA reallocated its costs between radio and television made possible the subsequent rental reduction. However, as the founding director of radio he was well acquainted with the quirks and qualities of the companies' chairmen and MDs. He recalls that: "I was much aware of several attempts by some

448 Nigel Lawson in his book *The View from Number 11* claimed that "had [Brittan] made public all he knew, she could not possibly have survived", although Lawson by 1992, when the book was published, had his own reasons for disliking Margaret Thatcher. If 1984-5 had been the high water mark of her premiership, Westland marked the start of the ebb, and AIRC's hopes risked being caught on an adverse tide – until *Death on the Rock* intervened (see Chapter 12).

449 IPA minutes 12 February 1985.

450 Richard Findlay interview op. cit.

companies to try and sidestep the IBA by going straight to ministers or the civil servants. In my view, this could be both unhelpful and confusing."[451]

However, although it may not have seemed so to the companies at that time, this was really an institutional question, not a personality issue. Findlay again: "The whole structure of the IBA was there really to limit and restrict and invent new regulations, rather than to enable. There were some members of the Authority who were enthusiastic about the regulatory aspect of what they did, rather than anything else."[452] AIRC was setting out to challenge the whole social and political settlement which had created ILR in the first place. The social liberalism of the seventies seemed increasingly out of place in the mid-eighties. The companies sensed their moment, and the IBA recoiled from the overturning of a settled order.

AIRC met John Whitney and John Thompson at the IBA on 4 July 1984, just 11 days after the Heathrow Conference, for the first of what turned out to be a series of eight meetings, lasting until November 1985. At this first session they agreed that AIRC would submit a 'shopping list' of proposals, which followed nine days later. Findlay, for AIRC, divided their issues under three headings: administrative, commercial and programming. He asserted that it was in the administrative area that the IBA had the greatest scope for independent action: "the right of companies to diversify as befits their commercial interest; the right to produce … publications; the market for companies' shares; share structure and geographical ownership; mergers, takeovers between and of ILR companies; the levy; copyright; the burden of regulatory costs – rental (primary and secondary), and the balance between radio and television; the high cost of the technical specifications; and ownership or leasing of premises".[453] Of these, the levy was a matter for the government, while copyright "is something the industry continues to try to resolve, and there may be something positive that the IBA could do to contribute to our efforts".

The commercial list featured copy clearance and the IBA's Code of Advertising Standards and Practice, separation of advertisements from programming, minutage rules, fixed rates for advertisements, sponsorship, broadcasts from named commercial premises, and contra-deals. Programming issues were controls on hours of broadcasting, balance within programmes rather than across the broadcasting day, mandating of programmes by the IBA, limits on competition prize values, prior approval of schedules, the amount of specialist programming required, splitting frequencies and the quality of transmissions. AIRC indicated also that it was exercised by the possibility of Independent National Radio, third tier radio and a whole raft of other legal and illegal, domestic and extra-territorial competition.

ILR chairmen had their own meeting with Lord Thomson on 19 July, but they largely left the campaign to their MDs. Their contribution, as well as being a catalyst for Heathrow, was to establish that companies should be prepared to challenge the IBA, notwithstanding any fear of losing their franchises, a key factor up until then in holding back any strong expressions of dissent. Some press support had been stirred up by AIRC. *The Times* acknowledged the case for some radical

451 Letter from John Thompson to the author, 19 January 2009.
452 Richard Findlay interview, op. cit.
453 Letter from Richard Findlay to John Whitney, 13 July 1984.

changes, moving away from basing ILR and BBC local radio "around a set of sub-Reithian corporatist policy ideas". It supported AIRC's wish for simpler administration with the resounding phrase: "What the government should offer them is freedom".[454]

The next set-piece encounter was an IBA/AIRC meeting on 4 September. For AIRC, the meeting did not go well, as Findlay's follow-up letter reveals. Having "entered into our discussions with you and your colleagues ... in a spirit of considerable hope", the company representatives "were deeply disappointed by your apparent lack of enthusiasm as regards the right of companies to diversify in their commercial interests without reference to the Authority", and only a little more encouraged by the response on publications and copy clearance issues.[455] A rather better meeting on 9 October saw the IBA concede the companies' wishes on involvement with publishing interests, freedom to diversify without prior approval, and to be allowed to experiment with presenters reading advertisements in their own programmes. The following day, George Thomson gave a speech in Birmingham which seemed to the companies to be conciliatory in tone and content. He announced that "the IBA is now going further and conducting a fundamental reappraisal of its own expenditures in relation to ILR ... the IBA does not believe that its existing practices are written in stone. It is determined that the ideal should not be made the enemy of the good. But it is equally convinced that good standards can still be maintained with a lighter touch and lower costs than at present."[456] This seemed to the ILR companies to be a good day. John Bradford told Duke Hussey, that "we seem to be making real strides with the IBA at last".[457]

While the country reeled from the IRA's attempt to assassinate the Conservative cabinet in the Brighton bombing, AIRC prepared to make the most it could out of the EIU report. But then the IBA detonated its own bombshell. In a statement in early November, not discussed in advance with the AIRC group, it announced a range of changes to its administration, engineering and regulation of ILR. Future development was to be limited to the 51 areas already established, with any new areas financed by 'forward funding', whereby the chosen company would meet the capital and running costs of the transmitters. The biennial 'rolls' of contracts were to be replaced with mid-term contract reviews, reducing the IBA's administrative load. That would allow it to pass responsibility to the companies for any diversification, advertising copy clearance, publications, hours of broadcasting and share structures, "subject to the companies' obligations to the IBA for comprehensive broadcasting standards".[458] Rentals would be reduced by 10 per cent from April 1985, and the government was to be urged to remove the exchequer levy on companies' profits. As for the mooted new services, "the IBA's first concern is that ILR should continue to flourish. The purpose of the IBA's new forward planning is to provide a constructive and flexible framework for future growth and continued public service of radio in the United Kingdom."

AIRC had little choice other than to welcome the proposals in public as "so

454 *The Times* 25 July 1984.
455 Letter from Richard Findlay to John Whitney, 7 September 1984.
456 Speech by Lord Thomson to the Royal Television Society in Birmingham 10 October 1984.
457 Memorandum: John Bradford to 'Duki' (Marmaduke Hussey), 12 October 1984.
458 IBA News Release "Independent Local Radio: The IBA's Forward Plans 12 November 1984.

far so good ... but the reliefs need to go further, once the act is updated".[459] Privately, though, the companies were caught off guard, and were furious at having been outflanked. The rental reduction of 10 per cent had been their original objective back in 1993, but it was no longer anything like enough for them. "All of the elements of the ... announcement arose from AIRC pressure", they believed, "but not all were especially high on our list of requirements".[460] Once they got wind of a statement in the offing, an EGM planned for 8 November was postponed "at the eleventh hour ... in the light of intimations from the IBA that a major announcement crucially affecting ILR was imminent". AIRC knew nothing of its contents, and had been offered a briefing only on the morning of 12 November.[461]

The EIU too was discomforted, as its unpublished report now seemed to be yesterday's news. It had already incurred extra costs taking on board AIRC's comments on its draft, and urged early publication "both for its publicity value and also so that we could recoup some of the additional costs we incurred".[462] In the event, the report was published by AIRC on 22 November, with a press release which noted that "the report does not propose a specific de-regulatory path"[463], a phrase which suggests more than a little disappointment with the EIU's findings. Certainly, the re-arranged AIRC EGM on 21 November was discouraging for those who had hoped that the EIU report would mark the way forward. "It was one of the less glorious debates in ILR history" one MD reported. "We should be approaching the government to argue our case over long-term policy development. But at the moment we don't have a case. Unless some attempt is made to put a composite view together, the industry will quickly polarize into deregulators versus regulators, and this would be very damaging."[464]

The IBA had paid back at least some of the taunts it had suffered. The EIU report also came under attack from the copyright societies for "a number of serious factual inaccuracies", and for disclosing information given to it in confidence.[465] One supportive ILR MD even expressed worry about eight major inaccuracies/omissions and "a number of minor [errors] which could only have an adverse effect if they led readers to think the research wasn't thorough, or it has been sloppily written/edited".[466] The broadcasting unions weighed in too, with a joint NUJ, ABS and ACTT circular to its members arguing that deregulation will lead eventually to the destruction of local news, information and creative programmes ... restriction rather than expansion of editorial independence, and a crude narrowing of the range of opinions and information broadcast."[467]

Prompted by AIRC, Angela Rumbold MP secured an adjournment debate on 30 November, arguing that "the notion that the government or their appointed bodies can continue closely to regulate local radio will survive for only as long as the idea that printing presses should be allowed to produce only the Bible and

459 AIRC Press Release undated.
460 Future of ILR Paper by Brian West 12 February 1985.
461 Memorandum from Brian West to ILR Managing Directors, 8 November 1984.
462 Letter from Ian Young, EIU, to Brian West, 13 November 1984.
463 AIRC Press Release Radio broadcasting in the UK, 22 November 1984.
464 Letter from Colin Walters, MD Piccadilly Radio to Richard Findlay, 23 November 1984.
465 Letter from M J Freegard, PRS to Brian West, 29 November 1984.
466 Letter from Bill MacDonald Radio Hallam to Brian West, 23 November 1984.
467 *UK Press Gazette* 18 February 1985.

approved texts ... ultimately, the market will dictate licensing requirements".[468] The analogy was not well chosen. However, Rumbold's odd example simply allowed David Mellor, then the parliamentary under secretary at the Home Office,[469] to make the orthodox case. "I do not think we can get away without various obligations being imposed for as far ahead as we can see". AIRC's moment seemed to be slipping away

The bi-lateral meetings between AIRC and the IBA continued, but into the new year of 1985 there were signs that each side was withdrawing to defensive lines. At the fifth meeting, in January 1985, the IBA – having declined to cancel the annual cost of living increase in primary rentals – was also, in effect, refusing to countenance a major reduction in expenditure on musicians. AIRC countered that "before a joint approach on community radio could be agreed, the present ILR system needed putting on a sound basis".[470] The companies felt increasing frustration at the lack of progress with the IBA. As chairman of AIRC for a second term, Richard Findlay had written to John Thompson about the IBA's November proposals. "I was certainly keen that your proposals should be viewed in a positive light by the industry", but "subsequent events have not supported that optimism".[471] To John Whitney he said that "the current high cost of IBA radio regulation (currently some £7m per annum) is not something that can be continually bounced from one meeting to another, the nettle must be grasped".[472]

AIRC decided to raise the stakes still further, since "the IBA announcement of 12 November really screwed things up" and "what was intended [by the IBA] did not go as far as the statement implied, and we had been lead to believe".[473] On 27 March the government announced the setting up of the Peacock Inquiry into the future funding of the BBC. On 28 March, AIRC went public with a request for a separate inquiry into radio structure and funding. "The government may not like the idea of two media inquiries at the same time, but our case for a radio inquiry ... is strengthened by the BBC decision ... 'sorting out' the BBC in isolation from the rest of UK broadcasting will only cause mayhem elsewhere."[474]

The government indeed had no time for this second inquiry, which would have complicated Douglas Hurd's elegant postponing of the issues around broadcasting. However, the tactical error of requesting it served to delay any potential resolution of the impasse over ILR, and actually took the pressure off both the IBA and the government. The existence of the Peacock Inquiry provided a potential forum to pursue the discussions about the future of ILR, but also moved the whole process from immediate decision into the middle distance. Both the IBA and AIRC now needed to turn their attention to the Peacock Inquiry, and the green paper

468 *Hansard* 30 November 1984 cols 1282–1288.

469 David Mellor rose rapidly under John Major's patronage to be the first secretary of state for national heritage, with a portfolio of responsibilities which gossips said was chosen to meet his specific personal enthusiasms. This was his first significant appearance on the stage of independent radio policy-making. He was generally a friend to independent radio throughout his time in office, most of all over the decision to deploy the national independent radio FM frequencies for a classical music service.

470 IBA note on Meeting between IBA and AIRC Representatives, 23 January 1985.

471 Letter from Richard Findlay to John Thompson, 21 January 1985.

472 Letter from Richard Findlay to John Whitney, 4 February 1985

473 Brian West paper op. cit.

474 AIRC Press Release 28 March 1985.

and eventual legislation which followed at the end of the decade. By then, John Thompson would have retired, and John Whitney would have escaped the toils of the IBA – but not before the Authority as a whole was sandbagged by *Death on the Rock*.[475]

The real impact of the Heathrow Conference was threefold. First, it focussed the strain and worries of the ILR companies on a challenge to the fundamental conception of ILR as *independent* rather than *commercial* radio. This was the pivotal shift in their collective perception and aspiration, which might otherwise not have happened so suddenly, or at that time. Second, it brought ILR as an industry into direct contact with the government for almost the first time, so that it became another of the players in the Great Game that was, and is, broadcasting policy in the UK – albeit a rather minor player.

Third, Heathrow changed at a stroke the relationship between the companies and their regulator. Suddenly, the companies were no longer content to be guided by the wisdom or lack of wisdom of the IBA. From that point onwards, the radio industry began to deal with its regulators in a spirit of confrontation, and in the language of confrontation. In the overall recasting of broadcasting structures in the UK, ILR did indeed become *commercial* radio, and by the start of the nineties it had gained almost everything it had wished for at Heathrow. However, the scarring of the relationship between regulator and the industry was not to be healed, and was a tainted legacy for the new regulator, the Radio Authority, and for its licensees.

475 See chapter 12.

Chapter 11

Left of the dial

The failure of community radio, 1965 – 1989

In the USA, where the whole raft of non-established radio stations – student, campus, community – is mostly found on the lower frequencies, they have been celebrated collectively as being 'left of the dial'. In the UK, the story of community radio from 1965 until 1989 was one in which those who advocated a politically-inspired concept were left high and dry. Community radio's supporters were doctrinaire to a striking extent in the seventies and eighties and failed almost wholly in their aims, even when they received unexpected support from the government. By contrast, ILR was a fudged compromise which successfully established a new medium with genuine local and demotic credentials. Only when ILR became commercial radio, and unwisely soft-pedalled on its local identity in the late nineties, was there unexpected space for a third tier of not-for-profit radio. This chapter tells the story of the first phase; chapter 23 tells how community radio eventually arrived in the new century.

The terms of the long, and often wearisome, debate over 'community radio' were determined by the arrival of ILR. Had the BBC retained the entire provision of local radio, it would quite properly have been challenged about the breadth of its community base. However, once advertising-funded local radio was established in 1972, the radical left took up the cause in terms of opposition to "local radio, private profit".[476] This consistently undermined the genuine case for community radio as a local social agent.

A definition of 'community radio' is important, since it often bedevilled the policy discussions. Almost every radio lobby tried to hijack the word 'community', to validate what they happened to be offering at the time. However, the definition of what comprises 'community radio', in the terms of the long debate, is fairly straightforward. It is a local service, run largely on a not-for-profit basis, involving a significant volunteer element, and concerned to deliver some social benefits beyond just the provision of an on-air radio service. The station also needs to be independent in the sense that it is not controlled by or beholden to mainstream media, and in most instances will also be stand-alone. It is on that basis that the term will be used in this chapter.

Much of the credit for an intellectual base for local radio in the UK belongs to Frank Gillard. His plan for BBC local radio represented more than just a defensive move against the growing commercial lobby. These stations were a genuine effort to take the Reithian high-mindedness of BBC radio in the mid-sixties down to a local level. Many of the elements of the definition of 'community

476 The title of a Local Radio Workshop pamphlet aimed at Capital Radio, 1983.

radio' were present. Funding from local authorities, chambers of commerce and so on certainly met the not-for-profit criterion, as did the extensive use of volunteers. Gillard described them as providing "a running serial of local life in all its aspects, involving a multitude of voices; what one might call the people's radio".[477] Yet, tellingly, they were part of the BBC machine, lacking the genuine independence of a community station.

The first documented use of the phrase "community radio" in the UK seems to have been in a paper by Rachel Powell in 1965, although Gillard's proposals to Pilkington pre-date that by several years. Powell argued for up to 250 not-for-profit local stations, financed partly by local government and partly from the BBC licence fee.[478] By the time of the discussions around the 1971 white paper, the tone of the community lobby was overwhelmingly one of doctrinal hostility to ILR. That is unfair to many of those – especially in Scotland – who had a genuine and non-partisan interest in radio, but in political circles at least a 'community radio activist' came to mean effectively someone from the loose coalition of interest which was the 'radical left' in the seventies. Taken together with trade union militancy, and the infiltration of the Labour Party by the Militant Tendency, they dominated the politics of the left. They were notably oppositional in style, including in their relationship with ILR.

Gillard's initiative, and the arrival of ILR, pretty much let the BBC off the hook in terms of the community radio lobby. Looking back over the years after 1972, it was fair for Professor Anthony Everitt to conclude that "despite a promising start, the BBC, as the country's publicly funded public service broadcaster, has not played a leading role in the development of community radio ... it has fallen to the regulator of commercial radio to promote its cause".[479] It is remarkable that, throughout what were at times very acrimonious debates, hardly a shot was fired in anger by the community radio activists in the direction of the BBC, which was the recipient of public money for the express purpose of providing public service radio, locally as well as nationally. Even in the early seventies, with BBC local radio wide open to input from the community lobby, they reserved their attention almost entirely for the new independent stations and their regulators, driven by what seemed a visceral hatred of the profit motive, misreading where the real opportunities existed to meet their objectives of grass roots radio.

The majority of the Annan Committee envisaged in March 1977 a single authority to oversee all local radio.[480] Their recommendations were shot through with the need for local radio of whatever stamp to be pre-eminently local, and in that sense they endorsed the concept of community radio. "Local radio has a quite different relationship with its community, and the community has, and should have, an almost proprietary feeling about its local station that it cannot have about a national network."[481] However, apart from a desire that coverage of all localities be achieved before competition be introduced to major areas,[482] that was as far as

477 *Connecting England, Local Radio, Local Television, Local Online* BBC English regions 2001.
478 Rachel Powell, *Possibilities for local radio*, Centre for Contemporary Cultural Studies, Birmingham University 1965.
479 *New Voices; an evaluation of 15 Access Radio projects* Professor Anthony Everitt for the Radio Authority March 2003 p 16.
480 See chapter 7.
481 *Report of the Committee on the Future of Broadcasting (Annan Report)* March 1977 Cmnd. 6753 p 206.

it went. Annan's wish was for a new type of Authority, which would be open and flexible, to accommodate the aspirations of the community radio lobby – but there was to be no specific third tier of radio. The report's main nod towards the community radio lobby was to urge the new authority to "break out of the present mould of financing broadcasting, and encouraging the growth of co-operative and other joint forms of financing to stimulate a direct involvement by the community in its own broadcasting services".[483]

In a marker for what was to happen in the mid-eighties, the Home Office flirted with direct licensing of 'community' radio services on cable in the aftermath of the Annan Report. It initiated two-year experimental licences to provide sound programmes over local cable services and had found a few takers. Thamesmead Insound, which began in January 1978, proved the most durable, becoming Radio Thamesmead and then Radio RTM, and living just on the right side of the edge of survival, but none of these stations presented any real competition to ILR. There was no progress towards the type of community radio which its true believers wanted.

At the forefront of the challenge to the profit motive in ILR was the Local Radio Workshop (LRW), a self-declared "collective". It was scathing about the efforts of the radio regulators within the IBA, citing the ILR companies' "cosy relationship with a none-too-critical IBA",[484] but had chosen its target poorly. Several of the staff within the IBA's Radio Division were sympathetic to the idea of not-for-profit radio. They had inserted into the IBA's response to the Annan Report in 1977 a clear indication of a willingness to implement "cross-subsidisation from more profitable stations"[485] for new services in non-viable areas. "We would also plan to carry out controlled, small-scale experiments with 'neighbourhood radio'." However, the signals were missed by the community radio lobby, and their stridently militant tone maintained the oppositional positions between lobbyists and regulator.

That lobby set out its aims in a Community Broadcasting Charter in 1979. This required community broadcasting to serve local communities; be non profit distributing; settle all its policies through a democratically elected general management committee; provide information, education and entertainment, and two-way participation in these; have mixed funding; recognise trade unions; deploy flexible working and volunteers; be an equal opportunities employer; provide training for local people; broadcast predominantly local material; and "have a programming policy which encourages the development of participatory democracy and which combats racism, sexism and other discriminatory attitudes".[486]

The IBA licensed one service intended to operate closely along these lines. Cardiff Broadcasting won the franchise for Cardiff, advertised before 1978 had ended, and awarded on 30 April 1979 by Lady Plowden's IBA in its most socially liberal mood. CBC failed by every measure, including the democratic when it was bought by Owen Oyston's emerging radio group. In Inverness, Moray Firth Radio began broadcasting on 23 February 1982, with a genuine community base but

482 Ibid. p 383.
483 Ibid. p 209.
484 *Public Radio; Private Profit*, Comedia Publishing, 1983, p 38.
485 *Independent Broadcasting* July 1977.
486 COMCOM charter, quoted in Partridge, *Not the BBC/IBA, the case for Community Radio*, 1982.

without the panoply of 'democratic' trappings that dominated CBC. That station, for many years the smallest ILR station in the UK, was a signal success, winning the radio industry's Sony award as 'small station of the year' with almost embarrassing frequency in the nineties, but it was a long way from the opinion makers in London. The disaster at Cardiff confirmed the IBA in its view that doctrinaire 'community radio' was something to be kept at arm's length, and a pretty duff notion into the bargain.[487] The lobby had lost its potential allies in Brompton Road for good.

The LRW produced a report in 1983 attacking Capital Radio, written by "our full time workers and our user groups: Women's Airwaves, Black Women's Radio Group and Rest of the News".[488] The report derides Capital's profit motive, identifies its social action programme as "public relations", and links ILR stations with the Conservative Party. The IBA is scorned as being dependent on the rental earned from the franchises.[489] The language seems anachronistic now, but it came absolutely from the mainstream of the radical left in the seventies. It was part of the confrontation seen at its starkest in the stand-off between two Conservative governments and the National Union of Mineworkers. In 1983, its partiality lent a political stridency to the lobbying of newly-formed Community Radio Association (CRA), which served its cause poorly.

Home secretary Willie Whitelaw seemed to confirm the government's lack of interest in community radio when announcing the prospect of frequencies for INR, in March 1983. "The development of community radio raises a number of difficult problems for broadcasting policy. Moreover, it would make substantial demands on resources in terms both of creating and operating a regulatory system, and of identifying the limited amount of spectrum which might be made available in the short term. Given the resource demands of other developments in the broadcasting field, and since the spectrum available in the longer term is not yet known, I have concluded that it would not be right to take matters further at present."[490]

Then, unpredictably, at the end of 1984, the Conservative government – in full Thatcherite mode by now – contemplated its own community radio experiment. This arose partly from concerns about land-based pirate radio output. George Thomson for the IBA had launched an "attack" on "government inaction" on the matter in June 1984,[491] with the *Listener* noting that "if numbers are anything to go by, the appeal of such stations seems considerable".[492] The Telecommunications Act, effective from July 1984, was aimed at preventing such piracy. The community radio lobby – never fond of the Thatcher government's laws against activists of any stripe – correctly saw this as counterproductive. Simon Partridge asserted that "simply to crack down without thinking about an alternative to what

487 Lady Plowden was to open the station formally in April 1980. In traditional form, she pulled the cord which opened a curtain in front of a carved stone block announcing that the station had been opened by Lady Plowden, DBE, Chair of the Independant (sic) Broadcasting Authority. When I visited the station several years later, the spelling mistake – which had seemed so striking in that era of careful officialdom - was still on display.

488 *Public Radio; Private Profit,* op. cit. p 4.

489 Ibid. p 106.

490 *Hansard*, House of Commons, 29 March 1983, col 90.

491 *The IBA and 'Pirate Radio'* statement issued 14 June 1984.

492 Stuart Simon in the *Listener* 19 July 1984 p 5.

some of the pirates are providing will be an invitation to more piracy. The problem is, there is just not room for everyone who wants to broadcast to do so, so there must be some kind of institutional solution."[493]

On 23 January 1985 Brittan announced his intention to "enable community radio to develop as soon as possible". That was at least partly at the urgings of officials in the Home Office, keen to try community media projects. Nevertheless, Brittan's personal enthusiasm shines through to a remarkable degree; presumably his libertarian Toryism made him warm to these potential social entrepreneurs, from however unlikely a stable. The language of the official statement is strikingly personal, especially for official documents of that period. "I have for some time been interested in the idea of community radio and am anxious to provide the opportunity for its development ... we now know what spectrum will be available to the UK, in what timescale, so that it will now be possible to establish what assignments could be devoted to community radio ... It is most commonly seen as representing a third tier of radio quite distinct from those services provided by the BBC and the IBA."[494] Even pirate radio operators would be encouraged to apply, which came as rather a shock to Graham Symonds of Sunshine Radio in Shropshire, who had only just handed over all his illegal equipment to the Department of Trade as a gesture of good faith.[495]

On 26 July, Brittan announced plans for "an initial" 21 community radio stations, ranging from the Shetland Isles to five stations in London, for what was to begin as a two-year experiment. Carol Thatcher, the PM's daughter, wrote in the *Daily Telegraph* that "it may be a long time before community radio achieves the status it has in Australia, where there are over 50 stations involving 20,000 volunteers, but the experimental licences are an exciting start".[496] The IBA, however, was furious. "The IBA is anxious that this new development is not introduced to the detriment of the existing Independent Local Radio system. In any case, the IBA believes that ILR already provides an effective and self-financing form of local community radio ... [the new stations] should be required to operate under similar financial and general obligations to the ILR companies; otherwise they will constitute unfair competition."[497] Radio Clyde's chairman, Ian Chapman, thought it quite simply "a mistake ... to announce an experiment in community radio without first examining exactly what that term meant and what effects it would have on the existing broadcasting system".[498]

It was indeed a remarkable departure. For the first time in British broadcasting history since the BBC was granted its Royal Charter to separate it from direct government control, there were to be free-to-air services licensed directly by the government with no intervening body, raising real issues of democratic concern. It had been axiomatic that governments acknowledged the democratic danger if they had direct control over broadcasters; the temptation to interfere would be irresistible, as it was in the several European countries where such arrangements applied. Whether it was a chartered BBC, or independent regulators, some inter-

493 Quoted in the *Listener* op. cit. p 7.
494 Home Office press statement 23 January 1985.
495 *Sunday Times* 24 March 1985.
496 *Daily Telegraph* 1 August 1985.
497 IBA Press Statement 25 July 1985.
498 *Radio Clyde Annual Report 1984–5*. Chairman's statement.

vening body was needed to keep elected politicians at arm's length. Actually, the Home Office intended that community radio services would be very largely unregulated, even in terms of their commercial revenues. The first signs that the idea might not have been fully thought through came with subsidiary technical regulations for the new stations, issued in September, which were "so strict that even the BBC or the IBA could not meet them" according to *Broadcast*.[499]

Nevertheless, the Home Office was deluged with 245 applications for the 21 licences by the revised closing date of 31 October, and set up a panel to select the first stations. There was some suspicion among the more radical applicants that this selection body was drawn overmuch from 'the great and the good', which might also have reflected a slight cooling of the government's feet. Still, the panel set to work, chaired by Stephen O'Brien, the chief executive of Business in the Community. They met regularly up to Christmas, submitting their recommendations to the home secretary early in 1986. Then there was silence, which continued to be deafening for months. Was the government unhappy with the panel's recommendations, which leaks said favoured ethnic minority services? Did the anguish of the ILR stations, which had only recently started to gain freedom from IBA regulations in a move strongly favoured by government, worry the free-marketeers in the cabinet?

AIRC director Brian West speculated that "there is concern over the public order aspect of community radio".[500] There had indeed been riots in Brixton, Toxteth, Peckham and Tottenham at the end of September and early October, which brought into sharp focus the simmering discontents within many minority communities, and the fragile state of law and order in England's cities, but these had preceded the start of the panel's work. It was rumoured that there was division within the cabinet, and opposition from the Foreign Office and the Northern Ireland Office, both concerned about broadcasting by 'enemies of the state'. Moreover, the panel's recommendations for licence awards showed the likelihood of licensing just the sort of 'marginal' groups that the hated GLC – poised for abolition that year – would have favoured.

How far the key determinant was personality, again, is open to debate. Leon Brittan had been succeeded as home secretary by the measured diplomat, Douglas Hurd, on 29 September 1985. Hurd may have had doubts from the start. The original application date for the licences was postponed as soon as he took office. Certainly, once the problems inherent in the experiment started to appear, the new home secretary would instinctively have sought an administrative resolution. There was one to hand. The Peacock Committee was about to issue its report (it actually appeared in July 1986), and there was growing government dissatisfaction with the institutions of broadcasting, and especially with the IBA's regime for ILR. It was time for a green paper, which followed in 1987. In the meantime, the community radio experiment could be safely postponed, until it could be considered within the broader picture. Hurd issued the announcement accordingly on 30 June. The community radio folk were outraged. John Gray, a BBC stalwart, later to embody small-scale radio in Scotland – and certainly not a political militant – spoke for them all. "As an example of insensitive incompetence, the indefinite

499 *Broadcast magazine* 13 September 1985 p 14.
500 *Mail on Sunday* 11 May 1986.

postponement of the experiment in community radio is a terrifying example of political ineptitude by the present government. To start a scheme, to encourage widespread participation and then, on dubious grounds, ditch the whole effort at the last moment is almost incredible."[501]

It was to be the last stand for the radical left in radio. By the time community radio came to fruition twenty years later, they and their passionate politics had left the stage. Peacock had a good deal to say about the fragile state of ILR finances,[502] and recommended pretty much all that the ILR companies were asking for in terms of looser regulation,[503] but said nothing at all about community radio. However, it had been enough to get Douglas Hurd off the hook on which Leon Brittan had hung the government. The Broadcasting Department of the Home Office found itself still rather uncomfortably awarding 'special event licences' for short-term services, which – in principle at least – had to be for localised radio coverage of actual events. It had also to find a home for a few cable radio experiments, and the hospital and student radio services mostly licensed on induction loop systems, but that needed to await the arrival of a new radio regulator.

As far as community radio was concerned, the discrediting of government policy left the field free for the regulators. Their response was threefold, and looked for a while likely to represent the total of what community radio in the UK might become: first, the new 'incremental' stations, launched by the outgoing IBA in 1989; second, the mostly short-term Restricted Service Licences (RSLs), which the Radio Authority developed in 1991 with such success from the Home Office Special Event licences; and third, small-scale ILR licences, especially those which were deployed on the expanded FM spectrum after 1994. Each of these is dealt with elsewhere in this history.

In the late eighties, the IBA – notably those within it who were to form the nucleus of the new Radio Authority – saw new 'incremental' licences as the clinching response to the community radio demand. When the IBA announced on 13 September 1988 a desire that there should be 20 new stations with a clear community remit – many of them within the service areas of existing ILR stations – it was "to the astonishment and horror of its larger radio contractors ... [that the IBA had] ... stirred itself from the torpor into which it has lately sunk".[504] Government approval was given formally on 2 November. Inviting expressions of interest, the IBA observed that "it is expected that a number of the contracts, within areas of large population, would be for 'community of interest' stations. The others will cover smaller geographic areas."[505] Paul Brown for the IBA was explicit that these were to be what the Authority considered community radio. The resulting stations included some of the land-based pirates whose illegitimacy had most troubled the authorities, including London Greek Radio and Kiss FM in London.

Measured in terms of a new foray into community radio – and by other measures too – many of the incremental stations failed. Those that survived became largely indistinguishable from mainstream ILRs, or were focussed upon specific ethnic minority communities where the dynamic was different. Ofcom

501 Letter to the *Scotsman* 3 July 1986.

502 *Report of the Committee on Financing the BBC (Peacock Committee)* July 1986 Cmnd 9824 p 19.

503 Ibid. p 140.

504 Nick Higham in *Broadcast* 23 September 1988.

505 IBA News Release 9 November 1988.

now dismissively characterises the incremental stations as a "false dawn" for community radio.[506] However, it was a response which, if nothing else, helped the government out of the hole it had dug for itself with the derisory 1985 experiment. The 1990 Broadcasting Act consequently placed no specific community radio obligations on the new regulator, the Radio Authority – no doubt to the satisfaction of both parties, but to the continued frustration of the CRA.

It looked at that point as if there would be no third tier of radio in the UK. The new regulator was almost as doctrinally opposed to the notion of separate community radio as were the ILR companies, and there was relatively little political support. The CRA lost heart, apart from a few enthusiasts. It renamed itself the Community Media Association, and set off to pursue the idea of local television. Once the political support for a leftist approach to media had been dissipated, by the twin forces of market economics and the new post-socialist political settlement, who remained with any real clout to argue the case?

Against all expectations, that was not to be the end of the story. A successful initiative came from the most unlikely source, the Radio Authority itself, which was to lead to the creation of a brand new, nationwide, and expanding third tier of not-for-profit radio in the UK. But that was not even to start happening until the next century, and is a tale for chapter 23.

506 *The Future of Radio*, Ofcom 17 April 2007 p 120.

Chapter 12

Changing the guard

1986 – 1989

During the last years of the eighties, the ILR companies generally enjoyed political supremacy in the radio argument. They undermined the IBA's long-held basic principles, and moved the medium firmly towards *commercial* radio. In 1987, programming assumptions were changed radically by Capital's new contemporary hits radio format, which gave ILR almost ten years' dominance of the centre ground of popular music radio. These years also saw the departure of the old guard of the IBA, notably John Thompson in the spring of 1987, and John Whitney just two years later. They were pivotal years, preparing the way for the shift from independent radio, as well as ushering in satellite television and the beginning of multi-channel broadcasting. Circumstances had seen off community radio, and it was clear to almost everyone that the IBA's stewardship of ILR was coming to an end. In these years, even the chairman of the Authority acknowledged that "it became fashionable to attack the IBA",[507] and at the end of the decade it would give way to two new regulators, the Independent Television Commission and the Radio Authority, while its prized transmission function would be privatised as National Transcommunications Limited (NTL).

Internationally, these were momentous times. The Berlin Wall fell in November 1989, bringing an end to the post-war era. Domestic politics were not quite as stable as the continued dominance by the Conservatives would suggest, with the Thatcher government shaken by the Westland Helicopters affair. However, as Labour wrestled to purge the Militant Tendency from its ranks, it was soon back to business as usual. Margaret Thatcher became the first prime minister for 160 years to win a successive third term, at the general election on 11 June 1987. A thumping majority of 101 seats marked her final and crowning electoral triumph. Privatisation remained the watchword, with airports, the steel industry, electricity, and eventually water all 'returned' to the private sector. Unsurprisingly, the ILR companies also yearned for greater freedom from regulation, to pursue their own commercial interests.

The impetus for large-scale expansion of ILR had largely dissipated by the mid-eighties, partly through the IBA's institutional loss of confidence, and from 1987 as an apparent consequence of imminent legislative change. The primary rental reductions which followed the Heathrow Conference, and the sharp fall in secondary rental yields, also placed severe limits on IBA capital expenditure and, therefore, its ability to fund new transmitters for ILR. The IBA changed tack, and

507 George Russell in *IBA Annual Report* 1988-89 p 5.

looked for a different kind of development, more modest than before. Attention shifted to expanding the coverage areas of existing franchises, either by adding transmitters on the edge of or within current areas, or through power increases made possible by the new Geneva Plan for frequency use. In March 1986, the Authority had noted that "ILR contractors are having a difficult time attracting the revenue to meet their costs. In many instances an additional area would also help them widen their opportunities."[508] As this approach was followed through, there were new transmitters to expand existing ILR franchises to cover Londonderry, Southampton, Basingstoke & Andover, Derby and Shrewsbury & Telford. In addition, there were seven entirely new FM-only franchises for Oxford/Banbury (Fox FM), Yeovil & Taunton (Orchard FM), Borders (Radio Borders), Dumfries and Galloway (South West Sound), Cambridge and Newmarket (Q103 FM), Milton Keynes (FM103 Horizon) and Eastbourne & Hastings (Arrow FM).

All the basic assumptions about private radio were starting to be recast. The radio companies were increasingly assertive in their dealings with the IBA. They were self-confident enough (or bloody-minded enough) angrily to confront the IBA's director of finance at the 1986 July RCC meeting, requiring DG John Whitney to intervene and suggest "that since agreement could not be reached on this issue [comparative rental levels] AIRC should, if they so wished, put their views and objections in writing".[509] Radio Wyvern's Norman Bilton "in an unprecedented display of kamikaze bravado [was] telling anyone who would listen that the IBA is an expensive bureaucracy living off the back of radio".[510] This was emphatically no longer 'master and servant'; the old deference had gone.

The government was tentatively changing some rules, too. In March 1986, it did away with the levy on radio company profits, which from 1981 had taken moneys out of the ILR system. Under encouragement from the Home Office, the IBA began to interpret the rules on programme sponsorship more flexibly, inventing the concept of 'co-funding', which allowed event-based programmes to be paid for by the sponsor of the event. Many stations simulcast a rock concert by Queen in July 1986 which was 'supported' by TDK. Everyone seemed to turn a blind eye to the evasion of previously sacrosanct principles. Selkirk's shareholdings in ILR, including a majority stake in LBC/IRN, were sold to an Australian company, Darling Downs Television (later to be Chalford Communications). It would have been unthinkable then for an overseas company to have controlled ITN. Torin Douglas asked the pertinent question whether "the IBA would have allowed 58 per cent of the shares in an ITV company to be sold to an unknown Australian company" and drew the conclusion that the different standards for ILR and ITV were an argument against radio and television being managed by the same regulator.[511]

The Peacock Committee reported on 29 May 1986. Set up the previous year, its start coincided with the low point in the prime minister's relations with the BBC, and many thought that Professor Alan Peacock's private brief was to recommend the abolition of the licence fee and the commercialisation of the Corporation. In the event, Peacock was something of a damp squib on television.

508 *IBA Paper* 25 (86) 12 March 1986.
509 *RCC Minutes* 51(86) of a meeting on 2 July 1986.
510 *Marketing Week* 25 April 1986.
511 *Marketing Week* op. cit.

Its proposals for that medium were limited to freeing Channel Four to sell its own advertising. However, muddled radicalism on radio made up for saying nothing interesting on television. The Peacock committee offered two conflicting recommendations. The first, rather strangely, that "the BBC should have the option to privatise Radio One, Two and local radio in whole or part".[512] Then, in case BBC turkeys decided not to vote for Christmas, a majority of the committee recommended that "Radio One and Radio Two should be privatised and financed by advertising ... [and] any further radio frequencies becoming available should be auctioned to the highest bidder". The dissenters were Judith Chalmers and Alastair Hetherington, both with strong BBC connections, who matched the stand taken by Marghanita Laski within the Annan Committee ten years previously.[513]

The IBA enjoyed no such protection, however, and there was no dissent from the recommendation that "IBA regulation of radio should be replaced by a looser regime",[514] although "it is vital not to destroy local services". Peacock quoted approvingly AIRC's 'shopping list' from the Heathrow Conference. Freedom for the ILR companies to operate without constraint, and thus to reduce programme content if they so chose, was thought acceptable since "the popularity of local news, local travel and weather, what's on and other features will secure continuation at least for these information programmes – the most important – if not for others". Bizarrely, Peacock proposed that the BBC should be allowed to "take over failing ILR stations" and "stronger ILR companies to buy out a BBC station if and when the BBC was willing to sell". The last (to date, anyway) of the long sequence of inquiries and royal commissions into broadcasting, dating back to Sykes in 1923, the Peacock Report was scarcely the most distinguished.

In ILR programming, something was happening during these years which changed the face of popular radio in the UK. Radio One continued to attract huge audiences with its rather outdated style of pop radio, ILR was doing nicely enough with its locally-flavoured imitation, and some more grown-up music, but without having moved forward to any single new approach since its original innovative mixed programming output. Then Richard Park arrived at Capital with a brief to re-invent the station. His version of contemporary hits radio (CHR) was ground-breaking, and was soon imitated across ILR. This was self-declared hit music radio, but it played good songs, old and new. The music was central, but it was also the vehicle for stunning competitions, personality presenters, contemporary spoken content, and all set off by a great jingle package. It had the energy of the best American radio, with a pronounced British accent.

The thoughtful Neil Fox, whose star career took off in that milieu, recalls that "it was happy pop, brilliant pop. It wasn't just the music, it was everything about the music. If you wanted to have access to the stars, listen to Capital. They are on our shows every day, you could come and meet them in the foyer. The radio station was the place to be. There was the job finder, the flat finder ... We are interested in your life. You can come here, find a job, meet the stars, meet the jocks, buy your

512 *Report of the Committee on Financing the BBC (*Peacock Report) *Cmnd* 9824 July 1986, p 140

513 Alastair Hetherington, for 20 years editor of the *Guardian*, was from 1975 to 1978 the controller of BBC Scotland, then manager of BBC Highland from 1979 to 1980. Judith Chalmers presented two major BBC radio programmes, *Family Favourites* and *Woman's Hour*, in the 1960s, and was also presenter of BBC Radio 2's mid-morning show from 1990–1992.

514 Peacock Report p 141.

T-shirt."[515] Taking the station out to its listeners was another element, in the road shows first used by Radio One in the seventies, and deployed by Capital to great effect around London. Audiences of 70,000 were commonplace, to hear the likes of Jason Donovan and Kylie Minogue live in Brixton or Romford. ILR stations flocked to follow Capital's lead, and they established a new and unique voice in British radio. This new style of UK popular radio was to drive the network weekly reach average from the 40 to the 50 per cents, and to lay the foundations for the commercial success that was down the road. Park's CHR version of Capital had kick-started an international radio revolution too. "Capital had become the model for a modern, big city, CHR radio station ... we had become the greatest CHR station in the world. To go to LA, and have people in the radio business coming up to you to say 'Are you Doctor Fox from Capital?' It was just amazing." It was also, as Fox implies, quintessentially the sounds of your life for listeners throughout London.

One of Fox's anecdotes explains how popular music and the immediacy of local appeal were woven together to such good effect in Park's CHR Capital. "Pete Waterman had my hot line in the studio. He phoned me up one day to say 'I've just finished this amazing record with Donna Summer. If I bike it over to you on tape, can you play it on the air, cos I'm sitting here with her and the engineer and we want to hear how it sounds on the radio before we mix it ... the bike arrives, reel to reel, I get it on the studer [tape recorder] and play it live, saying this record has just been cut this afternoon."[516] The song was *This Time I Know it's for Real* which stayed in the UK charts for 14 weeks at the start of 1989, and won a silver disc.

Financially, ILR was at last emerging from recession. In the year to September 1986, two thirds of the 40 ILR companies then on air reported profits, with Capital increasing its pre-tax profits by 80 per cent to £1.7m. Dividends were up too, which was starting to be important as the larger companies moved to acquire stock-market listings. Clyde and Piccadilly had followed City on to the USM, but the key move was that of Capital Radio, which gained a full stock-market listing and had a highly successful flotation early in 1987. Metro joined the USM in 1988. Such diversification as the IBA could be persuaded to permit, though, was not wholly successful. City's profits were badly hit over several years by the poor performance of its Beatles City subsidiary. Although radio advertising revenues for 1985/6 at £66.4m had still been disappointing, the following year showed a good recovery to a record of £93m, although this still remained stubbornly around 2 per cent of total UK display advertising expenditure.

A different kind of radio company was now coming to the fore. These were no longer the looser groupings which had initially held ILR franchises. The older notion of ownership by a consortium of mainly local shareholders, each with their other priorities, who came together for both the social and the commercial purposes of running a local radio station, was supplanted. The new breed were media companies, existing either to run radio stations or with a specific radio division amid other operational activities. This was a fundamental change, virtually unavoidable if stock-market quotations were wanted, and central to the shift away

515 Neil Fox in an interview with the author, 25 September 2008.
516 Ibid.

from *independent* to *commercial* radio. The companies understood better than their regulator that the previous mould of ILR ownership had now been fractured, and they appreciated its significance more quickly.

The IBA's attitude to mergers remained equivocal, although the 'rescues' continued. At the end of 1987, the IBA barred a merger between the three-station Red Rose and Yorkshire Radio groups, fearing that would create too large an entity, not in keeping with a local radio system. However, direct opposition was not going to be sustainable, and a merger between Radio 210 and 2CR was allowed in January 1988. Clyde acquired Northsound, and the IBA also permitted the merger of BRMB and Mercia Sound. The Authority therefore needed rules about ownership and takeovers, and in October 1988 set an ownership 'ceiling' of 15 per cent of the potential ILR audience, to prevent any one owner from becoming unduly dominant. In companies which they did not control, other stations could acquire from 5 per cent to 29.99 per cent of voting stock, provided that the companies concerned did do not cover more than 15 per cent of the total audience. However, 'control' was assumed to arise at 50 per cent, which was an unrealistically high threshold. In June 1989, the IBA needed to expand its rules to make clear that any new interest above 29.99 per cent would be counted as a controlling interest, although existing shareholders in the range from 29.99 to 49.99 per cent, or those already above the 15 per cent ceiling, would not have to sell down.[517] Inevitably, many companies built up stakes of 29.99 per cent, plus non-voting stock, in preparation for a more generous regime.

ILR began to celebrate renewed profits, thus giving the lie to Peacock's assertion that "under present regulations, there are no indications that ILR profitability will increase substantially in the future".[518] By the spring of 1987, buoyant advertising and cost-cutting, together with the IBA's forebearance on rental and the removal of the Treasury levy, saw the share prices of Piccadilly and Clyde double in a year, and Radio City's treble.[519] The IBA was persuaded in July 1987 to allow an extra two minutes of advertising per hour next to the morning news bulletins as part of the IRN service.[520] Sold as *Newslink*, this was to be a major revenue earner for the network, and would eventually turn IRN from a cost to a profit-earner. More Australian investors were arriving, with Paul Ramsay buying out the Canadians of Standard Broadcasting at the end of 1986. Chalford Communications became the Crown Communications Group in April 1988, adding to its 58 per cent shareholding of LBC and 56 per cent of Marcher substantial stakes in Forth, Beacon, Trent, Southern Sound and Mercury, as well as owning the sales house Independent Radio Sales. Its chairman was Christopher Chataway, the ministerial begetter of ILR.

According to Colin Walters, MD of Piccadilly, by late 1987 the radio industry "which a year ago felt itself the Cinderella of the media world – poor and unloved – has been feeling decidedly chipper".[521] For the year up to September 1988, advertising rose to a new record, up 27 per cent to £107m, and ILR was enjoying a genuine boom. Profits rose again, and the share prices of quoted ILR companies

517 IBA press release 8 June 1989.
518 Peacock report op. cit. p 20.
519 *Financial Times* 31 May 1987.
520 *UK Press Gazette* 20 July 1987.
521 *Guardian* 9 November 1987.

soared. Capital was very much regarded as the flagship, and a 239 per cent profit increase – to £9.3m – confirmed the general optimism. Radio stocks became fashionable, buoyed up by the expectation of a lighter regulatory regime and further stations, and the opportunities for expansion through takeovers. The *Independent* saw the "airwaves crackle with expectation".[522] The *Times* told investors "to prepare for a radio revolution". It believed that price-earnings ratios for radio stocks were "very conservative", and foresaw even for LBC's owner Crown "a stampede for the stock ... if advertising revenue stays at these (1988) levels".[523] The *Financial Times* provided the historical context. "Not so long ago, radio was perceived simply as television's poor relation. Now the medium is taking an increasing slice of a growing UK display advertising market ... compared with the early days of meagre revenues, the radio stations are currently walking on air."[524]

One takeover, however, was affected by factors outside the Authority's control. At the very beginning of 1989, *Investors Chronicle* reported that Midland Radio, which had resulted from the merger of BRMB and Mercia, was itself the subject of a bid from Piccadilly. The resulting group was to be the largest independent radio group outside London.[525] The cash and shares offer valued the purchaser, Piccadilly, at around £20m. All seemed to be proceeding smoothly until 6 February, when Piccadilly's shareholders were poised to agree the purchase. At that last moment, their company was itself subject to a bid from Owen Oyston's small group of stations, now called the Miss World Group since it had taken over the eponymous beauty competition. This bid valued Piccadilly at £35m, but was conditional on the Midlands deal not going ahead, since combining all those stations would breach the 15 per cent limit. Piccadilly's board described the offer as "wholly unwelcome and disruptive",[526] and urged shareholders to continue with the Midland purchase.

The battle dragged on until a much-adjourned extra-ordinary general meeting on 20 March. 50.33 per cent of the votes at that meeting went against the merger with Midland, and by the narrowest of margins this allowed Miss World's takeover of Piccadilly to proceed. However, this still required an amendment to Piccadilly's articles of association, which prohibited any one person from owning more than 15 per cent of its shares. That argument reached the High Court on 27 April, but the takeover proceeded. Piccadilly MD Colin Walters "walked out" in July,[527] and was replaced by Julian Allitt, a long-time associate of Owen Oyston. The resulting company changed its name again, to Trans World Communications. Following allegations, the Takeover Panel cleared the transaction in October, and again in February 1990. Independent radio had entered the new commercial dispensation with a vengeance.

The green paper *Radio: Choices and Opportunities* was published in February 1987. It was a response to the challenges raised by the ILR companies during and after the Heathrow Conference, to the radio issues in to which the Peacock Committee had strayed, to the policy vacuum left by the failed community radio

522 *Independent* 25 April 1988.
523 *The Times* 31 May 1988.
524 *Financial Times* 10 June 1988.
525 *Investors Chronicle* 6 January 1989.
526 *Financial Times* 7 February 1989.
527 *Broadcast* 14 July 1989.

experiment, and to the questions posed by the new international frequency plan authorised at the Geneva Conference in 1984. Policy-makers had been won over by the ILR companies' arguments about the parlous nature of their finances, and the green paper asserted that "unless changes are made in the regulatory framework for ILR, the quality and local character of the service may be increasingly at risk".[528] The failure of the community radio experiment is rather strangely presented as suggesting "a strong potential for services which appeal to very local or specialised interests, and in particular which meet the needs of ethnic minority communities". There is "a strong case for giving the BBC competition as a provider of national radio services". However, the privatisation of Radios One and Two, argued over by the Peacock Committee, was to be set aside in favour of new independent national radio (INR) services to provide competition, which would "not be required to fulfil all the obligations of public service broadcasting", and would run their own transmitters.

For local radio, the green paper was undecided about the case for community radio, the relevance of public service requirements on ILR and the nature and extent of regulation. The government favoured some community radio, with both that and ILR "operating under the same light regulatory regime as the new national services". [529] There was a case for a new radio authority for INR and ILR, which the green paper thought might be an expanded version of the Cable Authority, "which has experience of the sort of regulatory regime which would be appropriate to independent national and local radio". Whatever form it took, the new authority for radio would have much looser powers over ownership and shareholdings than the regime the IBA was currently required to enforce. It would also undertake its own frequency planning, in competition with the BBC, to ensure the most intensive use of the spectrum.

The green paper usefully rehearses the spectrum opportunities which conditioned the future development of both BBC and independent radio. As a guiding principle, it argued that broadcasting the same services on FM and AM should be phased out. It wished to take "a decisive step away from simulcasting"[530] by stopping that on national channels by 1995. Within the part of the VHF/FM spectrum known as Band II, there would be some limited possibilities for new services within 87.5–88 MHz immediately. 97.7–102 MHz would become available from 1990, for additional FM stations. The home secretary had already committed the sub-band from 97.7–99.8 MHz to the BBC, to provide Radio One on VHF/FM,[531] but 99.9–102 MHz "would remain available for an independent national radio service".[532] Looking further ahead, 105–108 MHz was expected to become available for radio use in the mid-nineties. Fitting in low-powered services within these bands was seen as "extremely difficult, but not impossible"; an unduly pessimistic forecast. On MF, there was room for a further 46 services in addition to the then current 50 ILR AM stations in the existing band, and perhaps double that number in the space made available by the Geneva Plan. The possible

528 *Radio: Choice and Opportunities* February 1987 Cm.92 p 38 (Green Paper).
529 Ibid. p 31.
530 Ibid. p 23.
531 *Hansard* House of Commons 23 October 1985 cols 147/8.
532 *Green paper*, op. cit., p 42.

re-allocation to INR of BBC national AM frequencies would clear the way for three INR services after 1990.[533]

The BBC had again turned to the McKinsey management consultancy to do its radio policy thinking for it. The result was a proposal for broadcasting Radios One, Two, Three and Four primarily on VHF, creating a sport and education network on one AM frequency and a public affairs network in place of Radio Four on long wave. There was a predictable outcry. The middle classes marched on Broadcasting House in April 1993. David Hendy's splendid history of Radio Four tells how "this was Middle England *par excellence* ... a column of polite individuals, holding balloons, marshalled by stewards eccentrically attired in pistachio green berets. No one shouted. Everyone spoke Standard English and 'within statutory decibel levels'."[534] An urban myth says that the marchers were chanting "What do we want, Radio Four; where do we want it, long wave; what do we say, please". Once again, outsiders had failed to understand the intensely close relationship listeners feel with their radio station, and even its frequency. Long wave may not even then have been a *modern* platform, but it was the one they knew and felt comfortable tuning to.

It had been axiomatic, since the radio engineers had begun to explore very high frequency (VHF) transmission in thirties America, that medium wave was the primary medium of listening, with VHF as a back-up of most interest to enthusiasts.[535] All ILR franchises, before the Basingstoke relay for Reading, were licensed for medium wave with a VHF back-up. Occasional splitting of frequencies was permitted by the IBA, but only to cover circumstances such as football commentaries, and then usually with some reluctance. The IBA and BBC had jointly won the right to experiment with split frequency local broadcasting in 1985, and those experiments allowed for up to 10 hours per week. Six ILR stations had done this. Capital ran a weekend 'beautiful music' service on FM (predating both Classic FM and Melody Radio). Viking, in Hull, put out rugby league commentary on AM only. The day of full-time splitting was edging closer.

Public use of radio listening technology in the UK had gradually changed over the 30 years since radio was first broadcast widely on VHF by the BBC from its Wrotham transmitter in 1954. Nobody really knew how common VHF listening was, since research conducted in the normal way was unreliable on this point, as it required listeners to know which technical platform they were tuned to; not something to be relied upon. The green paper thought that by 1983, although "there has been consumer resistance to VHF/FM listening", 85 per cent of

533 Ibid. p 39.

534 David Hendy, *Life on Air, a History of Radio Four*, Oxford University Press 2007, p 361.

535 For much of the twentieth century, it was popular usage to identify radio broadcasting bands, and the services on them, according to their wavelength, even if that upset the engineers. Thus services were broadcast on *medium wave* or *long wave*. The alternative nomenclature based on modulation characteristics of *amplitude modulation (AM)* or *frequency modulation (FM)* was not common outside technical publications. When the very high frequency bands began to be used for sound radio broadcasting, *VHF* was the normal identifier just as it was for television. Gradually however, the technical usage began to become common usage, so VHF became *frequency modulation* or *FM*. I have mostly tried to use the term which would have been current at the time of the events I am describing. That may upset the purists. At the establishment of Ofcom, some of the engineers who arrived from the Radiocommunications Agency insisted in all seriousness that what we know as radio services must be called "sound broadcasting" to distinguish them from the numerous other forms of radiocommunications services over mobile or fixed links. The BBC now has a director of audio, rather than a director of radio.

households had at least one set capable of receiving VHF/FM, and 29 per cent of car radios were equipped". It quoted research suggesting "inertia, and the absence of any strong incentive to use it ... [leading to] listening increasing gradually, by a few per cent a year".[536]

The companies were concerned that the IBA's response to the green paper had been institutional rather than innovative, on this point among others. Colin Walters argued late in 1987 that "the IBA, in spite of being rattled by the government's threat to rob it of control of radio, shows little sign of coming up with the sort of innovative thinking that would persuade the government of its ability to do the job in a new more liberal context".[537] The end of simulcasting therefore was driven chiefly by a shift in attitudes in the Home Office. The green paper had concluded that any new local stations should usually be allocated just one frequency, which carried the clear implication that those existing stations with two ought to be able to deploy them separately. For the first time, existing ILR stations were to be able, indeed required, to cease to be universal services and to segment their output for differing audiences. Companies must provide separate output on FM and AM, or contemplate being reduced to a single frequency only: 'use it or lose it'. There was some uncertainty within ILR. Would a station win more listeners and advertisers if it ran separate services? Terry Bate, with his transatlantic experience, was in no doubt. "You are being handed a free radio station on a plate ... advertising money expands to fill the media available."[538]

Radio Clyde hoped to be the pioneer, but its negotiations with PPL over needletime arrangements broke down in February 1988. The first all-week, all-day split, therefore, came from County Sound in Guildford, at the start of June 1988. It kept its normal output on FM, providing a 'golden oldies' station on AM: music from the fifties, sixties and seventies, plus news, weather and travel. That proved to be the generally adopted pattern. Piccadilly took the other option. At the beginning of September it left its normal output on AM, and launched Key 103 on FM "to win audiences which, by definition, chose not to listen to our traditional output".[539] Walters was keenly aware of the decline in the sales of single records and the growing importance of CDs, and wanted to avoid top forty music. His target was "quality – in artiste-image, in sound, production and lyric". This was quickly dubbed "Yuppie Radio", in contrast to "leaving the plebs to their pap" on medium wave.[540]

Piccadilly had actually gone off too soon with what proved to be the less favoured option, though not necessarily objectively the wrong choice. Almost all the others eventually offered their 'heritage' service on FM, and a version of a 'gold' format on AM. The most striking exception apart from Piccadilly was Downtown in Belfast, which offered and still offers Cool FM to supplement the more traditional mix of its heritage brand on AM. Ocean Sound, the successor to Victory, took advantage of its transmitter options to offer three separate services: Power FM for younger audiences from the Portsmouth and Winchester transmitter; Ocean Sound on FM from Southampton; and South Coast Radio on AM across the whole

536 Green Paper p 8.
537 *Guardian* 9 November 1987.
538 *Campaign* 1 April 1988.
539 *Music Week* 10 March 1989.
540 *Guardian* 19 September 1988.

area. Splitting transmissions was a key driver in introducing to ILR the so-called 'black box' technology. That would later permit extensive networking, the playing in of remote 'inserts' to add a supposed localness to shared output, and automation. All these were to be important battlegrounds between companies and their new regulator in the next decade.

The green paper endorsed the notion of national commercial radio, which became known as Independent National Radio (INR). This was not a new idea. In the debates around the offshore pirate services in the 1960s, there had been a moment when the BBC's 247m wavelength might have been put aside for a single national commercial pop music channel. As early as July 1976, the IBA was reflecting on whether it should seek a national radio channel. John Thompson had asserted earlier in the year that "it would be perfectly imaginable – both in programming and technical terms – that some national radio, like some national television, can be provided on a self-financing basis, along the lines of the present arrangements under the aegis of the IBA".[541] The frequency argument was less comfortable following the Geneva Conference in 1984, which meant that INR could only be broadcast on medium wave (or indeed long wave) if it could take over frequencies currently allocated to the BBC. On VHF, however, there was the prospect in the medium term of more VHF band being made available for radio broadcasting. Even at that stage, despite being fully taken up with the precarious nature of ILR, the IBA was contemplating the case that "Independent Radio can only be a full, credible and effective alternative to *total* BBC radio (as against local BBC services) if it has both local and national channels".[542]

With the ending of television use of what are called Bands I and II, as 405-line black-and-white gave way to 625-line colour, the government had commissioned, in 1982, a report from Dr J H H Merriman about the possible uses for this spectrum. He proposed that "Bands I and III be withdrawn from broadcasting use ... and used for ... land mobile services".[543] A concerted lobbying effort from the broadcasters, however, persuaded Willie Whitelaw to announce in March 1983 that the UK would seek international approval for the allocation of VHF frequencies for a new independent national radio network, along with a further BBC service on VHF. INR was now in play. Whitelaw told the House of Commons in March 1983 that "the IBA has proposed that one of the new national networks should be used to provide an independent national radio service. The government finds this proposal attractive, provided that satisfactory financing arrangements can be developed."[544] Thompson's hidden but elegant diplomacy had won another prize.

Nick Higham, then editing the radio pages of a trade magazine, was one of the few who was not caught off guard. "The announcement came as a surprise to a good many observers of broadcasting", he wrote, "even though the IBA (and in particular its director of radio, John Thompson) had been gently lobbying for such an allocation for at least two years, fearful lest the BBC walk off with all the extra VHF frequencies which become available at the end of the decade".[545] ILR companies were divided. David Pinnell of BRMB was a fervent supporter of INR,

541 John Thompson's IBA Lecture, 21 January 1976.
542 *IBA Paper* 194(76) 19 July 1976.
543 *Independent review of the Radio Spectrum (30-960MHz)* September 1982 Cmnd 8666.
544 *Hansard* House of Commons 29 March 1983 col 88.
545 *Broadcast* 6 Jun3 1983.

believing "strongly that a national service will increase radio's profile and prestige with advertisers ... and will lead them all to use radio, including ILR, more in the future". Hallam's Bill MacDonald thought the idea madness. "In going down this path, the IBA has lost sight of the whole dynamic of independent radio; its localness ... there's an obsession at the IBA about competing with the BBC." For almost the last time on a major matter of future radio policy, the IBA carried the day, persuading the Home Office to disregard the views of a divided radio industry. The Home Office even intended to legislate to allow the IBA to operate a national service, once parliamentary time could be found.[546] The application was duly made to the International Telecommunications Union, and approval given. The BBC was to have separate VHF networks for Radios One and Two, and an independent network was to be deployed.

The key to INR's prospects on medium wave was going to be which of the BBC-used AM frequencies could be prised from its grasp. That had been foreshadowed in IBA thinking back in 1976, but the matter was not decided until 1989, by which time the IBA had lost Thompson's subtly ruthless skills. The BBC and the IBA ferociously lobbied Douglas Hurd, then home secretary, but it was an unequal contest. The BBC was made only to give up 1053 and 1089 kHz, for one new INR service, and 1215 kHz for the other. This last had been Radio Three's medium wave channel, good enough for daytime in London (although not for classical music) when the bands were less crowded, but not much use outside. At the final, decisive meeting with Douglas Hurd and Quentin Thomas at the Home Office in 1989, the BBC negotiators, reportedly led by David Hatch, could scarcely believe their luck. Paul Brown – who became head of radio programming in the IBA's Radio Division in 1985 – takes up the story where the official record leaves off: "The BBC was under pressure from the government to offer spectrum to the commercial sector ... They argued that the commercial sector would only need medium wave, since audience figures clearly showed that nobody listened to radio after six o'clock at night. Douglas Hurd said 'that sounds OK to me'. When they left the office and got downstairs they could not believe they had pulled it off in a conversation of that kind, and they joined hands and danced down Queen Anne's Gate at half past seven at night."[547]

Meanwhile, John Thompson, the founding father of ILR, had taken his moment to hand over the radio job to his deputy, Peter Baldwin, in March 1987. Thompson had been at the helm of independent radio for almost two decades. He had designed the system, launched it, steered it through unexpected storms, at times narrowly avoiding shipwreck but moving forward against whatever adverse currents and tides circumstances engendered. He had laid the groundwork for INR, and even for the use of modified ILR licences to see off the community radio lobby, in a think-piece in 1985.[548] In 1977, he had expressed his deceptively simple vision of what this alternative service of radio broadcasting was all about. "It ... should be informal, radio in jeans as it were, but with informality combined with style, purpose and talent."[549] Independent radio had grown so used to him that it hardly realised how much it was losing with his departure. His achievement was

546 Letter from Bill Innes, Home Office, to John Thompson, 20 June 1983.
547 Paul Brown in an interview with the author, 15 October 2007.
548 *IBA Paper* 25(85) 15 February 1985.
549 *Listener* 7 April 1977.

monumental, but his stepping down passed with little ceremony. There was no John Tenniel cartoon to catch the moment of 'dropping the pilot', but its significance for independent radio was not unlike the departure of Bismarck from the empire he had conceived and nurtured.

The advertising of three new franchise areas in 1988 followed some considerable heart-searching by the Authority at an in-house conference. In the spring of that year, Peter Baldwin, a congenial and thoughtful former army officer, had reviewed the range of radio activity, and had urged the Authority members to permit some further advertisement of new areas. One reason he gave was the surprisingly direct aim of allowing Radio Division staff "to maximise the opportunity of winning places in the new regulatory regime".[550] By that July, there still remained ten locations from the list of 25, approved by the home secretary in July 1981, which had not yet been advertised, either as stand-alone stations or in some form of association with neighbouring franchises. Despite a Commons statement from Douglas Hurd at the start of the year, confirming that a Radio Authority was to be created, his minister of state Tim Renton told IBA DG John Whitney at a meeting on 25 May that a delay of at least twelve months in any legislation was likely. Renton "wondered if the IBA would wish to put forward new initiatives for extending ILR coverage, in which case the Home Office would welcome them".[551] At a private meeting between senior IBA and Home Office staff on 17 June, and again in a meeting between Renton and Whitney on 6 July, the government indicated that it would "treat future applications for relays and power increases more generously".

The end of 1988 was approaching, however, and there was still no white paper, nor a clear prospect of a broadcasting bill to usher in the expected new world. Seven ILR companies wrote to Tim Renton, complaining that the hold-up was de-stabilising their stations and their staffs. They were approaching the end of their franchises, with no clear idea how these would be extended or renewed. The Home Office response that "nothing has been decided yet. The options remained open on the transition from the IBA to the Radio Authority" [552] was hardly comforting.

The IBA itself seemed pretty much to have run aground, and to be stuck fast. However, Paul Brown and David Vick, the latter now responsible for licensing and development, came up with a wheeze, building on John Thompson's informed speculation about 'community radio' in 1985. They proposed that the IBA should advertise new services, "confined within existing ILR areas". That allowed for a more specialised or more local service, within the requirements of the act. Since they would be additional services, they were named 'incremental stations'. This initiative would "validate [the IBA's] claims regarding IBA expertise and suitability for a future radio regulatory role ... enhance listener choice and be in the public service ... be able to commence operation well in advance of expected new television competition ... offer the IBA a chance to show positive action ... demonstrate the Authority's imagination, flexibility and skill ... give [the IBA's] Engineering Division a head start in its involvement with new projects on a commercial basis ... [and] be attractive to all parties who seek to participate in the

550 *IBA Paper* 44(88) 18 March 1988.
551 *IBA Paper* 110(88) 11 July 1988.
552 *Campaign* 18 August 1988.

future development of localised radio; the more idealistic community radio enthusiasts ... aspiring operators of commercially profitable new local radio services ... and those existing ILR companies with an interest in involvement. All those bodies may lend their enthusiastic support to an IBA-led initiative."[553]

For students of how bureaucracies think when their existence is threatened, the hierarchy of arguments disclosed in that paper is particularly revealing. The new tranche of incremental stations was seen first as a way of re-invigorating the Authority in radio terms, either to protect its future – "Radio staff still believe it is unnecessary to give up all claims to regulate future non-BBC radio at this stage" – or to enhance the prospects of its staff in any new arrangements. Of course, there was a wish to bring new forms of local public service radio to listeners, but that is set out second, and there is no argument advanced at all around the impact on the existing radio industry. Although it was accepted that "some ILR contractors' first reaction could be against the proposal", it was said that "they did bring this general change on their own heads". Was this a reaction to all that the IBA's radio people had suffered since the Heathrow Conference?

Peter Baldwin was right about some companies' reaction. They were angered by the IBA's failure to consult them, before its application to the Home Office. "We are not best pleased" said Brian West for AIRC. "We were under the impression last time we met the Authority that they had not yet decided to do anything like this, and that we would be consulted before they did."[554] There is also, with hindsight, a strong sense of a rush to a decision. There seems to have been no research, no consultation and no chance to reflect on the impact on radio. The approval in principle from the Authority on 8 September led to advertisements being published on 13 and 14 November, inviting letters of intent to provide the new services. For the first time – except for the artificial division between LBC and Capital in London, and some local overlaps – ILR stations were to face direct competition within their franchise areas from other separate stations. For the first time also – again apart from LBC – contracts were to be advertised with a specific programming remit, being either "community of interest stations" or "small geographical/community stations".[555] And for the first time, advertisements envisaged ILR contractors "providing their own transmission facilities as the IBA's agents".

= The Radio Division was also right about the Home Office reaction. Delighted to have the community radio problem 'solved' for them, they gave quick approval to the new services, with little of the detailed scrutiny which had applied earlier in the decade to the more orderly expansion of ILR services, even though the long-awaited white paper was finally to emerge in November. A whole new concept of independent radio stations had got in under the wire. That was all the more remarkable since the IBA itself was less than wholehearted in support of its Radio Division's idea. David Vick recalls that "DG Shirley Littler was deeply unenthusiastic, believing that this initiative risked damaging the IBA's credibility overall, when it got the OK from members".[556]

553 *IBA Paper* 134(88) 1 September 1988.

554 *Broadcast* 16 September 1988.

555 *Incremental Local Radio Contracts – letters of intent* advertisements published by the IBA 13, 14 November 1988.

556 Note from David Vick to the author, 11 December 2008.

Early in December, the Authority approved the advertisement of 21 licences. Seven were designated as 'ethnic' stations, and one – Heathrow and Gatwick information radio – was a 'special project' to operate on low power. They were to be advertised in five blocks between 9 January and 15 May 1989, evaluated and recommended by a sub-committee of the Authority, with all the awards ratified by the summer or early autumn. Days before the Authority met, the list had been given to the Community Radio Association, which represented the main body of that sector; the Association of Broadcasting Development, a lobby group of mainly ex-pirate aspiring small-scale broadcasters, led by Paul Boon; and AIRC. A meeting with Home Office officials on 29 November had confirmed ministers' approval.[557] Never before, and not again until the launch of the Access Radio pilot stations in 2001/2, had there been such a wholesale introduction of new local services. Even then, the Access approach was a clear pilot experiment, without necessarily a long-term commitment beyond one-year licences. For the incrementals, all the pre-existing caution in the government and the Authority had been set aside, frequency engineers had moved with unprecedented despatch, industry reservations and the possibility of future problems had been set aside, all with barely a second thought.

The first of the new stations to begin broadcasting was Sunset Radio, an FM only service of black music for inner Manchester launched on 22 October 1989. Sunrise Radio, broadcasting from Southall for the Asian community of west London on medium wave, began on 5 November. Their names were uncomfortably prophetic. Sunset Radio lasted only until 1993.[558] Sunrise Radio was, and is, a considerable commercial success, although not without its own share of troubles. The first London incremental station to launch was Jazz FM, on 4 March 1990. A surprise choice by the IBA for the London-wide FM channel, the station had some original ideas about advertising, which included announcing that Jazz FM would not accept any advertisements which were not in keeping with the station's image.[559] Even so, the financial challenges inherent in the format soon emerged, as well as inevitable complaints that "it ain't proper jazz".

One of the stated intentions of the incremental stations was to allow illegal broadcasters the chance to become legitimate. Speaking to the Radio Academy in 1988, home secretary Douglas Hurd had announced that anyone convicted of illegal broadcasting committed after 1 January 1989 would be banned from holding an ILR or community licence for five years. This was, in effect, an amnesty for existing pirate broadcasters. In Bristol, black music station FTP had begun illegal broadcasting early in 1988, but stopped of its own volition on 31 December 1988 in order to qualify as an applicant for one of the IBA's new licences. It won a licence for Bristol, and planned to launch in September 1989. The day passed in silence on 97.2 MHz, as planning disputes had left the station without any studios. The actual launch took place on 21 April 1990, but by September the owners had turned to Chiltern Radio for a cash injection, leading to a full takeover in January 1991. On 27 January 1991, Galaxy Radio replaced FTP on air, offering commercial dance music in place of hip hop.

557 *IBA Paper* 179(88) 8 December 1988.
558 See chapter 16.
559 *Media Week* 13 October 1989.

Kiss FM in London, despite using a famous American station name, was the creation of the dreadlocked Gordon MacNamee. He had been a club DJ and pirate radio broadcaster since the age of 13, moving from base to base to avoid detection and the seizure of the transmitting equipment. He obeyed the rules to get off air by the end of 1988, and was rewarded with a licence to run a dance station in London. Gordon Mac and Kiss exactly caught the youth and musical mood of their era, when they began legal broadcasting in 1990. "Hip hooray!" said the London *Evening Standard* on its one trendy page. "KISS 100 finally hits the airwaves on 1 September. Unless you have been living in a small box, you will know that KISS is the first legal station to devote itself to playing the sort of dance music that has [us] strutting our funky stuff with complete abandon. And it's on 24 hours a day. Will we ever sleep?"[560]

At the other end of every imaginable scale, Melody Radio, owned by financier Lord Hanson and aimed at providing melodic music suitable for older listeners, launched in London on 9 July. At least it could rely on deep pockets from its backers. That was true also of London Greek Radio, which was supported by the leading members of the Greek diaspora in London. The station with which it – uniquely – shared a frequency, WNK, for African-Caribbean listeners, was less fortunate and survived only until 1993.

The boldest of all the experimental stations under the 'incremental' banner, in terms of community content, was Spectrum Radio. The IBA had licensed this as a multi-ethnic station for London. It was to operate rather as publisher, dividing the broadcasting day between different services, in a range of languages, broadcasting to different minority ethnic groups. Services came and went, some of them winning their own dedicated licences.[561] However, in the haste with which frequency selection and allocation had to be completed, the full implications of using 558 kHz had not been worked through. Although officially available, this had long been the frequency associated with the offshore pirate Radio Caroline, ("Caroline on five five nine" sounded so much better than the accurate wavelength) although usually in recent years it had operated only on low power. As Spectrum was preparing to broadcast its test transmissions, Caroline upped its power on this channel, effectively blocking the licensed station across most of London. In an echo of the early problems for LBC and Capital on medium wave, the IBA switched Spectrum to 990 kHz on 25 June 1990, this time using a temporary transmitter site at Fulham Football Club. Caroline promptly closed down. Spectrum threatened unprecedented legal action against the IBA, because of the delay and the changed circumstances of its launch. The IBA paid up, reportedly £175,000, although Spectrum's chairman John Kyriakides said he "was expecting a little more".[562] Spectrum simulcast on 558 kHz and 990 kHz until the end of March 1991, when it settled back to 558 kHz.

By the middle of 1992 Peter Baldwin was summarising the incremental experiment as successful. "There were 25 licences advertised, 23 of which came on air. Two have failed, and seven have experienced changes in ownership ... The incremental plan was an experiment. From our point of view it has been a

560 Metropolis in *Evening Standard* 24 August 1990.

561 By 2008, Spectrum was broadcasting to communities from 22 nationalities, from Irish to Somali, Chinese to Mauritian.

562 *Music Week* 13 October 1990.

considerably successful experiment."[563] Viewed from later, few apart from the 'ethnic' stations survived with their original remit for more than a few years, but that is an unsurprising characteristic of what were intended to be 'community' stations. Their historical significance is as new, single-waveband, themed services, presenting competition within previously monopoly areas.

The green paper had left open the future regulatory arrangements for radio, rehearsing three options but choosing none yet. John Whitney had envisaged in March 1987 that the IBA should "accept the philosophy in large part, take issue with some parts of it ... but ... argue that the IBA should be the Radio Authority".[564] That seemed at least worth arguing, especially given the close cooperation which was, by then, existing between Home Office officials and the Authority's radio staff, and the reluctance of the radio companies to press for a complete split. In November 1987 Radio City's Terry Smith weighed the IBA's stewardship of ILR. Despite "the totally unjustified amount of money they take from the stations", he thought that the IBA had administered a "helplessly out of date Broadcasting Act ... sensibly and reasonably, and in the past two or three years, quite liberally ... The IBA represent the best route for developing and expanding the system."[565] Even after the home secretary had stated on 19 January 1988 that he intended to create a new Radio Authority because the IBA "has major challenges ahead of it in the field of television",[566] many in the Autthority and elsewhere felt the matter of who runs radio was not yet finally settled. It was becoming clear that any legislation was going to be delayed for at least a year, perhaps longer. Perhaps the IBA had not yet lost the game.

Then, on Thursday 28 April 1988, ITV's current affairs series, *This Week*, carried a programme made by Thames Television entitled *Death on the Rock*. The die was cast for a wholesale change in the entire governance of independent broadcasting. The documentary, produced by Roger Bolton and presented by Jonathan Dimbleby, investigated 'Operation Flavius', an SAS mission in Gibraltar which ended on 6 March in the shooting dead of three Provisional IRA members: Danny McCann, Sean Savage and Mairéad Farrell. All three were subsequently found to be unarmed, but ingredients for a bomb, including 100 pounds of semtex, were later found in a car in Spain, identified by keys found in Farrell's handbag. The documentary interviewed witnesses, who alleged that the SAS had given no warning prior to shooting, and that the event had been carried out 'in cold blood'.

The foreign secretary, Sir Geoffrey Howe, attempted to stop the programme being broadcast, claiming that its transmission prior to the official inquest was an impediment to justice. Even the day before transmission, IBA chairman George Thomson, who had met Howe on other matters, "did not believe that the foreign secretary and the prime minister would react publicly in the way that they did".[567] *Death on the Rock* was broadcast on Thursday evening, and the roof fell in on the IBA. Howe had made a public statement condemning the broadcast even before the show was aired. The government briefed furiously against the IBA and Thames

563 *Broadcast* 27 March 1992.
564 *IBA Paper* 40(87) 19 March 1987.
565 Speech by Terry Smith, Radio City, at the annual Radio Conference of British Broadcasters, Advertisers and Agencies, Malta, 22 November 1987.
566 Quoted in *IBA Annual Report* 1987-88 p 21.
567 Letter from Lord Thomson, IBA chairman, to John Purvis, 12 May 1988.

Television, and the programme was heavily criticised by sections of the press, notably the *Sunday Times* and the *Sun*, both before and after transmission.

The government argued throughout that its concern was "first, that there should be no improper influence ... on the forthcoming inquest" and second, that it would constitute contempt of court, in practice even if not in law.[568] Thames Television's and the IBA's legal advice – the latter confirmed on the morning before transmission – was clear that the programme was not in contempt of court, but that was not really the issue. For all Howe's claims that "there is no question of the government seeking to 'muzzle the media'", that was exactly what was intended. Or rather, this was the heaviest pressure to make it clear that the IBA was expected to do the muzzling of its own accord. Howe's threat was unmistakable. "It is a matter of self-discipline by the broadcasters and of responsible judgement by the broadcasting authorities." Thomson's reply was equally direct. "The issues as we see them relate to free speech and free enquiry which underpin individual liberty in a democracy. The right of broadcasters and the press to examine events of major public concern is well established and should be preserved."[569] Five months later, on 19 October 1988, home secretary Douglas Hurd announced that organisations in Northern Ireland believed to support terrorism would be banned from directly broadcasting on the airwaves.[570]

Thames commissioned an inquiry undertaken by Lord Windlesham, an ex-government minister with experience as a managing director in television, and Richard Rampton QC, a barrister specialising in defamation and media law. In February 1989 Thames was cleared of the most serious criticisms,[571] and the IBA of any regulatory failure. John Whitney left the IBA in March to run Andrew Lloyd Webber's Really Useful Company, to be succeeded by the resolutely establishment Shirley Littler, all of whose papers for the Authority bear the daunting imprimatur, "Paper by Lady Littler". It is likely that – alone among directors general of the ITA or IBA – the IBA's resoluteness cost Whitney any formal public recognition, until his CBE in 2008.

Death on the Rock marked the breaking point with the old regulatory regime, at least so far as the independent sector was concerned. The benign autocracy of old-style regulation of independent broadcasting was anyway becoming obsolete, because the commercial media being regulated were changing, and the age of the market was belatedly about to arrive in television. The *Death on the Rock* controversy also highlighted the difficulty of having the body which licensed and regulated the programme companies being also, technically, the broadcaster. Most of all, it turned the withering eye of Thatcherism away from its previous target, the BBC, and towards ITV and the IBA. From then onwards, the Thatcher government

568 Letter from foreign secretary, Sir Geoffrey Howe to Lord Thomson, 4 May 1988.

569 Letter from Lord Thomson to Sir Geoffrey Howe, 12 May 1988.

570 Four years before, Margaret Thatcher herself had narrowly escaped assassination by the IRA in the Brighton bombing. She was outraged by *Death on the Rock*, which chimed also with her concerns about ITV's 'monopoly' in independent broadcasting. Just under two years previously, John Stalker, deputy chief constable of the Manchester Police, had begun but then been removed from an enquiry into allegations that in Northern Ireland a trained undercover RUC team known as the "Headquarters Mobile Support Unit" had carried out a "shoot-to-kill" policy. The report was never published, but government sensitivities over any such accusations were acute; and it was effectively that charge which *Death on the Rock* presented.

571 Windlesham, P, and R. Rampton, *The Windlesham/Rampton Report on 'Death on the Rock'*, Faber & Faber, 1989.

acquired not just an intellectual but a visceral determination to do away with the pattern of independent broadcasting overseen by the IBA, and especially with the IBA itself. For radio, its accidental effect was to be the end of public service local radio funded by advertising, *independent* radio, leaving the way clear for ILR and soon INR to be truly *commercial* radio. That might well have happened anyway. *Death on the Rock* made it a racing certainty.

The eventual white paper, *Broadcasting in the 90s*, was published in November 1988, and was truly radical. Its subtitle – "competition, choice and quality" – specifically replaced the Reithian injunction to "inform, educate and entertain". That shift, and especially the primacy given to competition and the proper operation of the market, is the most significant single change in the governance of UK broadcasting since 1927. Huge institutional changes followed from the new philosophy. The IBA and the Cable Authority were to be done away with, the former getting its come-uppance for supporting the old ITV system, replete with union-dominated restrictive practices, and for its temerity in allowing the transmission of *Death on the Rock*. The television responsibilities of both would be assumed by a new body, the Independent Television Commission (ITC). ITV franchises were to be replaced by licences given to the highest bidder (subject to a 'quality' threshold), and the ITC was to introduce a fifth terrestrial television service. Satellite transmissions were prefigured. The programme companies were to be responsible for their own transmission arrangements. The IBA's transmission arm was to be privatised, and was to become National Transcommunications Limited (NTL). The BBC escaped largely unscathed, although it was seen as "feasible and healthy to expose the BBC to competition in the provision of national radio services". Its own national radio services were to be enhanced, with separate FM frequencies for Radios One and Two, and permission was to be given to "complete its chain of local stations".[572]

Radical changes were proposed to non-BBC radio, although virtually contained on a single page. ILR was to be extended further, with as many new stations as the frequency planners could permit. To it was to be added Independent National Radio (INR), as the green paper had presaged, and these three licences were to be auctioned to the highest bidder, just like the new ITV licences. All were to be overseen by a new, lighter touch regulator, the Radio Authority, which was to be far more different from the IBA's approach to radio than the ITC's was to be for television. All the radio-specific public service obligations on ILR, which had underpinned *independent* radio, were to be removed. There would remain only the widely accepted rules governing taste and decency, due impartiality, and the avoidance of editorialising, plus a requirement to keep more or less to the plans set out in a successful licence application. "The new national services and independent local services will be subject to a lighter regulatory regime. They will not be required to comprise education, information and entertainment, although they may follow a public service pattern if they wish."

Each company was to provide and operate its own transmitters, which would usually mean contracting NTL or others to do so. ILR contracts were to segue into the replacement licence-based regime. All new ILR licences were still to be

572 *Broadcasting in the 90s*, 1988, Cmnd. 517, p 37; the same reference applies also to the quotations in the subsequent paragraphs.

awarded against statutory criteria – the so-called 'beauty parade' – but there was no check on the selling on of such licences once they had been awarded. The bases for awarding licences were also new, in both their specificity and the economy of expression. "At the local level licensing criteria will include financial viability, local audience demands and the extent to which the proposed services would enhance the range of programming, and the diversity of listener choice." Those four criteria were to pass in much the same language into the new legislation, and were the mantra for the Radio Authority's licensing process. "National services will be expected to provide a diverse programme service, calculated to appeal to a variety of tastes and interests and not limited to a single narrow format."

This was the culmination of the process started at the Heathrow Conference, four years previously. Encouraged by the demands of the radio industry, emboldened by what they had tried to do with community radio in 1984, and in keeping with their faith in the market economy operating benignly – even in sectors which had previously been dominated by social concerns – the Thatcher government was discarding the mould. It was for the first time wittingly and specifically abandoning the Reithian approach to broadcasting for free-to-air services using still-scarce public spectrum. Even though the relaxation was far from complete – as subsequent battles between the companies and the Radio Authority were to demonstrate – this was a massive shift, pre-figuring what was to happen to television a decade later. The ILR companies had won hands down in their challenge to the old regime of the IBA. It was to be some time before they awoke to their new freedom and its implications, but the hinge was about to swing. The 1990 Broadcasting Act opened the door to *commercial* radio, which would enter into its inheritance before too long.

Chapter 13

Copyright wars

The long battle over music copyright

Of all the struggles of the ILR companies collectively, none was as drawn out, as intense or as expensive, as the long war with the music copyright societies, and their usual ally the Musicians' Union (MU). It pre-dates even the establishment of the Association of Independent Radio Contractors (AIRC) in 1973, one of whose key purposes was to conduct and manage these relationships and to attempt to get progress by negotiations. There was, in effect, no final armistice to this struggle until the settlement which followed the judgement of the second Performing Right Tribunal in March 1993. With disputes continuing even now over rights to disseminate music within internet radio stations, the last battle may still lie in the future.

Music is the staple of popular radio, so music copyright, which has to be resolved before any commercially recorded music can be played by a radio station, was a central issue conditioning the fortunes of ILR. Its availability and cost – taking perhaps as much as a fifth of net advertising revenue – determined the commercial prospects of the companies. It was also a touchstone of expectations and attitudes about ILR in its earliest years. Some of the first radio operators, drawing on a good deal of overseas information, expected to get their music cheaply, and with little limitation on usage. They largely discounted the crucial influence of the BBC, which as a major patron of live music had significant influence on political expectations. They also misunderstood the nature of UK copyright legislation, and the realities of possible change. Here, as in other fields, they expected to be allowed to operate what was essentially *commercial* radio; but *independent* radio was not going to be able to ignore the British context. Some change was possible, as the political landscape itself changed, but even in a market-led environment the power of the copyright-owners in the UK means that commercial radio pays high royalties to this day.

It is inescapable also that copyright as a subject risks seeming a bit dry and legalistic. Yet it is such a pivotal issue that this chapter explores the skirmishes, set-piece encounters, parlays and treaties in some detail. It helps to be clear about the various parties who did battle over some thirty years. For the radio companies, most of their representations were conducted collectively by their trade association, AIRC/CRCA. Almost all of the major companies which produce recorded music for retail sale record companies are joined together in a single body, Phonographic Performance Limited (PPL). In more recent years, that has included video recordings as well. The oldest established music copyright society, and the one with the widest membership, is that representing the composers, performers and arrangers of music, the Performing Right Society (PRS). For a composer or performer, the arrival of their 'PRS cheque' is the moment when the bills can be paid. There exist also what are known as 'mechanical' rights, arising from dubbing or recording musical tapes, and in the importing of tapes from abroad. When ILR was set up, these rights were the business of the Mechanical Copyright Protection Society Ltd (MCPS), the Mechanical Rights Society Ltd (MRS) and the British Copyright Protection Association Ltd, although there were some subsequent mergers. Issues for any of these rights owners, which could not be settled by negotiation, would eventually fall to be disputed at and settled by the statutory Performing Right Tribunal, which later became the Copyright Tribunal.

The wild card in all of this was the Musicians' Union (MU), a trade union which wielded immense power during much of the period with which this history is concerned. Major issues of labour relations generally crop up throughout this history, especially in the early years. Not one of them, however, had the same power as the MU. It was in effect a condition of ILR coming into existence that the MU could be squared. They had at the start an innate, doctrinal opposition to commercial broadcasting, and the subtle and not so subtle ways in which they most effectively deployed their power made them a force to be reckoned with by governments of both persuasions. As events were to show, the MU had established a covert financial relationship with PPL, which showed how much the record companies needed the goodwill of the union.

At the heart of the long dispute is a paradox. Those producing and selling records – whether singles, EPs or LPs, vinyl, CDs or paid-for downloads from the internet – generally need radio stations to play the music on air, in order to generate demand for the mass sales on which the music industry depends. They therefore promote their products to the radio stations, offering them every encouragement and sometimes inducements to play them. On the other hand, those same owners of the copyright of that music are anxious that they should receive payment from the radio stations for playing it, because it forms such an essential part of radio output. Musicians also have an interest in making sure that the recording of live music for radio broadcast survives the wide availability of music recorded for retail sale, but is also used extensively by the radio stations. Much of this is a straightforward commercial argument, with the broadcasters keen to pay as little as possible and the owners of the music copyright seeking as much return as they can.

In many parts of the world, the balance between airplay and payment for rights is weighted quite strongly towards the former, with radio stations paying only relatively nominal sums for their music. It has long been argued that where the market for the sale of such music to broadcasters operates in an unfettered way, that is the natural outcome. In the UK however, the right to payment is one of the things protected by our legal and administrative structures, at least in part as a result of the special position attained by the Musicians' Union. Well before the arrival of ILR therefore, and with the legal support of tribunal decisions which are binding on both industries, substantial payments had to be made by the radio services to the music owners. In addition, broadcasters have been required to accept tight limits on what percentage of commercially recorded music they may play within their output (termed 'needletime'). The traditional patronage which the BBC has exercised as a supporter of major symphony orchestras and concert bands, and in arranging live music sessions in more popular genres, was therefore making a virtue of the necessity created by such arrangements.

Music copyright issues go back to the very start of broadcasting. PRS was first established in 1914, and the British Broadcasting Company had faced steady pressure on music copyright from its earliest days. The music publishers' representative told the Crawford Committee in 1925 that "it is the firm opinion of the owners of musical copyright that broadcasting has a deleterious effect upon the sale of music" while "paid musical performers were already suffering as orchestras in hotels were being replaced with wireless receiving equipment".[573] He claimed that 10 per cent of the takings of the BBC should be given over to payments of royalties. A legal challenge in 1928 led to a limit on the amount of any particular musical work which could be broadcast, which for example meant no complete Gilbert and Sullivan operas. When record sales dipped in the early thirties, the newly consolidated UK record industry "blamed ... their increasing use by hotels, restaurants and also by the BBC".[574] PPL was established by the record industry in 1934 to exploit the rights in its members' recordings established in the Cawardine case.[575] The so-called 'Tuppenny Bill' of 1929[576] had failed to reach the statute book, but it introduced the concept of a tribunal to set appropriate fees where the rights holder and the user could not agree. This formed a central part of the modern regime for music copyright in the Copyright Act 1956, which created the Performing Right Tribunal (PRT) "to determine disputes arising between licensing bodies and persons requiring licences, or organisations claiming to be representatives of such persons".[577] Under the 1988 Copyright Act, the PRT became the Copyright Tribunal, with a wider jurisdiction, and – crucially for broadcasting – there could be compulsory 'statutory' licences, giving broadcasters the right in certain circumstances to licences for the use of sound recordings whether or not they had the agreement of the copyright bodies.

The tribunal's first case concerned a long-standing dispute between PRS and halls promoting ballroom dancing, over how to calculate the fees to be paid for the public performance of copyright music. There followed judgements about music on juke boxes and in bingo halls, and a relatively minor issue for Southern Television. The first of the key cases affecting radio broadcasting was that of Manx Radio in 1965, and concerned the needletime restrictions which PPL sought to impose. In its judgement, the tribunal established a principle of permitting recorded music to be used for up to half the total broadcasting week. It concluded that a direct comparison with the BBC's entitlement was not a relevant factor. The tribunal set the level of copyright charges for this small commercial radio station at a percentage of revenue, on a sliding scale from 5 to 8 per cent, in this instance ruling that the level of the charges agreed with PRS – supposedly the 'major right' – were also not relevant to establishing the PPL entitlements.

573 A V Bradbury in evidence to the Crawford Committee 18 December 1925, quoted in Briggs op. cit. Vol.1 p 344.

574 Stephen Barnard, *On the Radio, Music Radio in Britain*, OUP 1989, p 26.

575 *Gramophone Co Ltd v Stephen Cawardine [1934]* Ch 450.

576 *Hansard*, House of Commons, 1 November 1929, col 476.

577 Copyright Act 1956 s 26.

In some instances, offshore pirate radio stations had made their own arrangements with the copyright bodies; in others they simply ignored them. Philip Birch's Radio London, determined to be a 'model citizen' with an eye to future legal commercial radio, actually tried to pay royalties to unwilling rights owners. Keith Skues, who worked as a disc jockey on both the Radio Caroline and Radio London ships, recalls that "we paid a Performing Rights copyright fee for the music. We offered to pay Phonographic Performance Limited ... but, in fact, they were not prepared to accept our money."[578] Skues confirms that the pirates understood that they were operating on the edge. "Gramophone records are copyright and, by law, their unrestricted use is illegal. The pirate broadcasters usurped the right to do so and without unrestricted use of gramophone records their programming would have been quite impossible." The point was not lost on those lobbying for legal UK commercial radio, who argued that although there needed to be arrangements in place to protect the livelihood of musicians "there should be unlimited 'needletime' and that the Copyright Act 1956 should be amended where it relates to copyright in sound recordings. The particular point at issue is not whether those who make recordings should be entitled to own copyright in their recordings; it is whether they should be allowed to use their copyright in a manner which constitutes a restrictive practice."[579]

It was a vain hope, entirely unrealistic given the tradition of copyright protection in the UK. Even the market-obsessed administrations of Margaret Thatcher and Tony Blair were to shy away from anything so radical. In the context of the early seventies it was unthinkable. That it was entertained at all shows the influence on the early thinking of the ILR companies of those whose experience had been primarily in overseas commercial radio. Both John Thompson and Jimmy Gordon speak disparagingly of BRMB founder David Pinnell's frequent citing of the copyright arrangements in Lourenco Marques, the colonial capital of Mozambique, where he had once run a commercial radio station.

Two key tribunal cases concerning the royalties charged by PRS to the BBC, in 1967 and 1972, established again that substantial fees should be paid by the broadcaster. The figure for the BBC to pay to PRS alone was set in 1967 at 1s 9½d per receiving licence, to be indexed against the retail price index. Following great dissatisfaction among PRS members, the issue was referred again to the tribunal in 1972, and it settled upon a formula of 2 per cent of the BBC's entire licence revenue, from both radio and television.

Government planning in 1970 was based on the realities of the existing UK music copyright regime, and with a keen awareness of the industrial power of the union. Christopher Chataway met with representatives from PPL that September, and was left in no doubt about their concerns both over their own commercial position and about the possible impact of the MU's disapproval, with the potential that strike action would put at risk £5 million of UK exports.[580] In October, meeting some consultants who were preparing radio franchise applications, Chataway encouraged early negotiations between the ITA, PPL and the MU, drawing upon the experience of the BBC and the Manx Radio cases. He was keen that the

578 *Pop went the Pirates* Keith Skues Lambs' Meadow Publications Sheffield 1998 p 227.

579 *The Shape of Independent Radio* Local Radio Association October 1970.

580 Government note of a meeting on 9 September 1979 between PPL and Post Office Department.

Authority should negotiate on behalf of the companies to avoid both delay and a situation in which "the larger companies might well settle for terms which would be impossible for the smaller stations. The Musicians' Union", he noted, "took the view that the service should offer opportunities for live employment and, provided this could be met, he did not think they would be too unreasonable, though they would play for time".[581] Any suggestions about amending the Copyright Act "would be contemplated only in the last resort; and even so might not prove possible".

When John Thompson was appointed in the autumn of 1971, one of his first and most critical tasks was to deliver the government's expectations that the future IBA would have a settled scheme for the use of music, agreed with the copyright societies and the MU, ready for the advertisement of the first franchises. A number of the parameters had already been set. There was strong UK copyright legislation, not to be changed, certainly in the short term. This was to be a collective agreement, applying to all the ILR stations. The BBC payment set by the tribunal equating to almost 5 per cent of radio revenue for PRS alone was the base from which the societies would expect to negotiate upwards for a 'commercial' network, guided by the 8 per cent PPL figure set for Manx Radio. In its Manx Radio decision, the tribunal had also established a 50 per cent needletime standard for non-BBC radio; the current BBC agreement was closer to 40 per cent. And even for a newly-empowered Conservative minister, squaring the MU was a *sine qua non*, in effect not just for the copyright deal, but for the future of ILR as a whole.

Thompson's early negotiations found the societies and the MU fully aware of the strength of their starting positions, with PPL "concerned not to upset the MU in any way by making concessions to ILR".[582] In addition, the MU was profoundly opposed in principle to the introduction of commercial radio, influenced by the damaging effects such systems had for musical employment in other parts of the world. The BBC, and especially its radio arm, was the UK's chief patron of music and the major source of employment for musicians in the UK. None of the parties was going to reach an agreement with ILR which would risk disturbing the arrangements with the BBC by seeming unduly favourable to the new service. The BBC at the time maintained eleven staff orchestras, employing nearly 600 musicians, and with a guaranteed level of casual engagements for musicians was spending nearly £3m a year on musical employment. BBC payments to PRS were to be £2.8m in 1973–74 and £3m in 1974–75, which equated to around 4.8 per cent of radio income. For its needletime, the BBC also paid PPL a sum similar to the PRS figure.[583]

For all that, Thompson found PPL at least to be "surprisingly flexible".[584] He met PPL general manager H G G Gilbert on 7 January and again on 1 February 1972. On the latter occasion, they spent more time discussing the position of the MU than that of PPL. Gilbert undertook to propose royalty rates, but these "would be conditional upon reasonable employment opportunities being offered to mu-

581 Government note of a meeting 6 October 1970 between Urwick Orr consultants and Post Office Department.

582 IBA Paper 108(72) Use of Musicians and Copyright on ILR 18 July 1972 – the seminal document for all that followed.

583 IBA Paper 108(72) op. cit. p 2.

584 John Thompson in an interview with the author, 8 November 2007.

sicians within the new medium".[585] Negotiations intensified, and at meetings on 11 and 14 April Thompson stuck firmly to his position that ILR should not have to face more than a 5 per cent charge on revenue to cover both copyright and musicians' costs. By 21 April, he was suggesting a phased introduction of royalties in view of the "very considerable problems" which the new stations were going to be facing for "this first period of about three or four years", with the intended review of broadcasting a likely "constitutional turning point" for ILR.[586] This formula was becoming the focus around which an agreement might crystallise. Thompson wrote to PPL board member and EMI director C B Dawson-Pane that "again and again it seems to me one comes back to the quintessential point – between us all we must try and get the pre-conditions for the first phase, i.e. roughly three years. It is on that basis that in the medium-term the new alternative system of radio should be able to flourish"[587] and, by implication, PPL might then expect its royalties to rise closer to the levels it was seeking. As the IBA's own deadline for franchise application loomed ever closer, the exchanges focussed down upon the actual figures. PRS proposed 6, 7 and 8 per cent for the first three years ... PPL then asked for 4, 5 and 6 per cent ... Thompson hoped PRS might come down to 4.5 per cent for that first year ... Thompson proposed 2.75, 3.5 and 4.25, followed by 5 and 6 per cent for years 4 and 5.

In the last week of July, Thompson reported progress to the full Authority. The MU was willing to co-operate with ILR if it was given a statement of principle at the outset that the system recognised its obligation to provide employment for musicians, a commitment of a percentage of net revenue being devoted to live music and a medium-term expectation of more substantial benefits from secondary rental. The copyright societies had moderated their demands to a degree, and by July there was a figure on the table of a total payment to PRS plus PPL of 10.75 per cent, rising over five years to 17 per cent. That would provide 50 per cent needletime over an 18 hour day, with some further scope for negotiation for 20 or 24-hour broadcasting. The Authority agreed this as a basis to move forward, noting presciently that the criticism for the companies would apply more to the arrangement with PPL, who would be benefiting from the airplay of their records, than to PRS, especially as most other countries did not allow record manufacturers the same rights. The Authority members were warned that some of the "applicants may argue, or imply, that they would have struck a better bargain with PPL than the IBA". However, he noted, "this is, under UK conditions as they exist, a complicated debate".[588]

There was time for a final exchange of letters, a last-minute reminder from PPL that their agreement was to 50 per cent needletime on an 18 hour day – thus a maximum of 9 hours at the agreed rate – and the deal was done.[589] The stations were to pay PPL 3 per cent of their net advertisement revenue in the first year of operation, 4 per cent in the second and 5 per cent in the third; although Thompson still hoped that the third year might "be expressed as being 4.75 per cent or 5 per

585 PPL's notes of meeting between J Thompson and H G G Gilbert at the ITA 1 February 1972.
586 Letter from John Thompson to H G G Gilbert, 21 April 1972.
587 Letter from John Thompson to C B Dawson-Pane, 26 April 1972.
588 IBA Paper 108(72) op. cit. p 4.
589 Letter from H G G Gilbert to John Thompson, 15 August 1972.

cent, subject to confirmation of either figure at a later stage".[590] For years four and five the figures were to be 6 per cent and 7 per cent, but that must "necessarily be subject to the circumstances which will then prevail and to the general review of broadcasting which will take place before July 1976". The MU received a guarantee that the new stations would have a contractual obligation to spend a minimum of 3 per cent of their advertising receipts on live music; PRS settled much more straightforwardly.

The eventual copyright and live music arrangements for the first tranche of ILR stations were set out in the application documents in October 1972. As Thompson had foreseen, the larger ILR companies instantly formed the view that they could have done better, but he remains robust: "We were really determined to get the figures down as far as possible, but granted the *actual* circumstances of that time and the *actual* copyright legislation there was very limited room for manoeuvre. And we were talking about the UK, not about North America, Australia, Canada – or Lourenco Marques."[591]

Where does the truth lie? It is relevant that the ITA/IBA had only nine months to conclude negotiations without which they would have no ILR system to regulate, a situation which placed them at an immediate disadvantage with highly-skilled negotiators from the societies, and not least the MU's John Morton. Typically, the BBC took around two years to agree renewals of existing arrangements. There was not going to be any help from the government, which was appalled by the prospect of a confrontation with the Musicians' Union to add to their wider troubles with the miners, railwaymen and others. Those factors alone probably meant that the IBA ended up settling for a less good deal that it might otherwise have achieved. Would the companies have done better? The deal had to be struck before any franchises were actually held and, given the failure of the commercial radio lobby to achieve pretty much any of its detailed aims over the years from 1970 to 1973, there is little reason to think that it would have been more successful in this "complicated debate" than the Authority. The frustrations the industry was to endure when it did take the matter on for itself further bears that out. ILR was saddled in 1973 with far higher charges than commercial radio in other parts of the world, especially in respect of PPL and the MU; but then it was not going to be like those other radio networks in very many other ways, either. It is hard, even with hindsight, to disparage the IBA settlement, however much the high costs added to ILR's problems in its early years.

As soon as they had the collective mechanism of their trade association, AIRC, the radio companies tried to challenge the settlement with PPL, which indeed seemed to them to be the central issue. In September 1973, Michael Flint[592] reported on behalf of AIRC to the IBA that they had held meetings with the MU, to try and alleviate the impact of the 3 per cent. They were also hopeful that they might gradually negotiate a better deal with Phonographic Performance Limited on needletime for those stations who wished to broadcast for 24 hours.[593] AIRC told the Performing Right Tribunal in 1979 that PPL had "made it clear that it was

590 Letter from John Thompson to L G Wood, EMI Records 8 September 1972.

591 John Thompson interview op. cit.

592 Michael Flint had been AIRC's first General Secretary, on an interim basis. He was an expert on copyright law, and the original author of one of the standard works, *A User's Guide to Copyright*.

593 RCC Minutes 26 September 1973.

not willing to make any alterations in the basic terms as to needletime and fees which had been arranged between it and the IBA".[594]

There was, though, at least one major opportunity to negotiate a new settlement. John Thompson recalls a private meeting with leading PPL board member, Len Woods of EMI, John Whitney and Jimmy Gordon, at which a really rather good deal was on the table. In his view, John Whitney and Jimmy Gordon who were present "grasped the point, but I have always understood that they simply couldn't sell it to their colleagues. I am absolutely confident, even with all the disadvantages of hindsight, that had that opportunity been grasped the radio companies would have saved themselves a great deal of money, had they been realistic about what was possible over copyright and the Musicians' Union arrangements." [595]

Jimmy Gordon confirms that such a meeting did indeed take place. It was partly to try and improve relationships between AIRC and PPL, which had been dogged by personality issues. "PPL's Herbert Gilbert, who had been a lifetime in the business, rightly or wrongly took umbrage at being lectured on copyright by a slip of a girl [Cecilia Garnett, AIRC's secretary]."[596] However, the influence of Len Woods was stronger. After a meeting over dinner in Scotts Restaurant in Mayfair in January 1980, Woods confirmed that PPL were prepared to settle for a flat rate royalty of four and a half per cent. As Thompson had thought, Whitney and Gordon were delighted. However, as Gordon recalls, "we took it back to AIRC; AIRC threw it out". Two tribunals, two decades later, after a slew of legal fees and twenty years of making much higher copyright payments, four and a half per cent still looks a good deal.

Collectively, the companies steadfastly refused to accept that what the IBA judged possible was a 'realistic' view of their prospects. On 18 July 1978, they came to meet Lady Plowden, Brian Young and John Thompson at the IBA, to explain what they planned to do. Thompson briefed his chairman and DG that of the two options – taking PPL to the tribunal or "using informal methods, e.g. social contact and persuasion" – he hoped that the latter might prevail. It did not. AIRC took its case to the Performing Right Tribunal in October 1978. As always, when embarking on major litigation, that led to huge amounts of work for the litigant, and corresponding fees for the lawyers. When I arrived at AIRC the following year it was to be greeted by a pile of plastic boxes containing cassettes, all examples of the output of the 18 ILR stations involved which were, by then, surplus to requirements. The tribunal did not need them after all, and we used them subsequently to dub tapes. That was not a good return for the work and cost involved in their original production, and is a gloomy paradigm for the progress of AIRC's whole case.

The tribunal sat under the chairmanship of H E Francis QC to hear preliminary legal issues for four days in March 1979; on the substantive issues for a costly total of 17 days in November and December that year; and then for a further 62 ruinously expensive days up to mid-May 1980. It issued its substantive judgement on 15 July 1980. In addition to the main litigants, AIRC and PPL, the MU was

594 Michael Freegard and Jack Black *The Decisions of the UK Performing Right and Copyright Tribunal*, Butterworths, London, p 118.

595 John Thompson interview op. cit.

596 Jimmy Gordon in an interview with the author, 25 April 2008.

formally included as an 'interested party'. AIRC pressed for a royalty for PPL reduced to between 1 and 1.5 per cent of net advertising revenue (NAR), resting its case on three comparisons – the BBC royalty, the PRS royalty and arrangements overseas – and the 'airplay' argument.

In respect of the BBC, AIRC's lawyers, Denton, Hall & Burgin, had calculated the hourly rate paid by the BBC in respect of Radios One, Two, Three and Four, and applied that to ILR's coverage of just two thirds of the UK. On that basis, they proposed a maximum total royalty payment of £385,639, compared with the actual payment of just over £2m for 1978/9. The tribunal did not accept that it was appropriate to link ILR and BBC payments, since the BBC was public service radio rather than a commercial enterprise, and it had since its inception done a great deal to provide employment for musicians and to promote audience appreciation of live and serious music.[597] AIRC's argument that PRS was the major right cut no ice with the tribunal, which was of the view that the rights of each were wholly distinct. As for the comparison with other English-speaking countries, just as the IBA had forecast, the tribunal was clear that the statutory position was very different there and could not be used to draw comparisons.

AIRC had argued that 'airplay' was of real importance to the record companies, since records being played on the radio promoted sales. The tribunal acknowledged that the record industry did indeed attach value to airplay, as a way of creating awareness and demand, but it thought that ILR was only one of the outlets, and not in its view the most important. Anyway, although airplay might increase sales of individual records, it did not believe that it increased total sales, which was the issue at the heart of the claims over the level of royalty payment. Since the selection of playlists was not done in order to promote sales, and given that ILR stations were often resistant to record pluggers' attentions, that confirmed the tribunal in its view that airplay was not such a major factor. Airplay might also encourage home taping, which cost the record companies money. The tribunal reckoned these factors cancelled each other out.

The tribunal's judgement was a body blow to ILR. It prescribed a rising scale of royalties payable to PPL from 4 per cent to 10 per cent of NAR, with the scales indexed to the retail price index at a time, remember, of very high inflation. New stations would benefit from lowered rates of 2 per cent and 3 per cent in each of their first two years. Needletime remained at 50 per cent to a daily maximum of nine hours. AIRC's effort and huge legal expenditure had resulted in a far worse outcome, especially for the major companies, than the much-abused settlement negotiated by the IBA. To make it even more galling, most of the tribunal's arguments echoed the warnings given by the IBA to the companies over the previous years. In the view of AIRC's own lawyers, "AIRC had sustained defeat on virtually all significant matters of principle".[598]

The ILR companies felt they had little option other than to appeal the decision of the Copyright Tribunal, and they were supported in their hopes of a better outcome by counsel's opinion. Accordingly, in October 1980 they lodged notice of a challenge to the tribunal's decisions by what is known as an "Appeal by Case

597 Performing Right Tribunal. Reference under section 25 between The Association of Independent radio Contractors Limited and Phonographic Performance Limited and Musicians' Union, Decision 15 July 1980.

598 Minutes of the AIRC PPL sub-committee, 21 July 1980.

Stated", where the High Court is asked to refer a case back to the tribunal to consider certain questions of law. In March 1981, Mr Justice Falconer referred 13 points of law back to the tribunal, which was now chaired by Leonard Bromley QC. There then followed what seemed to be interminable delay before AIRC could extract from the tribunal their Case Stated, the necessary preliminary before the High Court could consider the matter fully. It even required a separate application to the High Court to produce a fuller Case Stated than the reluctant tribunal originally offered, and this was not available until the start of October 1983, fully three years after the original notice of appeal.

Meanwhile, the other main copyright society, PRS, was becoming concerned about widespread under-reporting of music use. PRS used the 'logs' supplied by stations as the basis for the distribution of royalties to their members. AIRC's copyright sub-committee met PRS in December 1981 to work towards an agreement on this issue, and also to put in place an interim agreement while the PPL Tribunal case continued on its slow way. That was achieved in July, and with MCPS/MRS quiescent, AIRC was free to concentrate its efforts on the PPL situation. In the middle of October 1983, it faced a decision whether to continue to incur yet further costs involved in pursuing the Case Stated in detail in the High Court – estimated a year earlier to be £130,000 and now likely to be higher – or whether to drop the proceedings. Believing that if an out-of-court settlement were to be offered, it would need to be approached from an evident determination to proceed through the courts, AIRC determined "to pursue the PPL case into the High Court and raise from members of AIRC the sum necessary to fund the costs of the appeal".[599] Given that no offer had come from PPL, it was an inevitable decision.

It took until 16 January 1986 to get Mr Justice Harman to hear the appeal. He allowed himself some notably rude comments about the tribunal's supplementary submission, complaining that there was no legal authority to help him through "this turgid mass of material".[600] Having got that off his chest he delivered a complex judgement. There were, he said, some aspects of the tribunal's guidelines which betrayed errors at law. It was striking that Harman was not at all happy with the PRT's idea that the BBC had an 'ethos' which was different from the 'commercial' activities of the ILR companies. He concluded that the BBC's royalty payments were indeed relevant, and he therefore set aside the decision of the tribunal back in 1980, and ruled that the case should be reconsidered. However, it was not the overwhelming success that AIRC had hoped for, since the judge continued to reject AIRC's airplay arguments, once again concluding that the playing of particular records on air did not increase total record sales.

At least the volume of previous argument and counter-argument meant that the tribunal could re-consider its decision 'on the papers', without the delay and cost involved in further oral hearings. Under a third chairman, W Aldous QC, having taken into account the directions from the High Court, the tribunal delivered its revised decision on 23 October 1986. It was another disappointment to the ILR companies. The tribunal re-asserted its principle that the original rate equating to 7 per cent across the board should continue. However, "taking into

599 AIRC EGM Paper 13(83) 6 October 1983.
600 Quoted in Freegard and Black op. cit. p 118.

account the BBC comparison and matters relating to the Musicians' Union"[601] it concluded that there should be some modest relief from the 7 per cent figure, with a lower rate of 4 per cent applying to the first £650,000 of NAR. New stations would have only the relief in their first two years established in the 1980 decision.

Once more the ILR companies reeled away, bruised and disbelieving. Newish AIRC Director Brian West told the Department of Trade and Industry (DTI), which had responsibility for copyright issues, that they were "shell shocked".[602] Radio Hallam's Bill MacDonald thought the decision "outrageous".[603] The revised tribunal judgement had the extra damaging effect, from AIRC's point of view, of providing some relief to the largest ILR companies at the expense of significant extra royalty costs for the smaller and middle-sized stations, and the harmony of the trade association – never a given – was disturbed once again. AIRC's reaction might be characterised simply as 'denial' in face of the nature of UK copyright legislation and the position of the BBC as music patron, but the stations were determined to find other avenues to express their deep sense of injustice.

The ILR companies were feeling more politically self-confident as a result of the Heathrow Conference in the middle of 1984,[604] and the very real progress they were making in establishing direct channels of communication with the government. The 1987 green paper, which presaged further development of radio,[605] indicated that they were likely to win more of their arguments, as well as expressing disquiet over the high level of copyright charges. Meeting the secretary of state for trade and industry, David Young (Lord Young of Grafham), at a seminar on broadcasting chaired by Margaret Thatcher on 21 September 1987, Jimmy Gordon took the opportunity to talk about the relationship between the MU and PPL.

It had been disclosed in the first tribunal case that MU members were obliged, when making a record, to agree to assign their copyright on the public performance and broadcasting of the recording to PPL. PPL in turn gave 12.5 per cent of their net revenue from broadcasters to the Union. This looked for all the world like a restrictive practice. It therefore seemed to be a lever to help gain support from a potentially sympathetic government. More generally, as Gordon said in his follow-up letter, the ILR companies feared that "the record companies, through PPL, were effectively in a position to exercise a veto over the government's proposed expansion of radio as outlined in the green paper"[606] by raising radio companies' cost base to an un-economic level. "High copyright charges are a major inhibition on the growth and expansion of radio in the UK." Gordon proposed that the new copyright legislation should adopt the European formula of "equitable remuneration" as a basis for future copyright charges.

The new Copyright Act came into force in November 1988, but provided only a little comfort for ILR. It established the Copyright Tribunal, broader in scope than the Performing Right Tribunal which it replaced, and gave it a little more guidance about where it should look for comparators when assessing the continuing test of what was "reasonable in the circumstances".[607] Still, the govern-

601 Ibid. p 121.
602 Letter from Brian West to DTI, 20 October 1986.
603 Letter from Bill MacDonald to IBA, 19 November 1986.
604 See chapter 10.
605 See chapter 12.
606 Letter from Jimmy Gordon to Lord Young of Graffham, 24 September 1987.

ment was positioning itself as the champion of entrepreneurial activity against the restrictions of bureaucracy and 'the state', and as a self-proclaimed opponent of restrictive practices it should have been less open to influence from the Musicians' Union than any of its predecessors. David Young had clearly been convinced that there were issues which needed investigation, and in March 1988 referred "certain practices in the collective licensing of public performance and broadcasting rights in sound recordings" to the Monopolies and Mergers Commission (MMC).[608] Explaining his concerns in a reply to a carefully placed written question in the House of Lords, Young rehearsed the government's "general concern over uncompetitive practices", hoping for a report by October that year.[609]

When it arrived in December 1988 – after the new copyright legislation – the MMC report received a mixed reception from ILR. It endorsed the principle of collective licensing bodies, so long as they could be restrained from abusing their monopoly position, not least in order to protect smaller record companies who would otherwise have little bargaining power. Its main recommendations were aimed at avoiding any abuse of monopoly. Importantly, these included the concept of a 'statutory licence', whereby a new station could operate on the basis of self-assessed royalties pending a Copyright Tribunal order. The Copyright Tribunal should be strengthened, and the 'arrangement' between PPL and the MU should be discontinued in favour of 'equitable remuneration'. Crucially, AIRC won one and lost one of its main practical concerns. The MMC recommended that there should be no changes in PPL's current royalty rates; but also that needletime constraints should be abandoned.[610]

In his letter to Lord Young the previous year, Gordon had also floated the seemingly rather outré notion that, since there was little or no copyright payment made in the US by radio stations broadcasting records originated there, American records similarly should not attract any copyright charges in the UK. The MMC developed this idea, and reported in some detail on the concept of 'first fixation'. This would mean that foreign recordings would have copyright protection in the UK only if the recording company was a national of a state which was signed up to the Rome Convention for the Protection of Performers, Producers of Phonogram and Broadcasting Organisations; or if the "first fixation of the sound" – that is to say the original master recording – was made in a state signed up to that Convention.[611] The USA was not a signatory to the Rome Convention.

AIRC became very excited. Director Brian West told his members that "we regard this as very encouraging" while acknowledging that "we still have a long way to go" on the broader points.[612] Off went the trade association and its active members on a long pursuit of 'first fixation', persuading the DTI to commission a report on the notion from National Economic Research Associates (NERA). That suggested that the annual loss of revenue by PPL, and the corresponding benefit to radio and public users of records "would in the long run amount to

607 Copyright and Data Protection Act 1988 s 118(3).

608 Department of Trade and Industry Press Notice 30 March 1988.

609 *Hansard* House of Lords 30 March 1988 col 855WA.

610 *Monopolies and Mergers Commission Report on the Collective Licensing of Public Performance and Broadcasting Rights in Sound Recordings* 7 December 1988.

611 Ibid.

612 Memorandum from Brian West to ILR managing directors, 8 December 1988.

around £9m a year or nearly half PPL's current income".[613] Too much. The government used NERA's findings to rule out any change to present practice in this regard, and 'first fixation' departed the scene.

In other respects though, the government response in December 1989 was encouraging for AIRC. It announced its intention of legislating to introduce the MMC's proposals for a 'statutory licence' regime. The arrangement between the MU and PPL was a restrictive practice which should be abandoned, although the government would seek to achieve that by voluntary agreement rather than legislation. Better still, it endorsed the MMC proposals that PPL should abandon needletime altogether, and that the obligations for employing musicians for certain public performances should go too. John Redwood, then secretary of state for trade and industry, stated that "the proposed statutory licence procedure will prevent PPL from re-imposing needletime restrictions", which would avoid the need for any further statutory provisions. "As to the musicians' employment requirement I intend to ask PPL for a voluntary undertaking that the practice will not be reintroduced."[614] This was a fine outcome for AIRC, even though there was a clear flag that "no further changes are currently planned". It seemed an ideal time to try and reach a new negotiated settlement.

There had actually been negotiations through the previous year. AIRC and PRS had reached agreement in February 1988 for royalty rates of 5 per cent and 6 per cent, depending on station turnover, with Capital Radio paying just under 8 per cent, allowing 18 hours needletime.[615] Correspondence from February 1988 onwards, initially between Jimmy Gordon of Clyde and John Brooks, the chairman of PPL, addressed royalty rates with particular reference to the position of smaller stations which the tribunal's judgement had so disadvantaged; needletime; and the rates payable for split frequency broadcasts on FM and AM. However, the suspicion between the two bodies was manifest. There were spats over who had said what to whom, and over letters written to newspapers. At more than one stage, PPL went behind the backs of the AIRC negotiating team to address all the AIRC members. Gordon complained in May 1988 that "AIRC representation is a matter for us, just as PPL representation was left entirely to you. I hope you will recognise that we showed considerable goodwill in allowing the meeting to continue, despite the fact that you had unilaterally invited Manx Radio to attend, a company which is not in membership of AIRC and is not even subject to UK broadcasting legislation."[616]

These negotiations between AIRC and PPL continued during the three years after the government's endorsement of the MMC's report. However, although a settlement seemed within reach at many points, a deal could not be done. The two sides met yet again on 17 January 1991, but a follow-up meeting for the end of the month was postponed at short notice. PPL made new proposals on 12 February, but "the [AIRC] Copyright Committee regarded the proposal as ridiculous. It would mean stations paying more on current levels of usage, and many would pay

613 John Redwood secretary of state for trade and industry, in a House of Commons written answer, 20 December 1989 col 229W.

614 Ibid.

615 PRS press release 11 February 1988.

616 Letter from Jimmy Gordon to John Brooks, 11 May 1988.

substantially more."[617] After a critical final meeting with PPL on 13 February, AIRC concluded that it would invoke the 'statutory licence' procedure, applying to itself an arbitrary royalty rate of between 4 and 7 per cent depending on revenue, and then proceed to a new tribunal reference.

Some of the companies at least thought that PPL would only concede ground once a reference had been made. Brian West had told his members of an incident following the breakdown of the meeting on 13 February. "An extraordinary thing happened as we left the building. Trevor Faure, head of legal affairs [at PPL], ran after us and said ... 'look, whatever was said in there, PPL's members don't want to go to the tribunal; there is room for negotiation'."[618] Room perhaps, but not enough after such a long stand-off, and – even after the tribunal had begun – a last attempt to negotiate a settlement in July failed. The parties were close, but the prize of a negotiated agreement escaped them even when it had seemed to be within their grasp. Capital Radio's Richard Eyre sent a handwritten fax to Jimmy Gordon reporting on his conversations with Rupert Perry of EMI Records, and signing himself 'Despairing of Euston Tower'. "One step forward. One step back. I just spoke to Rupert Perry. In the course of our otherwise v. friendly conversation, he told me the backdated sum was still *not* agreed."[619]

AIRC made reference to what was now the Copyright Tribunal on 2 June 1992 on behalf of all its member stations apart from Radio Harmony in Coventry, which chose to be a separate applicant, and Classic FM. The INR stations, which began broadcasting during the hearings, raised similar proceedings with the tribunal which were settled later by agreement. The tribunal met on 30 separate days between 29 June and 9 July 1992, and then for 3 days in the autumn, giving its decision on 26 February 1993 and issuing a final decision and order on 30 March. In this instance the MU was not a party to the tribunal, but the BBC was an 'intervener'. The tribunal was chaired by a deputy chair of the Copyright Tribunals, Brian Gill QC. The AIRC team was led by Christopher Clarke QC, who came into prominence leading for the British government at the Bloody Sunday Inquiry from 2000 into the Derry shootings.

The arguments were not dissimilar to those in the 1979/80 tribunal, although in this 1992 case the changes brought about by the new regime of the 1988 Copyright Act and the 1990 Broadcasting Act were now relevant. Further, there were many more stations, including competition within localities, splitting of FM and AM services, and the abandonment of the public service obligations on the by now commercial radio stations. It was now some 20 years since the start of ILR, and much had changed. The industry's commercial prospects were different, and it now operated under very different legislation. The impact of the MMC Report, both in the ending of needletime and the introduction of statutory licences, allowed this tribunal to take a fresher look than had been appropriate for its predecessor. However, the two parties were no nearer agreement. AIRC proposed a flat rate of 3.5 per cent, PPL a rate of 15 per cent, the latter thought by the tribunal to be "preposterous" and thus not a valid starting point.[620]

617 Memorandum from Brian West to all AIRC members, 14 February 1991.

618 Ibid.

619 Fax from Richard Eyre to Jimmy Gordon, 6 July 1972.

620 Interim decision of the Copyright Tribunal, CT 9/91, The Association of Independent Radio Contractors Limited v Phonographic Performance Limited, Application under Section 135D, 26 February 1993.

The tribunal recorded the MMC's view that airplay in the end was of greater benefit to the radio stations than to the record companies, but decided that this might no longer hold good. Record companies continued to promote heavily to radio stations, even after the abolition of needletime, often of back catalogue material. Once again, the tribunal concluded that even in comparable European countries the way radio was organised there, and the different legislative context, ruled out making any valid comparisons.[621] Irrelevant also were the royalty payments for music videos broadcast on satellite television. However, the PRS agreement did have some relevance, and – crucially – PPL's agreement with the BBC was also relevant.

It is clear from the judgement that the 1993 tribunal was not persuaded by the logic of the PRT judgement back in 1986. It decided not to take that as the basis for its own decision, but to work out the respective benefits to AIRC and PPL from first principles, using the PRS rates as a rough guide.[622] The tribunal determined quite simply that the radio stations would in future pay a royalty to PPL of 5 per cent of net broadcasting revenue, with those broadcasting less than 15 per cent music paying just 1 per cent. Smaller stations would pay 2 or 3 per cent. Needletime, largely excluded anyway by the statutory licence regime in the 1990 Broadcasting Act, was not to be a factor. Stations would be able to reclaim any overpayment under the 1991 statutory licences, or would be required to make up for underpayments, by the middle of May 1993.

It was not quite all done, but it was nearly there. Jimmy Gordon, who had carried the weight of the work and worry, received congratulatory notes from around ILR, and a case of good red wine from the ever-thoughtful Neil Robinson at Metro Radio. Paul Brown, at that time head of regulation at the Radio Authority, spoke for all when he wrote to Gordon that "the success with the tribunal was owed in no small measure to your courage, tenacity, determination, hard work and clarity of vision and expresson".[623] AIRC's immediate task was to collect the legal costs from its members. The Association's total legal bill for this one tribunal was around £1.15m, of which PPL was expected to pay £425,000. The greatest winner, Capital Radio, was looked to for the major contribution, with companies contributing in proportion to one half of their 'winnings' from the decision. Capital had already funded the reference to the tribunal, underwriting the costs while the work was progressing.

That this was a genuine victory for the radio companies is confirmed by the concerned reaction of the music industry. PPL had also spent over £1m in legal fees, and although it had come out with a level of royalties which was one it would have settled for at an earlier stage in the proceedings, it was still less than it had expected. Of more significance was the tribunal judgement itself, which acknowledged for the first time that airplay was a factor of commercial value to the record industry. This was to weaken its future negotiating position with the new digital music platforms. The television companies preparing to do battle with PPL's sister organisation, Video Performance Limited (VPL), were delighted. Keith Macmillan, one of MTV's executive producers, noted that "there are many aspects of this

621 Ibid.
622 Ibid.
623 Letter from Paul Brown to Jimmy Gordon, 6 March 1993.

decision which are likely to have an impact on our own application to the tribunal".[624]

Once the bills were agreed and paid, the whole issue of music copyright could finally take a more subdued place among the concerns of the commercial radio sector. After over 20 years of struggle and strife, two major tribunals, with a return to the first after action in the High Court, an inquiry and report from the MMC, and two major pieces of legislation, ILR (and now INR too) had a deal it could live with without chewing the carpet. In the immediate glow of success, little else mattered. The copyright bill for the major companies was reduced, although it remained much more than negligible. Needletime had been seen off, and the Musicians' Union expenditure obligation had gone along with ILR's public service obligations in the 1990 Broadcasting Act.

Why had the ILR companies found it so difficult to achieve a better deal over those two long decades? Might they have achieved a deal with PPL similar to the eventual outcome, without the long and remarkably costly series of legal proceedings, had they not been working in a milieu in which their opponents were much more naturally at home? John Thompson's identification of the alternatives at that meeting in July 1978 was most apposite: talk or sue. The ILR companies approached PPL expecting it to be challenging and unresponsive, and perhaps provoked a less sympathetic response as a consequence. There was a sense around in the radio industry that the copyright societies would automatically be confrontational, and that may have hindered the chances of a negotiated settlement. That attitude found its way into some of the inter-personal difficulties which made negotiations so frustrating. Yet, during the mid-seventies there was a deal to be had, quite probably on terms even better than those imposed by the 1993 tribunal.

There is something admirable in the companies' persistence against repeated rebuffs, and certainly something heroic in the efforts of Jimmy Gordon. Yet AIRC collectively refused at almost every stage to understand the realities of the UK situation, the extent to which the BBC made it a different issue from anywhere else, and the nature and import of copyright legislation in the UK. It also – and this was a crucial point right up until the end of the eighties – wholly failed to appreciate the power of the MU, or to find a way to accommodate the musicians with a good grace. That let them down with almost all the movers and shakers in the critical early years. John Thompson captures the quintessential 'official' stance. "I shared the view up to a point of colleagues at Brompton Road that it was absolutely right and proper that the new radio companies should make a contribution to live music. And they were jolly lucky to have as small a commitment as they had."[625]

624 *Music Week*, 13 March 1993.
625 John Thompson interview op. cit.

IV. Victory of the commercial model 1990 – 2003

Chapter 14

Shadow and substance

1990

As the radio companies and the regulator prepared for a new world, there was more than a hint of change to come in the country at large. In December 1989 Margaret Thatcher faced the first challenge to her leadership. Disaffected Tories persuaded Sir Anthony Meyer to stand against her as a 'stalking horse', to see whether the party had any appetite for change. To the surprise of the commentariat, one in six Conservative MPs did not vote for her. Poll tax riots at the end of March 1990 showed how far her policies were stirring up unrest. She had lost her two key heavyweight supporters in the cabinet, Geoffrey Howe and Nigel Lawson. Howe's unexpectedly brutal resignation speech on 13 November 1990 – he famously said he felt so undermined by the PM that it was "rather like sending your opening batsmen to the crease only for them to find, the moment the first balls are bowled, that their bats have been broken before the game by the team captain"[626] – led directly to Thatcher's fall. She failed to win a wholehearted mandate from her cabinet colleagues while seeing off a leadership challenge from Michael Heseltine, and resigned tearfully on 22 November. Her largely unknown successor, John Major, was an uncertain heir to the throne of Thatcherism although, as a cricket enthusiast as much as a politician, he was one of those who would have appreciated Howe's resignation simile.

Private radio in the UK was impatient to start its second innings. The green paper, which seemed to offer the industry all that it could have wished, had appeared as far back as February 1987. The white paper did not emerge until November 1988, by which time it had become hugely concerned with television and with regulatory structures. That helped the ILR companies' aspirations, since it meant there would be little time or appetite to revisit radio policy which had already been established. However, it caused further delay, and the Broadcasting Act was not to receive royal assent until 1 November 1990, nearly four years after the green paper had seemed to answer their demands.

Meanwhile, the splitting of FM and AM services and greater consolidation of ownership had combined to sketch the pattern which commercial radio would adopt, predicated upon the greater freedom and commercial opportunities expected under the new dispensation. In August 1989, Ocean Sound and Southern Sound along the south coast of England merged to create a group of seven separate services, and started looking covetously towards Mercury and Invicta in Kent.[627]

626 Sir Geoffrey Howe 13 November 1990 in *Hansard* House of Commons col 464.
627 *Media Week* 4 August 1989.

The main shareholders included Michael Heseltine's Haymarket Publishing, to which he was able to devote more time once finally rebuffed in his premiership ambitions. A year later, Owen Oyston's Trans World Communications agreed to acquire the Yorkshire Radio Network. This was to bring Piccadilly, Red Rose, Aire and Red Dragon together with Hallam, Pennine and Viking, plus their Classic Gold AM services, thus running "from coast to coast in the same group [with] the same rate card."[628]

The provision of separate services on AM and FM continued apace, but now the nascent groups began to examine what benefits synergy between their licences might offer. Capital's launch of Capital Gold on AM and Capital FM in November 1988, with Park's innovative CHR format on FM as the younger competitor for Radio One, was a swift success. Not only did Capital's total audience increase, but advertising inventory for additional sales was doubled. In October 1989, LBC had followed suit, launching Crown FM (named for major shareholder, Crown Communications). This greatly increased its cost base, and the station was still unable to solve the challenge of "marrying sufficient news-gathering resources with the right anchor men and women".[629] Radio City attempted a short-lived AM talk service, launched in the same month. By the end of 1989, split services were being provided by Beacon, BRMB, Capital, Clyde, County, DevonAir, Essex, GWR in Bristol, Invicta, LBC, Leicester/Trent, Marcher, Metro/Tees, Ocean, Piccadilly and Yorkshire.

LBC apart, the cost increases were not huge, but the talent issue was. Music presenters "were like gold dust. Capital simply scooped up the cream of the sixties household names such as Tony Blackburn and Kenny Everett. Elsewhere, it was not quite so simple."[630] By the end of 1990, GWR had become a group covering five ILR franchises: GWR, Wiltshire Sound, Radio 210, 2CR and Plymouth. In a sign of things to come, it launched its own 'gold' AM network that September, with just two separate services (with local opt-outs) to cover the whole area. All the original ILR music-based stations except Moray Firth had split their frequencies by 1994, with Norman Bilton's Wyvern predictably the last to do so.

For a while, matters were looking healthy on the financial front, too. Shares in Capital Radio broke through the £10 barrier on 6 September 1989 and a leading city analyst opined that "£20 a share in two years time seems tenable".[631] The split had increased Capital's share of London listening from 17 to 28 per cent, and it anticipated profits for 1989/90 of £17m from remarkable margins of 30 per cent.[632] The radio sector's market value had risen from just £33m in 1987 to a whopping £400m, and although advertising still remained stubbornly around 2 per cent of the UK total, Barclays de Zoete Wedd were forecasting that it might rise to 4-5 per cent in the next five years.[633] Caution was suffused with the glow of hidden optimism. Crown's chairman Christopher Chataway observed that "the ones that survive" of the many new stations "will be those that can offer something identifiably different". For Peter Baldwin, "the winners will be the listeners, because of

628 Piccadilly MD Julian Allitt in *Music Week* 4 August 1990.
629 Radio critic Ken Garner in *Radio & Music* 5 December 1989.
630 *Broadcast* 29 September 1989.
631 Chris Akers of Citicorp in *Financial Times* 6 September 1989.
632 *Investors Chronicle* 22 June 1990.
633 *The Times* 11 October 1989.

the greater choice they will have ... even if all the new entrants don't succeed, the opening up of the market place will mean they will have had a chance". New frequencies, and independent national stations, were "likely to stimulate much more interest in ILR as an advertising medium".[634] The language of the market, of the inevitable benefits of competition, was nearly universal.

Many of those financial predictions were to prove correct over the course of the decade. However, as the nineties began along came another recession, which hit advertising expenditure and the national economy too. The mood around ILR turned suddenly gloomy as 1990 progressed. Crown announced redundancies, while Capital's financial director Patrick Taylor said it had "battened down the hatches ... we can ride out the storm".[635] Suddenly, competition seemed less certain to bring benefits. Capital was to face music radio rivals in Kiss, Melody and Jazz, and radio in London was broadening out with the split services from LBC and the unknown quantity of Spectrum. Analysts began to understand that the radio business in the UK was going to become more like that in America, as it became more commercial. "In New York's freely competitive radio market", observed the *Investors Chronicle*, "the biggest station in town gets around 5 per cent of the total audience. Capital's 28 per cent won't slump overnight, but in the long run there's only one way for it to go."[636]

The Radio Authority was established in 'shadow' form on 1 January 1990. It was to take up its substantive role at the end of the year.[637] Lord Chalfont, who as Alun Gwynne Jones had been a Labour minister from 1964 to 1970, had been deputy chairman of the IBA. An expert on military affairs, and military history, he was appointed chairman of the Radio Authority, with members Jill McIvor, John Grant, Richard Hooper and Ranjit Sondhi. Both Sondhi and McIvor, who became deputy chair on 1 June, had previously served as members of the IBA. Margaret Corrigan joined in June, and Michael Moriarty – who as a senior Home Office civil servant had been closely involved in broadcasting issues between 1981 and 1984 – became a member at the start of 1991. None of these appointments was through any open or competitive process. It was not until after the report of the Nolan Committee[638] in 1995 that Radio Authority members were chosen from among those who had actually applied for the job, rather than being just 'tapped on the shoulder' and invited to take on this public duty.

Staffing the Authority was similarly uncompetitive. IBA director of radio Peter Baldwin became chief executive of the Radio Authority, and senior IBA Radio Division figures Paul Brown and David Vick, joined by others such as Neil Romain and John Norrington, quietly 'crossed the floor' to radio; actually down to a lower floor at the IBA's Brompton Road offices, which was radio's new temporary home.

634 *Accountant* September 1990.
635 *Broadcast* 27 April 1990.
636 *Investors Chronicle* op. cit.
637 The starting date for the new pattern of regulation for commercial radio is usually given as January 1991. Strictly speaking, the Radio Authority came into existence on 11 December 1990, but took over from the IBA the responsibility for regulating Independent Radio in the UK from 1 January 1991. This is not just an academic point. The pre-existence of the Radio Authority allowed for the process of moving from contracts to licences. Thus when Alun Chalfont, Jill McIvor, John Grant, Richard Hooper and Ranjit Sondhi met on 19 December, they were already the Radio Authority, but the formal correspondence with Sunset Radio was to and from the IBA.
638 *Committee on Standards in Public Life* Cmnd 2850 I and II, May 1995.

There was a feeling among Radio Authority staff, when I arrived in 1995, that it had been the courageous and forward looking members of the IBA's staff who had opted for the 'adventure' of the new Authority, while those more cautious had stayed behind in the ITC. There is some truth in that, but it is also true that those who 'escaped' to the new body brought with them plenty of the attitudes and instincts of the IBA, and the aspiration for making things new would have been helped by more new blood at that time. One 'outsider', Marion Turner, was appointed as head of finance from the middle of 1990. She left to be succeeded in January 1991 by Neil Romain, her deputy and another former IBA radio officer. Mark Thomas, however, who joined from a commercial technical background as head of engineering in August 1990, became one of the Radio Authority stalwarts throughout its life.

The slow progress of the new broadcasting legislation left the IBA with some practical constitutional difficulties to resolve. Alun Chalfont continued as deputy chairman of the IBA through 1990, and as a member of the Shadow ITC as well. The Shadow Radio Authority (SRA) met for the first time on 21 February 1990 at Brompton Road. Thereafter, it was to meet monthly from 14 March on Wednesdays at 9.30 am in the IBA boardroom at Brompton Road, before the meeting of the IBA the following day. There was a huge amount to be done by the SRA, much of it widely different from the role and tasks of the outgoing IBA. As well as the practicalities of establishing the new organisation – finance, premises, staff appointments – it needed to have new licensing procedures in place for both ILR and the unprecedented auction of new national services; to work out how it was to undertake its own frequency planning; to devise detailed ownership rules; and to draw up codes to govern programming and advertising, and for allowing the newly permitted potential of sponsorship to be released in a controllable form.

At its first meeting in February 1990, the SRA had before it a paper from Peter Baldwin examining those issues which needed to be resolved at a fairly early stage.[639] First, there were matters where there might still be time to influence the final shape of the legislation: national news provision, religious output, education and social action broadcasting. Then came the sweeter topics: radio awards and EBU activities. Finally, the members were invited to discuss what, if anything, should replace the Local Advisory Committees, what type of research might be undertaken, and a range of issues around charitable fund-raising and commercial programming and marketing.

There was a debate about how far it was practical to urge a 'quality threshold' for the INR licences. This had been a late addition to the rules for the auction of the ITV licences, which nevertheless was to fail to save Thames Television. In the SRA, this was argued between the old social economists and the new market economists. John Grant, from his Labour/SDP perspective, made the case for a quality hurdle which applicants for the three national licences would have to pass. He was supported by Ranjit Sondhi and by Alun Chalfont. Richard Hooper doubted whether quality could be prescribed in licence terms. Paul Brown was of the view that "in a competitive market place in which the commercial system would have to cope with the high standards of the BBC, it was unlikely that low

639 *SRA Paper* 2(90) 6 February 1990.

quality services would survive".[640] Without consensus, the Authority let the matter be, and the auctions eventually proceeded with only viability as a pre-condition.

The SRA faced a formidable task, to be carried out by a small number of people, and against the background of continued slow progress with the legislation, which might well result in significant changes in the new Authority's powers, duties and limits. It was accomplished with real success and completeness, to the considerable credit of all those involved. Very little of the structure created under such pressure in 1990 needed to be revisited, except where the external, legislative context changed. Inevitably, policy evolved on the hoof. In April the SRA considered whether to press for a separate tier of community radio, with a fund to support it.[641] By June it had rejected that, but was drawing up guidance notes for local authorities to fund community-related activities via the commercial stations.[642] Even the nomenclature of the new system needed discussion. Should it be 'IR', independent radio, now favoured by AIRC, or would they stick with ILR and INR? David Vick and Peter Baldwin recommended the latter, hoped that the term 'incremental' would fade away, and proposed using 'community' only for non-profit-distributing, highly localised radio. As for 'commercial radio', that might "in the perception of much of the general public (including those who need to be wooed away from the BBC, as listeners) detract from the 'public service' elements of the role which many independent stations will continue to perform".[643]

As the broadcasting bill reached the radio sections at committee stage on 22 February, Alun Chalfont wrote to the minister, David Mellor, asking for the power to say which services each INR station should provide.[644] Chalfont had already gone on record as personally favouring using the FM frequencies for a classical music service, while the two AM licences should be for news and current affairs on one, and for popular music on the other.[645] The government accepted the argument about the relative weakness of the bill's requirement for diverse programme services between the INR channels. Mellor acknowledged the risk that "the economics of radio are such that this on its own might result in the emergence of three essentially pop-based stations with the minimum necessary diversity mixed in".[646] The bill was therefore to be amended to make sure each service was different from the other. Furthermore, "one of the new stations must be substantially speech-based, and the other must include a substantial proportion of music other than pop music".

Much of the fun in the radio sessions of the parliamentary debates on the broadcasting bill surrounded this last amendment, and the definition of what was or was not 'pop music'. Speaking in the House of Lords in June, minister of state Earl Ferrers revealed the government's less than subtle understanding of pop music genres. "The bill also now says that one of these stations must be a speech-based service and another must be devoted to music other than pop music – and I am sure that that will please some of your lordships. It will not all be 'Thump, thump,

640 *SRA Minutes* 1(90) of a meeting on 21 February 1990.
641 *SRA Paper* 15(90) 11 April 1990.
642 *SRA Paper* 25(90) 12 June 1990.
643 *SRA Paper* 20(90) 11 April 1990.
644 Letter from Lord Chalfont IBA to David Mellor, Home Office, 26 January 1990.
645 *Broadcast* 23 February 1990.
646 David Mellor speaking in London at a Campaign for Quality Television (sic) Forum 18 April 1990.

thump'!"[647] David Mellor was the speaker at dinner when the SRA met for a 'weekend conference' on 15 and 16 September, at the Swan Hotel in Streatley, and even at that late stage had indicated that there was still scope for changes in the bill. Notably, what was then Clause 84(2), which set out the external diversity test for the INR services, was to be "further amended so that the service other than pop music would be, in law, whatever the Radio Authority determined it to be".[648]

This Humpty Dumpty formula for 'pop music' went only part of the way.[649] The SRA members earnestly debated whether 'rock' music fell within 'pop' music. If not, then they might end up with two of the three INRs offering what the government, at least, would think of as pop. The minute of the conference discussion is a gem: "The chairman's solution was to declare that 'rock' music was 'pop' music. He said that the Collins dictionary which appeared to be the only recognised dictionary to include a definition of 'pop' music, supported this conclusion as 'generally characterised by a heavy rhythmic element and the use of electric amplifiers'. [He] put two propositions to members: (a) ... that the criterion for determining a non-pop station whatever frequency it was allocated should be that it was either composed of music that fitted the description of 'classical', 'jazz', 'easy listening', or other particular genres but that it did not include 'rock' or its equivalents; or that ... pop music meant 'chart hit singles'and (b) ... that the FM frequency should be advertised first as the non-pop frequency or that it should be unspecified. Members were divided on these issues, and it was agreed therefore to defer a decision until the next meeting."[650]

Peter Baldwin told the SRA in November that he "was confident that there was no widespread misconception about the scope of the 'non-pop' definition ... [it was] quite clear that 'non-pop' did not mean 'classical'".[651] John Grant asked for the minutes formally to record "his disquiet that the Authority may have been given selective advice previously about the definition of 'pop', maintaining that there were other dictionary definitions apart from the one quoted at the time". Richard Hooper went so far as to write to Peter Baldwin about being "increasingly concerned by the growing implication in the media ... that INR1 non-pop is classical music" and asking for "a firm and clear statement that non-pop includes classical music, easy listening (e.g. BBC Radio 2 and Melody), jazz, and country & western".[652]

The SRA's proposal for a 'points system' to govern ownership consolidation, devised by Neil Romain and initially raised formally with David Mellor in Alun Chalfont's letter of 26 January, was debated through the summer between the SRA and a sympathetic Home Office. The outcome was agreed between officials, approved by the Authority members at their June meeting,[653] and announced by David Mellor in a statement to the AIRC Congress on 27 June. The system apportioned all licences into one of four categories, on the basis of their population

647 *Hansard* House of Lords, 5 June 1990, col 1225.

648 *SRA Minutes* 8(90) of a meeting on 15/16 September 1990.

649 'When I use a word', Humpty Dumpty said, in rather a scornful tone, 'it means just what I choose it to mean – neither more nor less'. *Through the Looking-Glass*, Ch. VI. Lewis Carroll.

650 SRA Minutes 8(90) op. cit.

651 *SRA Minutes* 9(90) of a meeting on 22 November 1990.

652 Letter from Richard Hooper to Peter Baldwin 9 November 1990.

653 *SRA Minutes* 5(90) of a meeting on 20 June 1990.

coverage. Each A, B, C or D licence attracted a common number of points. Category A licences for example, covering Greater London, were worth 15 points, D licences with coverage of fewer than 400,000 adults in their measured coverage area just one point. At the suggestion of Richard Findlay, then AIRC Chair, INR was included in the scheme, with each INR licence attracting 25 points. All AM licences were then discounted by one third compared with FM. The total constituted the points universe at any one time, which had the considerable advantage, therefore, of being a moving total, able to cope with the steady expansion of the system. A company could own up to 15 per cent of that universe. Complex but clear, the system was widely accepted and, from the time of Mellor's speech to AIRC, formed the basis on which companies could plan their expansion and acquisitions for the new regime once the bill became law.

The Authority then formally considered representations from Owen Oyston of Trans World (TWC) for changes. His Piccadilly MD, Julian Allitt, had stated in August that the proposed takeover of Yorkshire Radio (YRN) by TWC would take the new group "to the limit of expansion" under the points system.[654] However, on 30 July a letter had already gone from Oyston to Alun Chalfont, asserting that the points scheme exhibited "a systematic bias in favour of the London-based companies and against the leading provincial companies, of which we are just one".[655] He proposed changes "as a constructive contribution to discussions on the present proposals". The letter was couched in general terms only. The request was put to the full Authority at their weekend conference in September. The Authority was told that Oyston's planned 'merger' could only be allowed under the proposed rules if TWC/YRN divested themselves of some stations, although once INR licences had been awarded, the problem would not arise since the points universe would increase. However, in the face of a 'hostile' bid for YRN from Metro, Trans World had a problem. The SRA paper suggested replacing the flexibility clause previously proposed with a 10 per cent tolerance "to avoid an unnecessary sell-down".[656] Aware that this might seem to be favouring one party in a dispute, the Authority had checked in advance with that year's AIRC chairman, Richard Findlay, and the paper noted that "the chairman of AIRC is content with the 10 per cent tolerance allowance ... and understands the reason for it". The Authority agreed, and the impediment to the merger was removed, allowing the flexibility for Trans World to complete its takeover.

A key issue to be resolved was the transition for the companies from being contractors operating a franchise, and providing programmes to be broadcast by IBA transmitters, to being licensees running their own transmission arrangements. This involved a complex contractual shift, and finding a basis for selling the IBA transmitters to the companies. The Broadcasting Act eventually allowed – as a fall-back – all the existing service providers to continue to operate under their contracts with the IBA, making exceptional provision for the Radio Authority to continue as the broadcaster from January 1991. However, those companies also had the statutory option – which they were earnestly enjoined to adopt – of

654 *Broadcast* 4 August 1990.
655 Letter from Owen Oyston, Chief Executive Trans World Communications plc to Lord Chalfont, Chairman Shadow Radio Authority 30 July 1990.
656 *SRA Paper* 44(90) 5 September 1990.

exchanging their contracts for a Radio Authority licence.[657] The wider deal involved companies buying their transmitters from the IBA at the turn of the year, together with an agreement which provided for them to be maintained by NTL for a period of two years.[658] A company exchanging a contract for a licence also gained an automatic extension of three years to the term of its licence, or to the end of December 1996, whichever was the earlier. (This did not apply to the 'incremental' licences, issued after the start of September 1989, all of which were to expire at the end of 1994.) It seemed to be a close-run thing at the time, but looking back now it is unsurprising that all the ILR companies agreed to move to licences, and did so with effect from 1 January 1991, which was also the formal starting date for the Radio Authority.

The licensing process for ILR, which the SRA established for the Radio Authority over the coming 13 years, and which was eventually completed by Ofcom, differed in five fundamental ways from the IBA processes which had preceded it. First, the Radio Authority (and the SRA) made its own selection of localities to license, and the timing of that was unconstrained by the government intervention which had been so significant in the earlier regime. Second, it undertook its own frequency planning, in a consciously competitive arrangement with the BBC. Together, these differences meant that the roll-out of local licences was no longer a matter for negotiation between the Corporation and the Authority, with the Home Office holding the ring. Third, since the stations were to own and operate their own transmitters, the Radio Authority was not limited by the availability of its own investment capital, as the IBA had been. Fourth, the Authority was given much more statutory direction about the selection criteria for local awards, while being permitted to use a less onerous regime for the awards themselves, which very largely did away with the set-piece interviews of applicant groups. Fifth, the new act did not require any local consultation before licence awards were made, so public meetings were to disappear, and the process was to be very largely paper-based.

The most significant of all the innovations were the four statutory selection criteria.[659] The Radio Authority's dogged insistence upon following them and virtually nothing else effectively conditioned the nature of new ILR stations, as well as the pre-existing stations in the process of re-advertisement. ILR was to change from being a broad 'alternative service of radio broadcasting', to one where each station existed because it had sufficient resources to provide the proposed service, catered for the tastes and interests of local people, broadened the range of local radio services and was supported by local people. Crucially, since the act did not provide for it, the Authority set its face against in any way prescribing the type of service it aspired to license in any locality. It was for the applicants to determine which type of service would best meet the statutory criteria and apply accordingly.[660] Equally, since there was no statutory need to consider any licence award

657 Broadcasting Act 1990 Schedule 11, parts IV and V.

658 *Radio Authority Annual Report* 1991.

659 *Broadcasting Act 1990* section 105.

660 It is hard to overstate the importance of section 105 of the new act. The SRA's initial inclination was to follow it to the letter, and that view was inherited by successive Radio Authority boards.
" Section 105: Special requirements relating to grant of local licences.
Where the Authority have published a notice under section 104(1), they shall, in determining whether, or to whom, to grant the local licence in question, have regard to the following matters,

in the overall context of a strategy for the whole radio industry, every award was on its own local terms.

The key decisions were taken at the SRA's September conference meeting.[661] Together with the advertisement of new INR licences, the expansion of ILR was from the start a major work stream, and licensing – both *de novo* and after re-advertisement – remained the Radio Authority's single major pre-occupation throughout its life. For all that time this effort was to be led by David Vick, who was the new Authority's head of development, and it is appropriate that his papers to the September conference should set the style for the following 13 years. The meeting agreed licensing procedures for ILR derived from that used by the IBA for the 'incremental' licences. These comprised initial analysis of each application by the Authority's staff, using common appraisal forms which 'scored' each against a wide range of criteria, derived from section 105. Those scores would then be weighted according to the Authority members' current assessment of importance.[662] One Authority member was deputed to work with staff on each award, to produce a short-list. The applications selected were debated extensively by the full Authority, which would then make the award.

National licence award procedures were even more fully specified in the act, which required that "the Authority shall, after considering all the cash bids submitted by the applicants for a national licence, award the licence to the applicant who submitted the highest bid".[663] There was considerable debate within the Authority over how far "exceptional circumstances" could be deployed, if the Authority wished to select a group which was not the highest bidder. On the basis of the staff recommendation,[664] the members concluded that the process would again be administrative, determined very largely by the highest bid, although "the right to interview should nevertheless be held in reserve". Paul Brown, by now head of regulation, noted that the Authority had "won a valuable concession from the government, which will ensure that each national service will be different ... [but] the precise nature of each national music service will be decided by the applicant who makes the highest bid".[665]

The final part of this group of seminal decisions was agreement on the selection of locations for the first phase of licensing. At the start of 1990, the SRA had trawled for interest through national and trade press advertisements, inviting

namely –
(a) the ability of each of the applicants for the licence to maintain, throughout the period for which the licence would be in force, the service which he proposes to provide;
(b) the extent to which any such proposed service would cater for the tastes and interests of persons living in the area or locality for which the service would be provided, and, where it is proposed to cater for any particular tastes and interests of such persons, the extent to which the service would cater for those tastes and interests;
(c) the extent to which any such proposed service would broaden the range of programmes available by way of local services to persons living in the area or locality for which it would be provided, and, in particular, the extent to which the service would cater for tastes and interests different from those already catered for by local services provided for that area or locality; and
(d) the extent to which any application for the licence is supported by persons living in that area or locality."

661 *SRA Minutes* 8(90) op. cit.
662 *SRA Paper* 43(90) 4 September 1990.
663 *Broadcasting Act 1990*, section 100.
664 *SRA Paper* 47(90) 10 September 1990.
665 *SRA Minutes* 8(90) op. cit.

'letters of intent' to run new ILR services, to identify the level of demand. By the closing date of 27 April 1990, it had received 850 letters. Once again extensively steered by David Vick's paper,[666] the SRA members concluded that the list of locations for which licences should be advertised in 1991 "should include a high proportion of areas currently unserved by ILR; few, if any, metropolitan areas until INR 1 & 2 were decided; and a number of locations suitable for small-scale 'community radio' stations or which could be appropriate for ethnic minorities".[667] The actual list would be subject to identifying and clearing the necessary frequencies, but the SRA agreed at the meeting a list of 27 areas from which it would select the localities to be advertised in 1991, and issued the press statement the following week.[668]

Next, the Authority had to establish its processes for the regulation of programming and advertising, placing the primary onus for delivery upon the licensees. The new regime swept away any prior approval of station programme schedules, which had been a concession to the Reithian prejudices of the early seventies, and replaced that with 'promises of performance'. Section 106 of the act, requiring the Authority to "secure the character" of the service as proposed at the time of application,[669] ranks alongside section 105 as the guiding texts of the new regulatory regime. The principle, which was to apply until a further loosening of the rules in 1999, was that the Authority would distil from each winning applicant's programming proposals the essence of what it had undertaken to provide. This 'promise of performance' would form part of the licence to broadcast, and be changed only in agreement with the Authority. The content of programming and of advertising were also to be self-regulated according to codes drawn up by the Authority and, along with the general law of the land, these formed the boundaries within which ILR and INR stations could operate unmolested by the new regulator. Most genuinely tried to work within the limits of the system, and there was a broad consensus that "to prevent a mass, commercially-driven migration towards lowest common denominator broadcasting, the set-up has to be regulated".[670]

All this may have seemed less radical later in the decade, but at the start of the nineties it was a real change compared with what had applied before. A notable liberalisation, approved by government at the urging of the SRA, was to remove any limit on the maximum number of minutes per hour of advertising on ILR and INR. ITV had been kept to a ceiling of six minutes, and when ILR was launched the Sound Broadcasting Act imposed a nine minutes limit for the new medium. David Vick recalls that "the Radio Authority's implementation of the 1990 act included doing away with such things as regulatory approval of the definition of marketing areas (TSAs) and of ratecards. Unlike the proto-ITC, we had persuaded

666 *SRA Paper* 46(90) 10 September 1990.

667 *SRA Minutes* 8(90) op. cit.

668 Annex D to *SRA Minutes* 8(90) op. cit.

669 Section 106: Requirements as to character and coverage of national and local services.
(1) A national or local licence shall include such conditions as appear to the Authority to be appropriate for securing that the character of the licensed service, as proposed by the licence holder when making his application, is maintained during the period for which the licence is in force, except to the extent that the Authority consent to any departure on the grounds—
(a) that it would not narrow the range of programmes available by way of independent radio services to persons living in the area or locality for which the service is licensed to be provided, or
(b) that it would not substantially alter the character of the service.

670 *The Times* 25 August 1990.

the government that licensees should be allowed to decided the appropriate volume of advertising minutage for themselves (on the premise that if they got it wrong, their resultant audience figures would produce the necessary correction)".[671]

The Radio Authority was to be obliged to remain part of the formal advisory process for advertising content, via the Advertising Advisory Committee. However, just as in the ITA years, it regarded this as an important mark of respectability for radio, a view largely shared by the stations. There was, in time, to be a separation of the copy clearance processes for radio advertising,[672] but at the start this remained a joint activity between AIRC and the ITCA. The other formal advisory committees ceased to have any involvement with commercial radio, although the Central Religious Advisory Committee still turned an ear in its direction from time to time. The SRA had decided not to have its own system of local advisory committees, nor to have any regional office structure. Routine monitoring of thousands of hours of programming each day was wholly out of the question, so the stations were left to their own devices to a degree which was striking at the time for its liberality. The statutory requirement in respect of monitoring was for each station to retain for 42 days a complete recording of all its output, which the regulator could then use to check retrospectively any complaints or concerns.[673] The system worked notably well. To avoid any doubt, though, the section of the act was specific that "nothing in this part shall be construed as requiring the Authority ... to listen to such programmes in advance of their being included in such services".

Later to be stigmatised as distorting the market, the new regime was mostly seen at the start of the nineties as a beacon of light touch regulation. The routine policing of promises of performance was done on an exception basis by the Radio Authority staff, was helped by public input, but was chiefly made effective by the stations themselves. Plenty of them were more than happy to draw the attention of the Authority's staff to breaches by their neighbours, especially if that neighbour's coverage area overlapped quite a bit with the reporting station. This was, after all, a newly competitive world for radio. I well recall the slightly furtive phone calls. "Tony, I don't want this to be formal, but I think you ought to be listening to Radio X. But you didn't hear this from me, of course." Later, as competition intensified, the informal tip-off was sometimes to be replaced by a formal and even quasi-legal objection to a rival's output. For a while yet, though, the personal relationships established in the shared task of *independent* radio were to hold good, into the new world of *commercial* radio.

671 Note from David Vick to the author, 8 July 2008.
672 See chapter 18.
673 *Broadcasting Act 1990*, section 95.

Chapter 15

Classic, Talk and Virgin

Independent National Radio 1991 – 1994

January 1991 marked the undoing of what was briefly thought to have been 'the end of history'. Once the Malta Summit in December 1989 between Gorbachev and Bush had formally declared an end to the cold war, Francis Fukuyama signalled the general euphoria by wondering whether this marked "the end point of mankind's ideological evolution and the universalisation of western liberal democracy as the final form of human government".[674] That was quite a claim, coming from the deputy director of the US State Department's policy planning staff. Gorbachev's resignation, and the effective end of the Soviet Union on Christmas Day 1990, marked its brief high point. 22 days later, the US unleashed Operation Desert Storm in response to Saddam Hussein's invasion and attempted annexation of Kuwait, and the 'new world order' started to look as brutal as the old.

Domestically, John Major was polishing his five 'great principles' for revitalising the Conservative government, to be announced in March 1991. These included the 'citizens charter', further privatisations including of the railways, and an attempt to strengthen the union of the home nations. This was an uncomfortable mix of the retrospective and the forward-looking. The prime minister still hankered after a Britain of "long shadows on cricket grounds, warm beer, invincible green suburbs, dog lovers and pools fillers, and old maids bicycling to holy communion through the morning mist".[675] Broadcasters and businessmen could be forgiven for feeling that the country was looking two ways at once.

This is the point at which the chronological narrative of this history of independent radio needs to separate again, into two streams. Between 1991 and 1994, national and local commercial radio developed along parallel tracks, with many of the same players and influences. However, there is so much information that to cover both together risks confusion and overload. This chapter will therefore deal with the progress of INR over those years; the next two will look at ILR. The two themes then come back together – they were never really apart in practice – when the development phase of INR was completed, in chapter 18.

674 Francis Fukuyama, *The End of History*, based on a lecture given at the University of Chicago Summer 1989. This was later expanded into a book *The End of History and the Last Man* (1992), but by the time of its publication the thesis had been bombed to hell on the Basra Road.

675 Anthony Seldon, *Major: A political life*, Weidenfield 1997, p 370.

Independent National Radio was at the top of the new Authority's task list. The SRA's decisions had excluded both rock and pop genres from the FM channel, so when the announcement was made on 30 October 1990, it was that the FM licence would be for a "non-pop music service".[676] This was greeted with relief by the ILR stations, who saw the disappearance of a long-feared potential competitor. Ron Coles, MD of the Midland Radio Group, thought it "very welcome and helpful. It will produce a service that will complement rather than compete with local stations."[677] Others were scathing. Citicorp media analyst Chris Akers opined that "a classical station just couldn't attract large enough audiences",[678] and that it would not add enough to enhance the overall sales proposition of commercial radio. Virgin spokesman Will Whitehorn stated that the company would have bid for a pop channel, but would back away from non-pop. "We have scoured the western world for a profitable classical station but we couldn't find one. Anyone who tries to set one up will be committing financial suicide."[679]

Undaunted, the Radio Authority advertised the licence in the national press on 11 January 1991. Potential bidders clustered round, with much attention on the expected consortium of Classic FM. This group was headed by David Astor, with David Maker as its chief executive. When the Authority announced in March that records released commercially as singles would jeopardise compliance with the non-pop format, Pat Falconer, MD of radio sales house IRS, announced that INR "will be an unmitigated disaster and will go broke".[680] It was the contemporaneous first Iraqi war which had given common currency to the term 'friendly fire', and here was radio's equivalent.

The Radio Authority was having problems. Several of the big hitters were shying away from applying, including the Classic FM consortium, which announced its withdrawal from the race.[681] At its April meeting, the Authority was contemplating the ghastly possibility that it might not receive any applications. The members decided to endorse a decision – which the timings of press comment indicate had already been taken by Alun Chalfont, and made known publicly – to postpone the closing date for one month, to 22 May.[682] They would not have been unhappy that press reports blamed the delay on proposals from PPL to increase music copyright charges.[683] Meanwhile, contingencies were considered, and David Vick and Neil Romain proposed either re-designating the licence as "open to 'all-comers' ... or using these valuable FM frequencies for a range of regional and local services instead of a national network".[684] By the Authority's meeting in early May, Peter Baldwin was reporting that "the position was not promising but there could still be bidders".[685] It must therefore have been a considerable relief to members, as they gathered at Brompton Road for their June meeting, to be given

676 *SRA Press Release* 30 October 1990.
677 *Broadcast* 2 November 1990.
678 *Marketing Week* 2 November 1990.
679 *Music Week* 10 November 1990.
680 *Media Week* 22 March 1991.
681 *Marketing Week* 29 March 1991.
682 *RA Minutes* 5(91) of a meeting on 4 April 1991.
683 *Financial Times* 26 March 1990.
684 *RA Paper* 33(91) 18 April 1991.
685 *RA Minutes* 6(91) of a meeting on 2 May 1991.

copies of three *bona fide*[686] applications – from First National Radio (FNR), UKFM and Classic FM.

FNR proposed 'Showtime'. The consortium had its origins in the Metropolitan Radio consortium, which had in 1983 unsuccessfully contested the re-advertised London franchise held by Capital. It was chaired by Sir Peter Parker, and led by Robert (no longer 'Bob') Kennedy, once of the IBA's Radio Division, who was the "business brain behind First National Radio", and was said to be unrepentant that the company had bid so much more than its rivals.[687] The Authority's initial impression was that its "programming proposals are rather vague",[688] with the linking theme of show business but not much specificity about the music other than an assurance that "music defined as 'not pop music' will form at least 75 per cent of all music broadcast".

UKFM was a joint application between the Radio Clyde Group, Lord Hanson whose huge company owned Melody Radio, the French radio group NRJ, Scottish Television and investment company Trevor Clark. It had John Whitney as a non-executive director, and proposed easy listening and light classical music, with rather more speech than the comparable Melody format. The initial staff analysis suggested that UKFM had already identified 70 per cent of the capital funding it would need, while "FNR seems to have no funding at all".

And Classic FM had, after all, come through with an application, even though its own funding seemed limited and "inconsistent".[689] The apparent prime movers were still David Astor and David Maker, but now its key partner was the GWR Group under Ralph Bernard. The staff briefing paper to members said that "it was the withdrawal of the group's original investors, rather than the reasons given publicly, that led to Classic FM's temporary retirement". There was more to it than that.

Two groups had been separately pursuing ambitions to apply to run classical music on INR1. GWR's Brunel Radio had carried a successful three hour afternoon show on AM only when it split its frequencies in 1988, devised by Michael Bukht. Bernard had convinced his board that this format would translate to the national scene. Meanwhile, the Golden Rose group led by Astor and Maker had applied, without success, for a London FM incremental contract for a similar service. When it was reported that Golden Rose was not proceeding with a bid for INR1, Bernard spoke with Maker and the two agreed to join forces. It was an uneasy alliance, but the application, written by Bernard, GWR board member Nicholas Tresilian and Deanna Hallett – and driven by Hallett's partner Jonathan Arendt at improbable speed across London – reached the Radio Authority just before the 5 pm deadline. This was seat-of-the-pants stuff. Bernard recalls that "everyone thought we had deliberately put our application in late to shock. The reality is we got it in so late because we cocked it up."[690]

The Radio Authority staff set about examining the three applications to see whether any 'exceptional circumstances' applied, while its financial advisers,

686 A fourth 'application' was received, backed by an unsigned cheque for the £10,000 application fee, and lacking any cash bid. It was therefore disqualified.

687 *Financial Times* 5 July 1991.

688 *RA Paper* 46(91) 30 May 1991.

689 Ibid.

690 Ralph Bernard in an interview with the author, 19 November 2007.

Barclays der Zoete Wedd (BZW), evaluated the financial viability of the proposals. The highest bid had come from FNR/Showtime with £1.75m. Classic FM was next with £0.67m, and UKFM third with £0.30m. We may guess that nobody at the Authority liked the Showtime bid. Alun Chalfont and Peter Baldwin, to say nothing of David Mellor, had not gone out on a limb to keep the INR pop music channel away from FM, only to see it become 'songs from the shows'. However, there was nothing in its application to inhibit a licence offer, and INR1 was duly awarded to FNR on 4 July, subject to it securing its funding by midnight on Friday 16 August. The day before the deadline, FNR wrote to the Authority requesting an extension of four weeks to allow it to complete a new funding structure as "a joint venture ... between Hutchinson (Telecommunications) and a group of UK and EC investors".[691] The Authority met in emergency session the following Monday, and "decided not to grant this extension".[692] The offer to FNR was therefore withdrawn. The group had needed to raise £12m, given their ambitious business plan. Had they been more prudent in their proposals, it is wholly possible that the almost £10m they actually raised would have been enough to gain the licence.

The Authority next turned to Classic FM, with much the same offer as it had made to FNR – show us that your full funding is in place, and we will award the INR licence to you – but with a shorter deadline. However, there were complications. Golden Rose and GWR, suddenly given renewed hope, rushed off on what turned out to be separate tracks. GWR made over 100 presentations, "to the City, to trade investors, overseas, to other radio companies" aiming for the necessary £6m.[693] Golden Rose, meanwhile, had privately decided on a target of raising £10m – not just the £6m it needed to confirm its right to the INR licence – which would have allowed it to acquire local radio stations in addition to securing the national licence. This widened the rift between GWR and Maker's Golden Rose Group. In a meeting at the Dorchester Hotel, the assembled group of investors rebuffed the hopes of Golden Rose that the new company would take over Jazz and Buzz FM.[694] Once again at the eleventh hour, the group that Ralph Bernard had assembled – GWR itself, Time-Warner, Associated Newspapers and a last- minute investment from Sir Peter Michael – was able to deliver the necessary documents to the Authority at 11.40 am on 30 September, beating the deadline by just 20 minutes.[695] The funding was confirmed as being in place, and the Radio Authority awarded the first national commercial licence to Classic FM. Alun Chalfont and Peter Baldwin, supported by the now secretary of state David Mellor, had their FM classical station. Maker, who had been MD-designate of Classic FM in its original guise, chose instead to stay with Jazz FM, and it fell to Bernard to set up the station. When Classic FM launched almost a year later, it was to change the face of commercial radio.

The same claims could not be made for INR2, the AM licence open to all formats and therefore almost certainly destined for pop music. When advertised on 29 October 1991, it was accompanied by an acceptance from Alun Chalfont

691 Letter from Secretary of First National Radio to Peter Baldwin, Radio Authority, 15 August 1991.
692 *Radio Authority News Release* 19 August 1991.
693 Ralph Bernard interview op. cit.
694 *Media Week* 4 October 1991.
695 *Financial Times* 1 October 1991.

that the Authority would need "much more solid" financial information than it had gathered first time round.[696] Anyone who shared passions for radio and cricket would have known that the frequency was going to be a problem. This had been the medium wave back-up for BBC Radio Three, which the Corporation had surrendered with such private glee.[697] The Radio Authority forecast 85 per cent coverage of the UK in daytime, reducing to just 40 per cent as soon as darkness began to arrive. This was better than the skywave-afflicted Radio Luxembourg service on 208, but it shared some of those disadvantages. When the frequency had been used to broadcast Radio Three, it had been highly inadequate for classical music, especially as the wavebands became gradually more crowded and interference increased. The BBC's *Test Match Special* was a summer daytime programme, but even so it had some problems being heard adequately in many parts of the UK.

Once again a national licence was not after all awarded to the highest bidder. The Independent National Broadcasting Company (INBC) bid an over-the-top £4.101m. The Radio Authority understandably decided that the company would be unable to sustain the service over the eight-year licence period. On 2 April 1992, it awarded the licence instead to Independent Music Radio (IMR), a consortium jointly owned by TVAM plc and Virgin Communications Ltd, with the expectation that it would come on air late in that year. However, IMR almost at once asked to delay the launch. Bruce Gyngell wrote to Peter Baldwin in May to request "a flexible start date between 1st October 1992 and 1st May 1993".[698] He cited the large number of start-ups that year – Classic FM, Carlton TV and Meridian, GM-TV – the need for more audience pre-research, and transmitter problems. Gyngell even asked for a delay in actually issuing the licence so that the company could avoid paying the bid figure of £640,000 until they were ready to start properly. That went against the senior staff recommendation that the licence should be issued as originally planned,[699] but, following a meeting between Baldwin and Gyngell on 11 May, the Authority acceded. The INR 2 station, which was soon called Virgin Radio, began broadcasting on 30 April 1993.

Classic FM began broadcasting from its studios in newly fashionable Camden Town on 7 September 1992, and it was the making of commercial radio. This supposedly elitist station, introduced in the face of predictions of doom from all those who flattered themselves that they knew what was what, did more to refresh UK radio than anything since the start of ILR itself. Freed from any public service obligations other than public expectation, it had the chance to show what the market could do once set free in radio. It was going to prove its promoters right, and rout the doubters. It was also going to put commercial radio on to the map, helping it at last to break through the 2 per cent ceiling of advertising share and attain, for a while, real commercial success.

Classic FM's chief executive was John Spearman, a branding expert faced with what was for commercial radio an unique marketing task, while Michael Bukht was the station's first programme controller. As the launch PC for Capital Radio, he had shown his great skill in setting up radio stations, and had a particular passion for this format, which he shared with Ralph Bernard. Bukht asserted that "when

696 *Guardian* 30 October 1991.
697 See chapter 12.
698 Letter from Bruce Gyngell TV AM for IMR to Peter Baldwin, 12 May 1992.
699 *RA Paper* 32(92) March 1992.

I was at Capital we did four hours a week of classical music programming and we got bigger figures in London than Radio Three got nationally".[700] Bukht was brought in by GWR to make sure there was a format which would work, and he was in his element. Robin Ray was the music consultant – although there was to be a falling out and legal actions further down the road – but for Bukht "Classic FM was Capital come again but with slighter smarter clothes".[701] Easy and popular 'tunes' were given high rotation during the day; virtue was practised in the evenings with more substantial programmes, opera coverage and concerts, and news was a priority. The station quoted its achievements against Radio Three; in truth, its rivals were Radio Two and Radio Four, and it was soon easy to spot an archetypal group of listeners, identified as 'Radio Four plus Classic FM'. Best of all for commercial radio, where previously it was hard to find senior advertisers who listened regularly to commercial radio, Classic FM now became respectable and fashionable. If the chairman's wife was a fan, the chairman was that much more likely to direct some of his advertising budget towards radio. Listening to Classic FM became "the guilty secret of the chattering classes".[702]

In conscious homage to the launch of Capital Radio in 1973, Bukht opened Classic FM with Handel's anthem *God save the King*, from *Zadok the Priest*, followed by Weber's rondo *Invitation to the Dance*. This was Classic's equivalent of *Bridge Over Troubled Water* with which Capital had begun, and a clear statement that this was – during daytime at least – a pop station with classical tunes. A heavyweight line-up of presenters helped the station to establish a good profile among the quality press and its readers. Simon Jenkins was the political correspondent. Margaret Howard had been a BBC stalwart, as had Susannah Simons. Henry Kelly's Irish charm sounded the right note – even if his pronunciation sometimes caused ripples – while Paul Gambaccini was an outstanding and knowledgeable presenter of most types of music.

The BBC called it wrong. Challenged as it seemed in its heartland by new national competition, it decided to slug it out toe-to-toe with the new station in cheerful vulgarity, but the odds were always impossible. When Classic breakfast presenter Nick Bailey announced *Jesu Joy of Man's Desiring* and played instead the prelude from Previn's *Invisible Drummer*, Bukht and Ray simply laughed.[703] Radio Three listeners would have been apoplectic. The BBC, meanwhile, went "from the esoteric to the populist extreme ... weather reports, plugs for BBC programmes, interviews and fragmented music hardly to be distinguished from Classic FM",[704] but all the counter-cultural impulses were just too strong for success to be gained on these terms.

Classic won a remarkable four and a half million listeners in just a few months, while Radio Three dropped to two and a half million.[705] Classic FM was dubbed "an instant classic",[706] firmly established in the UK psyche, especially among opinion leaders. It had at once become "the sound of middle England, the voice

700 *Daily Telegraph* 6 June 1992.
701 Michael Bukht in conversation with the author, 2 February 2009.
702 *Sunday Telegraph* 25 October 1992.
703 *Observer* 20 September 1992.
704 Samuel Brittan in the *Financial Times* 11 January 1993.
705 *RAJAR* quarter 1 1993.
706 *Sunday Telegraph* 7 February 1993.

of the heartland". When "George calling from his car on the M4" won a magnum of champagne in a competition to find the link between Beethoven, Haydn and Strauss,[707] many were amused but few were surprised when the caller turned out to be housing minister Sir George Young.[708]

Middle England was in need of such comfort. Since 'Black Wednesday', 16 September 1992, when the pound fell ignominiously out of the European Exchange Rate Mechanism (ERM), the Conservative government was – in the words of sacked Chancellor, Norman Lamont, himself largely responsible for the situation – giving "the impression of being in office but not in power".[709] A new spirit was emerging in the country, as 'mondeo man' found himself courted by those who would do things in a 'modern' way. Classic FM, "Mozart for the masses",[710] caught this arriving mood perfectly. The new classless populace loved having the cachet of being classical music listeners without all the bother of understanding what classical music was really all about. 'The tunes without the hard work' might be a metaphor for how 'New Labour' was to be sold to a disenchanted electorate after the middle of 1994. Classic FM argues that exposing people to bits of music they like encourages mass listening to substantial classical music; as if its success somehow needed intellectual justification. It was as show business radio that Classic was terrific, and one of the media success stories of the age.

National radio elsewhere was going less well. Virgin AM had already delayed its launch, and the Radio Authority now determined to move more slowly than planned to advertise the third INR service, the speech-based network. That was partly a consequence of the continuing recession, and partly to allow what Alun Chalfont called "a kind of orderly progress in the development of the radio regime".[711] AIRC's chairman Stewart Francis was displeased. With AIRC now seemingly converted to the notion that "national stations offer a chance of attracting genuine new money to radio", he complained of the Authority that "if you have a plan you should stick to it". One benefit for the BBC was that they were able to re-launch struggling Radio Five as Five Live, ahead of its anticipated speech-based rival.

Virgin itself was now firmly Richard Branson's baby. He had bought out TVAM, and brought in David Campbell as chief executive, over the head of the previous MD John Aumonier. Campbell, a long serving Virgin executive who had previously run the group's television facilities company, and respected sales director John Pearson, were to be at the helm of Virgin throughout its early years. This was to be a rock channel, with an impressive line-up of presenters: Graham Dene from ILR, Richard Skinner from BBC London, Chris Evans from breakfast television and Tommy Vance from Radio One. It even offered Emperor Rosko doing two shows from Los Angeles, one of them live on Friday evenings. Its news was to come not from IRN but from a new supplier, Network News, which was emerging from the local Chiltern Radio Group based in Dunstable. It offered only modest audience guarantees to advertisers – wisely as it turned out – promising to deliver just 3.3 million adults.

707 The correct answer was 'emperor'.

708 *The Times* 30 December 1992.

709 *Hansard* Commons 9 June 1993 col 285.

710 Anthony Thorncroft in the *Financial Times* 29 August 1992.

711 *Financial Times* 16 September 1992.

Virgin, with its own transmissions on a questionable AM frequency, was up against Radio One on FM. Radio Luxembourg had finally shut down its English language medium wave transmissions in December 1992. You might have thought that the long courtship between European-based commercial radio and the UK, dating back to Radio Normandy in 1930, had ended at last with the arrival of a UK-based national popular music station. Not so. An unexpected competitor had begun broadcasting in September 1989 from the Republic of Ireland. Transmitted from a giant mast in County Meath, with studios in Trim, Atlantic 252 was a considerable success, despite being on long wave which most thought hopelessly old-fashioned.

Atlantic 252 offered a music format of high-rotation mainstream pop and rock music, with influences borrowed heavily from American radio. It mixed the best songs from the last few years along with the best songs from the top 40, in a format which it called 'today's best music variety'. Atlantic held a UK audience of between three and four and a half million adults right through until 1998, despite being almost inaudible in London and south east England. A legal station, licensed by the Irish government, it adopted voluntarily many of the rules which applied to ILR and INR in the UK – notably in terms of advertising content controls – joined the RAJAR research, and generally conducted itself as part of the legitimate commercial radio community. AIRC was happy enough to add its audience figures to those of INR and ILR, to demonstrate that commercial radio was overtaking the BBC in nationwide audience share. The reviving of Radio One, and the repositioning of Radio Two, eventually saw it off in 2002.

The precedent of Atlantic 252 on AM should have provided encouragement for Virgin when it launched on 1215 kHz on 30 April 1993. It had the benefit of a strong brand, a good line up, and a distinguishing music policy in its rock format. The first hour featured music from INXS – opening the station with *Born to be Wild* – the Cure, the Beatles, Dylan, Queen, Simply Red, U2, Genesis, Steve Miller and Elvis Costello; hardly underground, but not mainstream pop as offered by Radio One or ILR either. Virgin quickly built an audience closely in line with its original pessimistic predictions, hitting 2.6 million adults by the autumn of 1993.[712] That rose to just short of four million by the end of the following year, but the station thereafter stayed between around 2.8 and 3.5 million for the rest of the decade. A continuing feature was its chairman's public rubbishing of its frequency. The day before the launch, Richard Branson announced that he was meeting John Birt to discuss swapping 1215 kHz for the Radio Four FM frequency around 93.5 MHz. Branson's argument was that such a deal would save him bidding for a privatised Radio One, and held the unstated implication that he might otherwise use his influence to have it so privatised.[713] Not only was this baying at the moon, but it left the station's listeners and advertisers with the feeling that even Branson thought that what they were getting was second best.

Undeterred, Branson met Birt and Radio MD Liz Forgan in May 1993. When rebuffed by them, he took his case to the newspapers – who then, as now, found him the provider of good copy – and to the House of Commons National Heritage Select Committee in early July. The Radio Authority stirred itself into action on

712 *RAJAR* quarter 3 1993.
713 *Times* 29 April 1993.

29 June, with Alun Chalfont branding the proposal "unacceptable"; which, of course, it was.[714] The Authority had advertised a licence on AM, with governmental and parliamentary authority. That had been bid for in an open auction, and awarded to Branson and Gyngell's IMR. It was, or should have been, unthinkable to change the terms of the award retrospectively, leaving Virgin with a hugely valuable national FM popular music licence in flat contradiction of the legislation and public policy. Nevertheless, when David Frost became a major shareholder in Virgin Radio in July, buying the stake of venture capitalists Apax, there existed a formidable potential lobby, well positioned to press its case when the Radio Authority came to award new FM licences in London in 1994.

Meanwhile, another possibility arose. Additional FM spectrum in the 105 - 108 MHz sub-band was at last being released to the Radio Authority, and it was at its own discretion how this should be deployed. One option could be a further national service, an INR4 on FM. Inevitably, Branson and Virgin eyed up this opportunity, salivated, and made assumptions. They were tremendously keen to acquire this licence, but even at the hypothetical stage – before the wide consultation on the use of 105–108, and well ahead of any consequent decision – Virgin already found itself in a legal bind. The Authority in its shadow form had successfully worked to persuade the government that any INR licences, while awarded by highest bid, should have formats distinct from each other. This was a waveband-neutral requirement. As a consequence Virgin, which already held a rock music licence, could not bid for another.

The best idea the Authority's lawyers and those for Virgin could come up with was that Virgin, in advance of the INR4 auction, would have to commit to relinquish its AM licence whatever the outcome of the contest; that, unsurprisingly, it was reluctant to do.[715] Branson, however, was not averse to dangling a threat, if he thought it would help his case. He told a Coopers & Lybrand seminar that Virgin would not survive a national rival on FM, and was reported as saying that he would "consider handing back the AM licence". Together with an apparently unguarded comment that he "no longer listens to Virgin 1215" this seemed to be a gaffe-prone occasion.[716] Yet it served Virgin's purpose in ratcheting up the pressure on the Radio Authority very well, given the range of press stories generated by the Virgin public relations machine immediately afterwards. In the event, the new spectrum was deployed for regional and local radio only.[717]

The third INR licence, postponed amid the pressures of recession and ILR licensing, was finally advertised on 30 November 1993. The timing was still awkward. ITN, Capital and Classic were all said to have decided not to bid because of the uncertainty over INR4, a real loss to commercial radio.[718] INR 3 was to be 'predominantly speech', which the Authority interpreted as a minimum 51 per cent speech in any three hour period. To ensure the required diversity of format, the advertisement also forbade applications devoting more than 10 per cent of their non-speech time to either rock or classical music. On 15 March, the Authority received six bids, more than for any of the previous INR licences. Talk Radio UK

714 *Daily Telegraph* 30 June 1993.
715 *RA Paper* 16(94).
716 *Broadcast* 18 February 1994.
717 See chapter 18.
718 *Observer* 20 March 1994.

bid £3.82m; Newstalk UK £2.75m; Apollo Radio (the only application to specify music content – contemporary adult music) £2.27m; LBC £2m; First National Entertainment Radio £1.54m; and Jim Black Broadcasting, proposing speech-based programmes for 10–24 year olds, £1.04m.

Talk Radio was backed by US group Emmis Broadcasting, with radio stations in Los Angeles, New York and Chicago, and Australian broadcasting group Prime Television. Brought together by media venture capitalist Christopher Turner, it proposed mainly speech, with talk-back and phone-ins. Its MD John Aumonier had been squeezed out of Virgin by Richard Branson, and returned for INR3. Gillian Reynolds thought that Talk Radio offered "big money, but possibly too big to be serious".[719] Many commentators thought that the Radio Authority had a real choice to make; after all, INR1 and INR2 had both been awarded to the second highest bidder. That was wide of the mark. Both Showtime and INBC had been obviously flawed. Matt Baker's prediction was the most accurate: "Rejecting Talk Radio's bid would make a mockery of a bidding system based on the principle that highest wins".[720] The Authority duly decided in June that there was no impediment to the Talk Radio bid, and awarded the company the licence.

Talk Radio officially began broadcasting on St Valentine's Day, 14 February 1995. In a measure of what it was to offer, however, the first live broadcast happened from 10 pm the previous evening, a phone-in show hosted by Chris Ryder under his radio name 'Caesar the Geezer'. Along with Wild Al Kelly, Ryder was one of the 'shock jocks' which Talk Radio's overseas owners mistakenly copied from their home jurisdictions. Unfortunately for them, the broadcasting legislation still outlawed editorialising, and frowned heavily upon causing offence to public taste and decency. Fifteen complaints were upheld against the station between 14 February and 4 March.[721] It was twice fined by the Authority during 1995, £2,000 in July for three failures to provide logging tapes, and a further £5,000 in October for an offensive and blasphemous broadcast. Thereafter, the output was usually within the rules, but this meant neutralising the initial proposition of the station, which had been to shock and provoke. After a year, its audience settled at under 2 million, compared with almost 5 million for Five Live. Even the growing stream of advertisers returning to radio tended to fight shy of the low-audience, down-market environment offered by Talk Radio, and it faced considerable losses compounded by its high bid for the licence. The British radio public, brought up on the BBC Home Service/BBC Radio Four, and nourished by the genuine aspirations of ILR and then Classic FM, still knew better than to be fooled by the 'shock jocks'.

The Radio Authority decided against deploying 105–108 MHz for INR4, effectively meaning that there could be no further national commercial radio stations, by terrestrial transmission at least, until the arrival of the national digital radio multiplex in November 1999. However, there is a curious postscript. One of the frequencies allocated to the BBC was 225 metres long wave, which had been used for a while as a back-up to the long wave Home Service transmissions in the fifties. The Corporation decided it could let 225 go from its hoard, and it was

719 *Daily Telegraph* 19 March 1994.
720 *Broadcast* 25 March 1994.
721 *RA Paper* 14(95), 27 July 1995.

offered by the Department of National Heritage to the Radio Authority in November 1996. In February 1997, the Authority consulted the industry about the potential development of an INR4 service on long wave, but the outcome was clear. No one wanted it, or at least no one thought they could carry the high cost of transmission and electrical power, nor find suitable back-up frequencies to fill the gaps.

As 1995 dawned, INR meant Classic, Virgin, and the current version of Talk Radio. In Northern Ireland, IRA and Loyalist ceasefires in August and October 1994 marked the beginning of the end of the Troubles, although there was still some way to go. Sunday trading for retailers had arrived in England in August of that year, confirming the final secularisation of society. The National Lottery institutionalised the British fancy for a flutter, and started to transform the funding of culture and charities across the UK. Few seemed to care that it might take money from the poor to fund the pleasures of the affluent, but then this was a society which had largely shed its collective social conscience. This was Thatcher's Britain, her *apotheosis in absentia.* Commercial radio, shorn of many of its public service obligations, seemed to have little to fear from the new free market, and everything to gain. The ILR stations were to realise that too, and it is to their increasing success in the mid-nineties that this history now turns.

Chapter 16

Glad confident morning

ILR 1991 – 1993

While Independent National Radio was taking its first steps, ILR was now a lusty post-adolescent. When the bright commercial morning arrived on 1 January 1991, the network already included 126 ILR licences. During 1991, 19 new local licences were advertised, and even though the consideration of applications could not begin until June, one station actually began (legal) broadcasting before the end of 1991. Shetland Island Broadcasting Company had been the UK's most tolerated land-based pirate service. Travel news involved looking out of the window on to the harbour, to see if the weekday ferry from Jamieson's Quay Aberdeen had yet arrived. SIBC was run entirely by Inga Walterson and Ian Anderson, in a Norse version of what the Americans called 'mom & pop radio'.[722] When it began broadcasting as a properly licensed station in October 1991 it was as if it had never been away.

Seven more ILR licences were awarded in the first year of the new dispensation: for Lincoln (Lincs FM), Pitlochry & Aberfeldy (Heartland FM), Cornwall (Pirate FM), York (Minster FM), Kings Lynn (KL-FM), Blackpool (Radio Wave) and Salisbury (Spire FM). If their modest localness seems a rather mild way to enter the supposed free-market utopia of commercial radio, that was because many of the assumptions which had driven the IBA's Radio Division continued among the largely unchanged personnel who made up the Radio Authority. What was more, pre-existing ILR stations were not enjoying the commercial climate, and many of the new incremental stations were experiencing great difficulties. Often with little commercial experience, they were up against BBC local services in many areas, and frequently competing with established ILR stations. Gillian Reynolds as usual had the well-phrased measure of the times, in reflecting on the plight of the incremental companies. "Too many looked at the fat cat commercial radio had become, and thought there was going to be plenty of cream left in the saucer. Wrong. These are alley cat times in radio, the same as everywhere else."[723]

Operating an ILR station changed little to begin with, apart from companies having now to take responsibility for their own transmissions. However, the

722 This is not a derogatory term in the US, except perhaps among executives in huge radio groups. Jerry del Coliano, once professor of music at USC, argues that "The radio business used to be a conglomeration of mom and pop stores run by some of the most colorful people you could wish to work for ... mom and pop radio operators were good enough to build the radio sector into the kind of assets that eventually sold for hundreds of millions of dollars a piece ... They sold high. Now the consolidators are owning – low." (*Inside Music Media* 30 May 1998).

723 *Music & Media* 3 November 1990.

regulatory processes changed overnight, none more so that the evaluation of licence applications, which also revealed the underlying assumptions about what the Radio Authority wanted ILR to become. There was a new scoring system, evaluating five areas: general, including as sub-criteria the abilities of the applicant's chairman, directors, their history, staffing plans and management calibre; programming; audience projections and evidence of local support; finance, including "adequacy of income, credibility of outgoings ... plausibility of forecasts", plus funding structure and commitment of investors; and engineering.[724] Staff marked a number of sub-criteria under each main heading on a scale one to five, with the specific wording of the criteria and the weightings reviewed regularly, and changed at times to reflect the current priorities of the Authority members, who made their decisions after scrutiny of the applications themselves and the staff analysis, including the scoring.

This was a good 'rule-of-thumb' way to produce a short list for more detailed consideration, but it was in no sense a merely mechanical system for selecting licensees. From the start, David Vick expected that "a group scoring 83 per cent should [not] necessarily beat one scoring 81 per cent when a licence award decision comes to be made. All the scoring system is telling ... is that there are two applicants which, overall, deserve serious consideration as the potential licence-winner. It would then be for members to decide whether one applicant group has particularly attractive qualities which make it the preferred choice."[725] In practice, the Radio Authority's award of ILR licences was only in part more 'scientific' than the IBA's, although in my experience the staff marking was done scrupulously and with a keen sense of fairness pervading the process. However, the members kept to themselves the duty and right to make the final decisions, not infrequently surprising the permanent staff, and occasionally to their dismay. The top scoring applicant was by no means always the winner; on occasions the successful applicant had not even been scored among the top three. Over time, the perceived unpredictability of certain award decisions threatened seriously to undermine the credibility of the regulator.

The Radio Authority's secrecy about the details of its process made this worse. In December 1992, it considered in whether to make public in more detail the method it used for assessing applications, in the context of Alun Chalfont's personal recommendation to be no more explicit than the information already included in the licence specifications.[726] The legal advice from top city firm Allen & Overy – who had been the IBA's lawyers and the ITA's before that – was firmly against giving any detail or reasons. Allen & Overy's job was to keep its clients out of court, so far as possible, and it judged this was the best way to do that. Peter Gibbings was persuaded to lift the curtain slightly, after the *Newsnight* episode in 1999,[727] but it was not until Richard Hooper's chairmanship that the Authority's approach changed significantly. However, by that time its reputation as a secretive and supposedly capricious licensor was already established, and disappointed applicants spoke of the award process as a 'lottery'.

724 *RA Paper* 41(91).
725 Ibid.
726 *RA Paper* 1 December 1992.
727 See chapter 19.

The new style of programming regulation *post hoc* made itself felt quickly. The system of Programme and Advertising Codes, and promises of performance, operated as had been intended on an exception basis, usually reliant on complaints to trigger investigations. In its first year, the Radio Authority received 346 complaints dealing with programming and advertising. It upheld 33 programming complaints and seven advertising complaints, and imposed three fines for programming offences. Modest in amount, they shocked the companies not least as a new aspect of regulation, and served *pour encourager les autres*.[728]

The Asian station, Bradford City Radio, was fined £2,500 in March for breaching its promise of performance, by refusing to broadcast Afro-Caribbean or Afro-American music, amid concerns that it "was carrying programmes from Sunrise Radio, Southall, unabridged, including London traffic reports".[729] Mellow 1557 was fined in May for breaching its promise of performance. This followed a complaint from a former volunteer broadcaster on the station that local speech content, including news and community information, was being eroded. Radio Authority staff monitored a full day's output, which involved listening to the logging tapes requisitioned from the station, and found that Mellow was "4 minutes (or 33 per cent) per hour short of its speech commitment".[730] In a move designed "to give confidence to the community radio lobby that the Authority regarded community radio's contribution ... as important and worth protecting",[731] the Authority imposed a penalty of £1,250. Galaxy Radio in Bristol was fined £800 in September for failure to maintain and provide recordings of its output. Under the 1990 act,[732] each station was required to keep tapes of all its output for the past 42 days, an essential element in a self-regulating system, and the Authority quickly realised that it had to treat the failure to provide tapes in the face of a complaint as a breach in its own right. "Failure to maintain recording of output makes it impossible for the Authority to regulate licensees retrospectively on the basis of complaint."[733]

This was not the type of new, 'light touch' regulation which the radio industry had been hoping for, and it showed the Radio Authority still bound by the statutory requirements interpreted through the prism of its public service past. Some Authority members, especially those who had come from the IBA, also found it hard at times to let go of their prior assumptions about appropriate programme content. When the Authority met in December 1992, deputy chair Jill McIvor – who had been a member of the IBA – expressed concern over remarks made by the chair of judges, John Timpson, in a religious broadcasting awards ceremony that entries from independent radio did not match previous years in number or

728 The court martial and execution of Admiral Byng after the British lost the Battle of Minorca in May 1756 is referred to in Voltaire's novel *Candide* with the line "*Dans ce pays-ci, il est bon de tuer de temps en temps un amiral pour encourager les autres*". ("In this country, it is wise to kill an admiral from time to time to encourage the others.") That was very much the Radio Authority's approach to applying sanctions for breaches of the Programming or Advertising Codes. It could probably have found dozens of station misdemeanours equivalent to any particular one under investigation, but used the making of an example to keep order across the system. This worked surprisingly well, helped by the closeness still between those working at different stations.

729 *RA Minutes* 4(91) of a meeting on 19 March 1991.

730 *RA Paper* 45(91) 30 May 1991.

731 *RA Minutes* 7(91) of a meeting on 11 June 1991.

732 Broadcasting Act 1990, section 95 (1)(2)(a).

733 *RA Paper* 60(91) 16 August 1991

quality.[734] Had the new regulator felt comfortable with its different role, it would have said simply 'so what?' However, Paul Brown, now deputy chief executive as well as head of regulation, took issue with Timpson privately, and quickly researched the religious output on ILR. His findings were remarkable for how much, rather than how little, of such output was included. Seventeen per cent of stations had daily religious programming, while a further 38 per cent broadcast weekly religious output.[735] That might have reflected their sense that this was popular programming; more likely, many of the companies had also not yet discarded this aspect of their own public service past.

With the removal of other of the IBA's minor prohibitions, such as on competition prize limits, many had expected American-style commercial radio from the very start. One newspaper predicted that "competitions with prizes worth millions of pounds will soon be taking local radio by storm. Cars, yachts and even houses will all be up for grabs as commercial stations fight for listeners."[736] Given the impact of recession and high interest rates, Capital's Richard Park was less sure: "I don't think British radio will go completely down the American track. I think it will be done with fun and intelligence." Nevertheless, Capital offered its listeners £25,000 in a Birthday Bonanza promotion. Clyde offered a house, while BRMB MD Ian Rufus thought his station's offer of a £53,000 detached residence "was more expensive and really very nice".[737] Jazz FM of all stations – struggling to survive and ever unrealistic – launched a breakfast-time competition, *Hot Notes,* with an ultimate prize of £1m.

Jazz FM was the most tendentious of the IBA's incremental stations. Launched in March 1990, it found the combination of recession and elusive advertising revenue more than challenging. The station struggled to build what would in the terms of that time have been a credible audience, and suffered inevitable attacks from self-appointed jazz experts. The company went through several rounds of re-financing before being acquired by David Maker's Golden Rose Broadcasting in October 1991. That involved a significant shift in format away from 'proper' jazz, and the station lost founder Dave Lee, leading jazz DJ Peter Young, and most of its credibility with the jazz buffs. Perhaps that did not matter. David Maker thought not. "The promise of performance is not just about jazz ... the problem is, what is jazz?"[738] The Radio Authority also determined that the station was not operating outside its licence, leaving those who regarded blues and soul music as anathema with no further recourse.

The most successful of the incremental stations were mostly those which were for ethnic minorities, and ran themselves as effective businesses, too. Sunrise Radio in West London and Choice FM in Brixton quickly made themselves into legends for their ethnic minority audiences, and more widely within ILR. They demonstrated that there was a particular type of community; people with shared values and distinct from those around them. This became known as a 'community of interest', in contrast with a community in a single locality. Sunrise was to be a

734 The Sandford St Martin Trust, endowed by the Wills family and administered by the Church of England, has run regular awards for religious output on both television and radio since 1978.

735 *RA Paper* 95(92) 30 December 1992.

736 *Today* 29 October 1990.

737 *Independent* 17 April 1991.

738 *Evening Standard* 10 October 1991.

popular channel on satellite radio just as much as in Southall, and it greatly extended its territorial coverage by winning an AM licence in 1993. Choice took its Brixton formula into north London also, when it eventually acquired a second licence to extend its coverage in 1999.

Small-scale stations which managed to survive the shock of setting up in reasonable shape, such as Mellow Radio in Tendring, Isle of Wight Radio and KCBC in Kettering, were starting to show how new models of commercial radio might be made to work; all three, notably, on AM rather than FM for many years. Mellow 1557 ran on five full-time staff, and took a sustaining service from its parent, Kent-based ILR Invicta, from 10 pm to 6 am. KCBC ran a classified advertising spot three or four times a day, charging just £15 for an advertisement. This was closer to the community radio model which the IBA had envisaged for its incremental stations, but it was still avowedly commercial. Multi-tasking was the order of the day, with presenters doubling as sales people and everyone able to manage increasingly solid-state digital studio equipment. Even the smallest of the original ILR stations had full time engineering cover; Radio 210, during my time there in the early eighties, employed three full-time engineers. The new small stations, relying on new solid state equipment, just used call-out technical support.

The incremental stations were consciously thought of as experimental, and some of the experiments were bold enough not to succeed. One of the most challenging problems for the new Radio Authority was what to do with the Haringey frequency on which London Greek Radio and WNK were both broadcast, in an uneasy pattern of alternation. The former station was underwritten by the wealthier members of the Greek diaspora in London.[739] The latter, eventually owned by the Midlands Radio Group, was losing its shirt (despite the efforts of its hugely committed founder, Joe Douglas). The Authority was presented with a demand for a second frequency, to allow each station full-time transmission, but none was able to be cleared and allocation without advertisement could raise many issues. The Authority had hoped to broker a tripartite meeting, to re-arrange the hours of broadcasting in an attempt to help rescue WNK, but Midlands Radio refused. Group MD Ron Coles wrote to Peter Baldwin that it was no longer prepared "to struggle on with WNK minimising the losses until the licence comes up for renewal ... we have opted for immediate advertisement, recognising the risks".[740] The Authority decided in July, nevertheless, not to offer a second frequency[741] and WNK eventually resolved the dilemma by closing just before the re-advertisement of the licence.

WNK was not the only station near the brink of extinction. FTP had changed to become "the harmless, empty-sounding Galaxy",[742] when bought out by Chiltern Radio, and Sunset Radio was in its long descent. Other incremental stations were failing too. The licence for East End Radio in Glasgow was revoked on 30 August 1991, when it transferred ownership to a third party without the Authority's

739 Some years later, when I visited London Greek Radio in my capacity as chief executive of the Radio Authority, I was welcomed to a lunch with all the directors of the company, who seemed to represent every significant Greek business in London. The excellent buffet comprised all the Greek-style foods supplied by a company owned by one of the directors to the UK's major supermarkets; humous, dolmades, taramasalata, olives, pita bread and the like.

740 Letter from Ron Coles, Midlands Radio to Peter Baldwin, Radio Authority, 14 June 1991.

741 *RA Minutes* 8(91) of a meeting on 4 July 1991.

742 *The Voice* 19 February 1991.

approval. There was enough outrage, real and synthetic, in Glasgow for future first minister Alex Salmond to argue fruitlessly that "notwithstanding technical breaches in applying for the licence, of which I am sure there were some, the key thing is that the station remains on air".[743] Within six months, on 7 February 1992, Airport Information Radio, which provided travel information for Heathrow and Gatwick airports, ceased broadcasting, and handed back its licence. In a fierce recession, and given the experimental context, none of these failures were major problems in themselves, but they showed how much of the Authority's attention – like that of the IBA's Radio Division in its latter years – was being devoted to the small-scale services. The new programme for licensing emphasised that still further. The Radio Authority was in charge of commercial radio, but throughout its life it kept its inherited public service assumptions close to the front of its collective mind.

Meanwhile, on 10 April 1992, John Major confounded opinion poll expectations to win another Conservative general election victory. His majority of 21 seats looked tricky to manage – and it was made almost impossible by party divisions over European policy and membership – but Major managed to hold on for a full five-year term. For commercial radio, paradoxically, the effect of having a government unable to attempt major legislative reforms was to preserve unchallenged the market-directed changes contained in the 1990 act, and even to leave legislative time for some further tweaking upwards of the ownership limits.

The mainstream 'heritage' ILR stations were still dominant within the radio industry. Capital Radio plc made a profit of nearly £16m in 1990, but the going was tough for many as the advertising recession continued. Some public radio companies were looking around for reasons other than the recession to explain to their poor profit performance. In a speech in Manchester, Radio City MD Terry Smith attacked the introduction of promises of performance, limiting the established ILRs, while the incremental stations were being allowed to shift as economic pressures threatened their survival. He cited "black ethnic stations, for example, becoming straightforward top 20 operators for most of the day".[744] On a falling turnover, City's profits were well down, and in March 1991 it replaced its failing City Talk service with music on AM, and cut fourteen journalist posts. The failure of Smith's efforts to provide talk radio on AM, in imitation of the successful trans-Atlantic pattern, was sad for innovation in ILR; for the stock market, the cuts were not enough to save the whole company's independence. Emap, previously most known for regional newspapers as East Midland Allied Press, but also the publisher of *Smash Hits* and *Q* magazines, paid £10.7m for the radio company, acquiring in the process a 50 per cent stake in the radio sales house BMS. Emap already owned minority stakes in Kiss, East Anglia Radio, and 16 per cent of Trans World Communications, so it was well placed to make the purchase of Radio City "the cornerstone for further expansion into radio".[745] Consolidation was seen as a solution by others, and Capital and the GWR Group merged their interests in DevonAir and Plymouth Sound in July 1991.

The recession got steadily worse for ILR. In the year to September 1991 its

743 *Scotland on Sunday* 1 September 1991.
744 *Media Week* 26 January 1990.
745 *Investors Chronicle* 17 May 1991.

net advertising revenue fell by 10 per cent, to just over £112m.That was in line with advertising trends as a whole in the UK, but radio was still stuck at around 2 per cent of national display spend. There was some improvement at the start of the following year, but the third and fourth quarters saw national sales tumble again, falling by 10 per cent between April and June alone. Capital's profits for 1992 were down by a third over 1991, but they were still nine times as big as the rest of ILR's aggregate profit. Only two other companies, Clyde and Metro, made more than £1m profit before tax in 1992. Several of the larger players, including Midlands, GWR, Chiltern and Southern, were said to have expanded too quickly on the expectation of a swifter recovery.[746] Many stations were in the red, with only Fox and Wyvern bucking the trend among the smaller stations. Most diversifications had failed; Radio City's Beatles Museum had been a major factor in its downfall. Only West Sound in Ayr stood out as an exception. It had invested in nursing homes, which soon dwarfed the earnings possible from one small ILR licence. Anxious for savings, AIRC once again took up the fight against PPL, making reference to the Copyright Tribunal in June 1992.[747]

In 1992 also, LBC's troubles occupied centre stage once more. The split between FM and AM in 1989 had not worked well for LBC, losing it audience all round. It was only with the impact in 1992 of new MD Charlie Cox, an engaging Australian motor racing enthusiast, that the station had recovered its previous listening levels, and began to reverse the down-market trend in its audience composition which had been unattractive to potential advertisers. The news service and the radio station both operated from Crown's premises in Hammersmith, with the former paying £500,000 annually for the privilege, which served to prop up LBC. There was tension between the ILR companies and IRN, which at one stage threatened the future of IRN. AIRC Chairman Stewart Francis warned on 17 February 1992 that "in addition to Capital, one or two other stations/groups are thinking of pulling out of Newslink and taking an alternative news service". [748] In March 1992 agreement was reached for Crown Communications to sell all but 10 per cent of its stake in IRN, and for LBC and IRN to be decoupled. However, by July Crown was in serious financial trouble, dragged down by a costly involvement in a French radio network, RFM, to the point where chairman Christopher Chataway was talking of a "critical problem".[749] When the French broadcasting regulator blocked the sale of RFM to rival group NRJ, Crown reached the end of the road.

Crown went into receivership in January with debts of £16m.[750] LBC had been in play for months, with talk of ex-financial presenter Douglas Moffitt mounting a takeover with backing from the City. In the event, the LBC licence passed to Chelverton Investments, run by Matthew Cartissier and John Porter. Chataway remained chairman of LBC for the time being, but was succeeded by Lady Porter in January 1993. By now, the re-advertisement of the London news licence was imminent, and the stage was set for the next ILR drama. The re-advertisement of LBC's licences is covered in the next chapter.

746 William Phillips in *Broadcast* 16 October 1992.
747 See chapter 13.
748 *Broadcast* 13 March 1992.
749 *Broadcast* 3 July 1992.
750 *Financial Times* 29 January 1993.

There were plenty of problems in national politics and the economy in 1992. The UK faced a fully-blown sterling crisis, and on 16 September, 'Black Wednesday', surrendered to the currency speculators. Major's authority never fully recovered during the remaining four and a half years of his second premiership. The year 1992 also began the years of scandal and sleaze for Major's government. The most prominent casualty was broadcasting secretary of state, David Mellor, who was by then heading up the Department of National Heritage, some said specially created for him by his friend John Major. Mellor was the subject of a tabloid exposé on the basis of an alleged adulterous affair. Allegations about a holiday as the guest of a Palestinian Liberation Organisation official's daughter forced his resignation in September 1992.

A combination of press prurience and public appetite for undoing establishment figures prevailed through these years. In 1991, right at the start of the Radio Authority, its chairman, Alun Chalfont, took the step of resigning as a member of the ITC, when allegations arose surrounding his position as a non-executive director of the public relations group Shandwick plc, which had entered into a contract with an applicant for an ITV licence. In a statement to the Radio Authority on 11 June, he explained that he had been unaware of that at the time, and also "unaware that a subsidiary of Shandwick ... had entered into a contract with Spectrum Radio to provide public relations services".[751] Spectrum found another public relations adviser, and the matter was therefore resolved. This miasma of veiled allegation was to return to shroud the Radio Authority on occasions throughout the decade, as it was to bedevil the Major government to the end of its days and beyond.

Meanwhile, back in ILR, a number of the incremental stations continued to give the Authority problems, but none at this time more than Sunset Radio. The tortuous process by which the Authority sought to regulate the company illustrates the amount of work and complexity which was generated by even the smallest companies when problems arose. There had been difficulties surrounding Sunset from the start, and its board had been summoned to meet the Authority in December 1990, even before the licence was granted. In December 1992 the Authority was considering whether the company's directors were fit and proper persons, triggered by a series of accusations made by a freelance journalist.

The Authority held a special meeting on 12 January 1993, and met again on 26 January when it considered whether to revoke the licence. By February, John Norrington, the secretary to the Authority, was of the view that "Sunset Radio appears to be intent on provoking the Authority" and recommended that a fine should be imposed for failure to provide a declaration, with a clear statement that non-payment would lead to the revocation of the licence.[752] Emotions were running high around the regulators' table. Deputy chair Jill McIvor asked for her dissent from one decision to be recorded formally, while Richard Hooper "expressed his disquiet".[753] Sunset resolved the matter eventually when its electricity supply was disconnected on 27 August, and the company was formally wound up

751 *RA Minutes* 7(91) op. cit.

752 *RA Paper* 16(93) 24 February 1993.

753 *RA Minutes* 7(93) of a meeting on 6 May 1993.

on 3 September. A minor station, but it had dominated a whole series of Authority meetings, tied up staff time and divided the members.

Regional licences represented the next step in introducing competition for the heritage ILR stations. This was intended by David Vick to produce a three-level approach to ILR: small scale local stations; the traditional ILR stations (now often called 'heritage' stations); and wider regional licences. The opportunity to do this at an early stage had come about by chance. Since the Radio Authority was required to do its own frequency planning once separated off from the IBA, it decided initially to contract this out to NTL, and thus to its former colleagues within the IBA. A small team headed by Andrew Woodger did this work. When Vick and Woodger were chatting over the Shadow arrangements, Woodger asked him "do you realise how much you could do?" – meaning, within the 2.2 MHz of FM spectrum allocated for INR1.[754] This was because Classic FM coverage was not statutorily required to be 'universal' – the BBC standard – but only to reach areas where transmission would be economic, leaving space to fit in some quite wide coverage local stations. As a consequence, the Radio Authority was able to announce in September 1982 its intention to advertise five regional local licences, designed to come on air during 1984, each transmitted on frequencies 'left over' from the Classic FM sub-band. These would encompass the territory of two or three conventional ILR services, and be the largest-coverage licences outside Greater London.

Regional licences were duly advertised for the Severn Estuary, the North East, the North West and the Midlands in England, and for Central Scotland. The focus here was to be on broadening choice, following the direction taken in offering incremental licences, and the assessment of applications was weighted accordingly. The Authority made clear that it would not regard favourably an applicant proposing to hold all five licences, together with a Greater London licence, broadcasting a common programme format. It was keen to avoid a quasi-national service being set up, which would not have to incur the same obligations, such as payment of a cash bid, as the INR services. The steer was unambiguous; these were to be specialist services, not mainstream pop, offering diversity of output and providing plurality of ownership.

The implication that new entrants would be welcome was also clear, certainly once the awards started to be made. They attracted plenty of interest – 13 applications were received for the West Midlands and 11 for the North West – and the award decisions bore out the original intentions. Galaxy Radio was to provide dance music to the cities and towns around the Severn. A second Jazz FM service was licensed out of Manchester for the North West. Two services promised a high proportion of speech output amid a different music style: Century Radio, with easy-listening and country music in the North East; and Scot FM, offering adult contemporary music and a substantial proportion of speech in Scotland. With Heart FM providing soft adult-orientated music for the West Midlands, this clutch of stations introduced relatively new owners to ILR. After its success with Century, Border Television seemed for a while to be a possible major radio player, but lacked sufficient expertise. With Heart FM, though, Chrysalis took its first steps towards a significant position within ILR.

754 David Vick in an interview with the author, op. cit.

The new regional tier was unpopular with many of the heritage ILR companies. Jimmy Gordon had asked the Authority to delay their introduction,[755] while AIRC chair Stewart Francis, in a heated debate with David Vick and the BBC's Mark Byford on Radio Four, described it as "a tier too far".[756] Brian West for AIRC worried that the Radio Authority "could end up destablising the whole industry" by introducing competition. Douglas McArthur at the new Radio Advertising Bureau (RAB), on the other hand, saw national and regional stations as a chance to attract new advertising sources. The new stations enjoyed varying fortunes, often tempted to drift further than the Authority would permit from their 'specialist' programme remit. Century was fined £2,000 in November 1994 for failure to meet its promise of performance, specifically a shortfall in speech content, while Galaxy was to find that the decline in the fashion for 'dance' music presented problems in keeping to the licence terms. However, all the stations apart from Jazz – a not unfamiliar refrain following the London experience – built sizeable audiences and looked to be highly viable, at least once the advertising market recovered.

One effect of adding new, large-coverage stations was to open up once again the question of the ownership rules, just after they seemed to have been settled by the 'points' system. For example, should a company holding a local licence be allowed to hold an overlapping regional licence? And how could the Authority prevent regional local licences using common formats coming under single ownership, and possibly creating a quasi-national network by the back door? The Authority concluded that it could not set non-statutory limits on the ownership of regional licences, or on their programme formats. It thus inadvertently demonstrated that the ownership rules were by no means a fixed quantity, but could be amended by legislation, or tuned on the advice of clever lawyers, opening up a running issue for the rest of the Authority's existence.

Other unanticipated consequences flowed from the introduction of specialist regional licences. They took away the ground from under the feet of those in the 'heritage' stations who might have maintained voluntarily the old public service model of *independent* radio for rather longer. Losing their monopoly status, and facing new commercial competition, they had to cope with the freeing up of the radio market, which they themselves had set in train at the Heathrow Conference. The Radio Authority, needing to distinguish between different genres of popular music, had also opened a box which was not to be closed again. This had regulatory consequences, since it became necessary to ensure that stations were fully observing the music obligations in their promises of performance; and market implications, as companies increasingly flirted with music-based branding, taking them further towards an ever more commercial model.

The process of advertising and awarding new ILR licences, mainly designed to fill in the 'white space' on the radio map, proceeded apace. This rarely made national headlines outside London, but here and there generated fierce competition and local controversy. The events surrounding the four new Greater London licences, and the London re-advertisements, are covered in the next chapter. Even without those controversies, however, the workload for the Authority staff and

755 Letter from Jimmy Gordon, Radio Clyde, to Alun Chalfont, Radio Authority, 31 August 1992.
756 *The Radio Programme* BBC Radio Four 28 June 1992.

members was formidable, and the licensing achievement remarkable. In 1992 alone, the Authority awarded 16 licences: for Ceredigion (Radio Ceredigion); Paisley (96.3QFM); Cheltenham (Cheltenham Radio); Barnstaple (Lantern FM); Ludlow (Sunshine 855); Montgomeryshire (Radio Maldwyn); Jersey (Channel 103); Guernsey (Island FM); Morecambe Bay (The Bay); Harlow (Ten17); Slough, Maidenhead & Windsor (Star FM); Alton (Wey Valley Radio); Colchester (SGR); Carlisle (CFM); the North Wales Coast (Marcher Coast FM); and Weymouth & Dorchester (Wessex FM).

When Spire FM won the Salisbury licence in December 1991, it became a model for a new breed of ILR stations, 'small scale' but commercially robust. Spire's licensed area covered just 55,000 adults, and was backed by the local community with venture capital group 3i providing almost half the necessary capital. Run by Chris Carnegy, who had just resigned as programme controller of 2CR in Bournemouth, it set much-imitated standards for low staffing, multi-tasking and making the most of technology. It remained closer to the *independent* radio ideal than most of the much richer and better resourced *commercial* stations. In doing so, it raised the sights of other small-scale operators, including those trying to make a precarious living in less favourable localities.

The two Channel Island licences took the Radio Authority literally into new territory: the States of Jersey, and the States of Guernsey & Alderney. The Authority found itself in effect working for, and with, two new parliaments. It was one of the pleasanter duties of regulation to fly to visit the island authorities, to be met with a warm welcome, some searching questions, and – if time luckily allowed – lunch at the yacht club in St Helier or St Peter Port. Less cheerfully, in both Glenrothes & Kirkcaldy and Pembrokeshire none of the applicants sufficiently met the requirements of section 105 of the Act, and no awards were made, as happened also for Dunfermline the following year.

The most improbable localities seemed to generate the fiercest competition for licences. Five groups applied for Cheltenham and four for Paisley. In west Wales, the two rivals for Ceredigion (Cardigan Bay) represented what was to become a familiar dichotomy for small scale licensing. Blaca FM was a subsidiary of Swansea Sound, a well established ILR company looking to expand. The board of its opponent, Radio Ceredigion, read like a local *Who's Who*, including Ifan Edwards, who was chair of the Wales Tourist Board, and Elvey MacDonald, who headed the National Eisteddfod. The Authority strictly followed the totality of the section 105 criteria, without giving any undue weight to the financial prospects. It awarded the licence to Radio Ceredigion, which was hard pressed to make a go of it. The established company felt rejected, and that its chances of expansion were being limited.

The Radio Authority was now fully into its stride. In May 1992 it had moved out of the old IBA building in Brompton Road, opposite Harrods, to new premises within Holbrook House between Covent Garden and Holborn. The Authority was encouraged by the high number of applications for new ILR licences. Despite objections from the lobbying group, the Voice of the Listener and Viewer (VLV), that this was an "insane rush to license new competition"[757] for the BBC, 14 new

757 Jocelyn Hay, Voice of the Viewer and Listener in a letter to *Broadcast* 30 October 1992.

services were licensed in 1993. Along with the five regionals, they ranged from Londonderry to Inverurie, and from Scarborough to High Wycombe.

Among the new local stations was the first self-styled Christian station, in High Wycombe. Starting as Eleven Seventy in December 1993, it was chaired by comedian Roy Castle. Station manager Mark Austin claimed it was "a revolutionary idea but we are operating on sound principles – and we have got God on our side".[758] The station was to have 12 full-time employees "of a Christian ethos" and to be non-profit-distributing. Ten other new licences followed in 1994, and a further six in 1995. All in all, in its first five years the Radio Authority increased the number of ILR licences in issue from the 126 inherited from the IBA – just 79 if FM and AM services are not counted separately – to 177 when the process paused briefly for breath in 1996. Given the increasing role for short-term licences[759] and other non-commercial services, this was a time of tremendous expansion for non-BBC radio.

As ILR reached its twentieth birthday in October 1993, across the radio industry as a whole things were starting to look up at last. The Radio Authority celebrated with a lecture on the therapeutic effects of music at the Merchant Taylor's Hall in the City of London, given by Professor Robertson and attended by Prince Edward, and then a Grosvenor House dinner graced by a speech from Lord Attenborough. If in both events the hand of Capital's founding fathers was undisguised, it was, after all, Capital's birthday too. It had won probably the largest audience of any metropolitan commercial radio station in the world.[760] Audience numbers were rising, up 2 million in the year for all commercial radio, to a weekly reach of more than 26 million.[761] Revenue was also starting to recover, as the advertising recession at last reached its end. The nationwide display advertising market was already recovering, with national display advertising rising by 10 per cent to £5,794m in 1994.[762] The impact of Classic FM was opening up revenue opportunities for ILR as well, to an unprecedented degree. The introduction of the new RAJAR audience research in the autumn of 1992 – shared with and endorsed by the BBC – gave radio further credibility among advertising agencies.[763]

The last and crucial element in the improving fortunes for commercial radio was put in place with the establishment of the Radio Advertising Bureau (RAB) in 1992. AIRC had various attempts at getting a widely-supported marketing arm to promote radio, going back to the Radio Marketing Bureau in 1982, but none had succeeded. This time, the major companies, brought together by the quiet persistence of Metro's Neil Robinson, made sure that the necessary commitment was there, and not undermined by the national sales agencies or others. The RAB's highly intellectual Scottish managing director, Douglas MacArthur, whose appointment had been championed by Jimmy Gordon, told the Radio Authority in October 1992 that, through the efforts of RAB, radio could expect to double its national advertising revenue by 1996.[764] National advertising revenue for 1992 was

758 *Daily Telegraph* 8 January 1993.
759 See chapter 23.
760 *Funding Universe* www.fundinguniverse.com
761 *RAJAR.*
762 *Radio Authority/RAB.*
763 see chapter 6.

£67m. In 1996, it exceeded £172m. MacArthur's approach was to promote the UK commercial radio network to the large number of national advertisers who were still not including it in their advertising schedules, by extensive use of research data, case histories and general marketing information, promoting the medium rather than any sectional interests.

As a symbol of its new style and success, in July 1993, commercial radio launched a new network chart show. Named for its new sponsor, the *Pepsi Chart Show* was hosted by Neil 'Doctor' Fox from Capital's London studios. At different stages, between 80 and 110 ILR stations carried the network show, scheduled against Radio One's flagship programme. Produced by Unique Productions, with a mixture of chart music and a fun style of presentation, the *Pepsi Chart* became the most listened-to chart show in the UK, enjoying audiences of 3.6 million – at its peak almost a million more than Radio One's *Chart Show*. It was now all going so well. Everything at last seemed set fair for a glad, confident morning for commercial radio.

764 *RA Minutes* 22(92) of a meeting on 1 October 1992.1

Chapter 17

Awards and re-awards

Licensing in London and beyond, 1993 – 1994

In the first quarter of 1993, the UK economy just escaped out of formal recession, with a rise in GDP of 0.2 per cent. The public were allowed to tour Buckingham Palace that summer for a fee of £8. Yitzhak Rabin and Yasser Arafat shook hands on a peace deal for Israel/Palestine in September, sharing the Nobel Peace Prize the following year but little else, and London and Dublin concluded a pact to bring peace to Northern Ireland. Three national commercial radio channels and some 170 local commercial stations were offering listeners nearly 30,000 hours of programming a week. The audience for commercial radio had reached 26.4 million adults that year, with commercial radio taking nearly 43 per cent of all radio listening in the UK.

For the radio historian, however, these were the years of licence awards, re-awards and non-awards, and their accompanying controversies. There are four separate elements in this part of the tale: the re-advertisement between 1992 and 1995 of the existing licences held by ILR companies; the process of re-advertisement of the London licences held by LBC, Capital, Jazz, Melody and Kiss, in two sets in the Autumn of 1993; and the four new Greater London licences awarded in October 1994. A continuing theme throughout these and later years was what all this meant for the licence award process of the 1990 Broadcasting Act, and the position of the Radio Authority as the licensing body.

Re-advertisement of ILR licences

In October 1992, shortly after Classic FM launched to the sound of Handelian trumpets, the Radio Authority began the process of re-advertising all the existing ILR licences. 126 licences had to be re-advertised, and either re-awarded or replaced before they expired on dates up to 31 December 1996, a formidable task for regulator and industry alike.

'Advertisement' meant exactly that. The Authority placed press advertisements in national, trade and relevant local publications, inviting companies who wished to apply for existing licences, including of course the existing ILR franchisee, to submit detailed application documents. Where the existing franchisee operated both an AM and an FM station, these were advertised simultaneously but as two separate licences, which might nevertheless remain jointly owned.[765] The Authority had taken the view that it was not empowered either to prescribe or proscribe programme formats. The awards would be made on the four criteria in

765 *RA Paper* 14(92) 24 February 1992.

section 105 of the act – viability, catering for local tastes and interests, broadening choice, and local support – plus, in theory, the broader requirements of section 85 regarding quality.[766]

The scoring system took on greater significance. The Authority decided that it would not consider new applicants scoring less than 60 per cent unless there was a particularly weak incumbent,[767] and would – all else being equal – expect to re-appoint incumbents where their competitors failed to exceed 70 per cent scores for one or both wavebands.[768] Interviews would be the rare exception, not the rule, and in practice were only used when the decision was unusually close, and where an incumbent was under threat. For the re-advertised licences, as for new licensing, one Authority member was designated to work with the staff to oversee the process and help agree a shortlist.[769]

Much of the re-advertisement process was uneventful. It began with the re-advertisement of North Sound's FM and AM licences for Aberdeen in October 1992, which were retained by the company, and ended with the re-award of Horizon's FM licence for Milton Keynes in February 1996. Most of the incumbents were unopposed, and all but nine licences were re-awarded to the existing operators. Much of this was straightforward. However, in instances where licensees were not re-appointed, the consequent work for the Authority and the impact on each local radio service was substantial.

In November 1993, the two Exeter/Torbay licences held previously held by DevonAir were awarded to Gemini Radio. The Authority saw this award as turning primarily on the issue of broadening choice between separate FM and AM services. DevonAir's original re-application proposed largely the same output on both wavebands; the new applicant Gemini offered distinct services. Realising its mistake too late, DevonAir tried to offer new proposals after the closing date, but all legal advice showed that the Authority could not accept that. It was, therefore, on "a rigorous interpretation of the statutory criteria" a straightforward decision.[770]

Already off the air, Sunset Radio lost its licence to Faze FM. In 1994, Buzz FM lost its Birmingham licence to Choice FM, which was to provide a service for black listeners in the 'second city' following its notable successes in London. The Leicester licence catering for the Asian communities went from Sunrise East Midlands to Sabras Sound. In Bradford, the Authority summoned for interviews in London both the existing station – also Sunrise – and its challenger, Rainbow FM, before deciding narrowly to leave the licence with the incumbent.[771] In 1995, the two licences held by Allied Radio for Guildford were lost to Surrey and North East Hampshire Radio, which ran them as The Eagle and County Sound.

Doncaster presented a new slant on this process. Although a separate licence area, the locality was served by shared programming from Radio Hallam in Sheffield and Rotherham. Lincs FM applied against the incumbent to run a service with stand-alone output for Doncaster, and the Authority had to weigh this against the popularity of the existing shared South Yorkshire service. When its bid failed

766 *RA Minutes* 21(92) of its strategy conference on 4 September 1992.
767 Ibid.
768 *RA Minutes* 24(92) of a meeting on 3 December 1992.
769 *RA Minutes* 1(93)of a meeting on 7 January 1993.
770 David Vick in an interview with the author, 24 June 2008.
771 *Radio Authority Annual Report* 1994 p 12.

in 1995, Lincs unsuccessfully sought a judicial review of the Authority's decision.[772] Not least of the odd features of this episode was that the long and expensive judicial process, which required disclosure of documents needed by the courts, did not have the effect the Authority had feared of making public cherished secrets about the award process.

London licences round one – LBC

In June 1993, the Radio Authority invited applications for eight London-wide licences, four on FM and four on AM. Six of these were held by existing licensees: LBC with both FM and AM; Jazz; Kiss; Melody on FM; and Spectrum on AM. Capital Radio's two licences, FM and AM, only expired in 1995. The whole process attracted a great deal of media interest, with the *Guardian* talking about a "radioactive war of the airwaves".[773] To judge from the media journalists, there was some real appetite for change, summed up by Torin Douglas: "If the RA cannot afford to take risks in London, where listeners have two dozen stations to choose from, where can it?"

None of the licensing outcomes, before or subsequently, had anything like the depth-charge effect of the Authority's decision on the London News licences. LBC and IRN had been at the heart of the ILR project since its inception. IRN had enjoyed some notable successes, starting with its leading role in the parliamentary broadcasting experiment in the summer of 1975, and LBC had built a significant audience (not only, but extensively, among London's taxi drivers, from whom it won wide word-of-mouth recommendation). It was the model for what was to become the most successful demotic UK national radio speech station – BBC Five Live. Despite that, and the extensively favoured treatment it had received from the IBA, LBC's 20 years as an ILR station had been a series of disasters: managerial, commercial and financial, and most of all in its industrial relations. When its FM and AM licences were advertised in June 1993, the company had only one year's new track record since it emerged from the wreck of Crown Communications, and was now owned largely by private investment vehicles Chelverton and Wildhaven, and the Bank of Scotland.[774]

LBC had been showing signs of improvement, but MD Charlie Cox, credited with starting to turn the station around, left just before the application date and re-surfaced as part of a rival bidder for the licence a month later. Shirley Porter[775] was an enthusiastic high profile chair for a company which had gathered as directors David Dimbleby, Tony Banks MP, Baroness Jay and financier Ernest Burton. It could boast the best weather man in the business, Philip Eden, along with national names such as Angela Rippon and Richard Littlejohn. LBC's audience reach had declined from 24 per cent in the 1986 JICRAR research to 14 per cent in the 1993 RAJAR study; on the other hand its average listening hours had

772 See chapter 19.

773 *Guardian* 16 June 1993.

774 *RA Paper* 64(93) 25 August 1993.

775 Dame Shirley Porter, daughter of Tesco founder Sir Jack Cohen, had been a controversial leader of Westminster City Council between 1983 and 1991. Following allegations, and an investigation started in 1989, the district auditor accused her and five others of "wilful misconduct" and "disgraceful and improper gerrymandering", between 1987 and 1989. In December 2001 the House of Lords ordered Dame Shirley Porter to pay a surcharge of £27m (plus interest and costs). In 2004 she paid £12m to Westminster Council in a mediated settlement.

risen steadily, so that it maintained its share of London listening at 9.5 per cent, not far from its peak of 11 per cent. LBC was clearly playing well to its own base, but that base was shrinking. Its Achilles' heel – beyond its damaged history – was its programming proposals. Despite seeming to offer different strands of 'newstalk' on FM and 'talkback' on AM, these did not sufficiently address the weight the Authority was going to give to the statutory criterion of broadening choice. There were head-to-head phone-ins on FM and AM each weekday morning and afternoon, and although the application proposed reducing the amount of simulcasting by around 30 hours a week, that still left 60 hours of joint output.

LBC faced two competitors. The London Radio Company (LRC) was a vehicle for Associated Newspapers, GWR – in which Associated was a major investor – ITN, Reuters and SelecTV. Chaired by newspaper heavyweight Bert Hardy, this was a talented but not necessarily cohesive group. It proposed FM services similar to those offered by LBC, but shifting AM much more towards comedy and sport. London News Radio (LNR) was, in effect, the 'alternative LBC' application, offering rolling news on FM, as well as the prospect of bringing back to LBC pre-Crown favourites such as Brian Hayes, Ed Boyle and Philip Hodson, under the management of Peter Thornton. LNR was chaired by John Tusa, seen by the Authority's analysts as a "major acquisition for ILR ... if the group won".[776] Financial muscle was intended to come from Guinness Mahon's Bruce Fireman, along with the Independent Newspaper Group and Allied Radio.

Meeting on the first Thursday in September 1993, the Authority members – including recent appointees Jennifer Francis and Michael Reupke – had to consider whether to award one of the groups both the FM and AM licences, or to mix and match. The scoring in the staff paper was pretty even for each of the three groups, although there was an unusually clear indication that if one group were to be awarded both licences that should be LNR, since it did the most to broaden choice.[777] The members also had to assess the potential for rolling news in the UK, which had embroiled Radio Four in recent controversy.[778] Paul Brown's advice was that this "was a public campaign to save Radio Four, rather than oppose plans for a BBC rolling news service. The campaign alleged that 'only' 25 per cent of its research sample wanted a rolling news service. A London licensee would be pleased to obtain this level of appeal."[779] That was good enough for the members, who decided to award both licences to London News Radio.

At that point the sky fell in. LBC presenter Mike Carlton thought that "our jobs have been snatched from us by one of those ridiculous and unaccountable quangos which infest every corner of British life and multiply like amoebas in a pond".[780] Authority member John Grant unwisely hit back at Carlton in a letter to the *Guardian* saying "I don't see why I should be on the end of a pompous lecture by a middle-aged imported Aussie, overpaid, over-bearing and over here".[781] That did little to elevate the Authority above the battle. The Radio Authority received letters by the thousand – it had to recruit extra secretarial staff to deal with them –

776 RA Paper 64(93) op. cit.
777 Ibid.
778 See chapter 12.
779 *RA Minutes* 11(93) of a meeting on 2 September 1993.
780 *Guardian* 6 September 1993.
781 *Guardian* 7 September 1994

telephone abuse, and some withering media attention.

The award to LNR was as controversial as the decision not to award the licences back to LBC. The winning group was alleged to be supported by only a small equity investment, and "an underwriting of the £4m financing by Guinness Mahon, a wholly-owned subsidiary of the Bank of Yokohama".[782] Authority member Richard Hooper was making his own unwelcome headlines too. An independent media consultant, he had declared at the key September meeting that "he had undertaken an assignment for ITN, completed at the end of July, on its international television news strategy. He declared also that he had agreed to serve on the advisory panel of the Guinness Mahon Global Rights Development Fund."[783] The Authority formally resolved that he could take part in the discussion about the London licences, since neither assignment concerned UK radio development. That was not how the *Sun* saw it, especially its enraged columnist Richard Littlejohn, who stood to lose his radio job. His colleague John Kay ran an exclusive story, headlined "chief who helped axe LBC linked to rival bidders",[784] and Hooper was on the front pages of the tabloids for most of the week. Littlejohn used his show on LBC the same day to encourage a critical discussion of the decision, despite a warning that presenters were risking breaking impartiality rules.

Even the Authority's former friends were taking pot-shots now. David Mellor himself said that "there was a strong case to be made for LBC, and I personally regret its demise". LBC, he thought, had lost "more for the sins of their forefathers than for any deficiencies of the present management",[785] undermining the Authority's continued assertion that its decision was a straightforward assessment against the four statutory criteria for licence awards. Thinking it "a bit hard on Shirley Porter", Mellor joined the chorus of those excoriating Alun Chalfont for his rather lofty refusal to give any reasons for the licensing decision. He regretted, with the benefit of hindsight, that the Broadcasting Act – his act – had not made provision for the giving of reasons. LBC backed down from seeking a judicial review of the award decision, not least since the company was in serious financial trouble. Having laboured on with increasing difficulty after the award, in a 'lame duck' period towards the end of its licence, LBC eventually went into receivership on 29 March 1994. One month later, it was acquired from the receivers by LNR to prevent the station falling silent, with the Authority's formal approval.[786]

To make matters worse, LNR had not been able to raise its financing through a proposed private placement. Fireman, the MD designate of LNR, returned to Guinness Mahon which was anyway disqualified from owning an ILR licence because of its non-EU parentage. By June 1994, the Authority had to face squarely the embarrassment of seeing the licence sold to Reuters, a member of the unsuccessful applicant LRC. The June meeting was a troubled one. The members had in front of them a cutting from that day's *Financial Times*, which reported renewed criticism of the Reuters purchase.[787] They concluded however that they had no powers to prevent the sale, since the new owners were not a disqualified company,

782 *Financial Times* 5 October 1993.
783 *RA Minutes* 11(93) op. cit.
784 *Sun* 7 September 1993.
785 *Guardian* 10 September 1993.
786 *RA Minutes* 5(94) of a meeting on 5 May 1994.
787 *Financial Times*, 2 June 1994.

and "took consolation from the fact ... that the Authority was not being asked to agree that control could be passed to a company that would not have been acceptable to the Authority in the first place".[788]

It was a sorry ending to the tale, and one anticipated by the Shadow Radio Authority in 1990 when it pressed the government to include in the Broadcasting Act some sort of moratorium on the selling on of recently awarded licences. The government refused then, and continued to do so in the face of renewed requests through the rest of the Radio Authority's existence, and even into the new Communications Act of 2003. The argument put to me by ministers and civil servants, when I pressed the matter repeatedly after 1995, was that any such rule would interfere with the proper workings of the market. Until the credit crunch of 2008, few in the official world truly appreciated what damage unregulated markets can do. The whole ILR licence award system, the so-called 'beauty parade', as opposed to a simple auction, was predicated on the Authority making a judgment about who would best deliver against the statutory criteria. By refusing the regulator the power to keep its chosen operator in place, even for a year or so after the licence award, the government knowingly made a mockery of the process it had itself designed, opening the way to selling-on which would discredit the regulatory institution it had created, which it then left to swing in the wind. The LBC episode seriously and unfairly damaged the credibility of the Radio Authority.

London licences round two – Jazz, Kiss, Melody and Sunrise

It was not only the LBC's London licences which were up for consideration that autumn. All eight of the FM and AM Greater London licences (Capital's licences excepted) were to be considered at the same time. This was a sensible attempt to ensure that the Authority considered the wider implications of any individual decisions, but was also a source of pressure which put in some jeopardy the amount of attention to, and therefore the quality of, each individual decision.

At the same September 2003 meeting, therefore, the Authority members turned their attention to the licences then held on FM by Jazz, Kiss and Melody. When these had been re-advertised, they had attracted 48 applications, the most ever received from a single advertisement. Working with three 'nominated' Authority members, the staff, led by David Vick, reduced those to 24 for fuller consideration, and then down to 10 for a final decision at the meeting. Three of those were the FM incumbents, and on the indicative points system they were among the top four scoring applications in the staff analysis. Sunrise, which already held an AM licence giving modest coverage of West London, actually outscored them all. Two other challengers, both offering adult contemporary music – Crystal FM and Music FM – also scored well. Choice FM was seeking a London-wide licence, to expand from its Brixton base, while Xfm was surfing the growing wave of indie music with another credible bid. Of the short-listed groups, only the comedy proposals of Radio Barking (yes, really!) and London Country Radio lagged a bit behind in the staff marking.[789] Two of the decisions on these FM licences found staff and members in ready agreement, and Jazz and Kiss were re-appointed without much disagreement. There were "widely shared

788 *RA Minutes* 7(94) of a meeting on 2 June 1994.
789 *RA Paper* 63(93) 17 August 1993.

reservations" about Melody, but in the absence of agreement on an alternative, and given its Hanson-backed service for older listeners not otherwise catered for, Melody survived.[790]

Was there to be no end to this meeting? Having started at 9.30 am, the Authority members were now faced with awarding four AM licences. The decision to replace LBC with LNR led, inescapably, to awarding one of the AM licences to LNR as well. There were, though, still 10 single-waveband applications for the remaining three licences. Spectrum, the unusual and experimental 'publisher' of a wide range of services targeting ethnic minority groups in London, was seeking to retain its existing licence. The other bidders were Radio Asia, London AM, London Business Radio, London Christian Radio, the London Sports Radio Consortium, Metropolis AM and a proposed women's station, Viva. Many of the scores were again very close. In the light of the earlier decisions on FM, the award of a London-wide AM frequency to Sunrise Radio was agreed without dissent, and Spectrum's AM licence was similarly re-awarded.

By this stage, those at the meeting could be forgiven for starting to flag. Agreement on the fourth AM licence was proving much more difficult to achieve, with London Christian Radio splitting the Authority, and so the members broke for lunch. Returning refreshed, they awarded the final licence to London Country Radio; drew a deep breath; and turned their attention to agreeing a change to the promise of performance of Q96 FM in Paisley. Alun Chalfont, Peter Baldwin and Authority secretary John Norrington met LBC's executive director Matthew Cartisser after the meeting, at 5.55 pm that evening, to tell him of the decision not to re-appoint LBC. Norrington's note reports that "the discussion concluded at 6.00 pm without acrimony". As already discussed, the acrimony was to follow all too soon.

London licence round three – Capital and beyond

A year later came the next, bruising round. The Authority advertised six further Greater London ILR licences in March 1994 – three on FM and three on AM – and the award of one of these was to lead to more trouble.

Richard Branson – who had run a formidable lobby to try and get an FM frequency for his Virgin INR only to be rebuffed nationally – had turned his attention to securing a London FM licence for Virgin. With an FM outlet in London, the station would be much better placed to cope with the disadvantages of offering a music station on AM. The AM frequencies were notably poor in central London, and the company argued its case assiduously. Branson and David Frost – who joined the Virgin board in July 1993 – were guest speakers at the Radio Authority's September 1993 conference in Streatley-on-Thames at the Swan Hotel. On 2 March 1994, the Authority's management team visited the Virgin Radio studios and once more met Branson, who "again pressed the issue of an FM frequency for Virgin".[791]

At the Authority meeting on 6 October 1994, thirteen months after the LBC non-award and the first round of Greater London awards, this next set of licensing had to be concluded. The members faced decisions on whether to re-award Capital its two licences, and on the award of a further two FM and two AM licences for

790 *RA Minutes* 11(93) op. cit.
791 David Vick an interview with the author, 24 June 2008.

Greater London. With Margaret Corrigan unable to attend, only eight of the nine members were present. The Capital re-award was reasonably straightforward, although in re-appointing the flagship company to its FM and AM licences the members asked Alun Chalfont to "convey the Authority's disquiet about Capital's complacency".[792] The meeting had been told that there had been "a certain amount of arrogance" in Capital's approach, and noted – presciently as it turned out – that the strong adult contemporary applications "represented a challenge to Capital's dominance".

There was to be little else that was straightforward on that day. The meeting had begun with something of an apology from Alun Chalfont, that Peter Baldwin's intention to retire as chief executive had suddenly been made public, without telling the other members in advance. There had been a wish not to risk public misinterpretation, if it were announced too close to the current licensing decisions or to the impending announcement of Chalfont's own successor as chairman. Chalfont then formally stressed the need for confidentiality, since "a member of the press had been claiming that he had obtained guidance from a member of the Authority about the licence award decisions due to be taken that day".[793] It was not a comfortable start. To help the members negotiate their way through the various applications distinguished by their music genres, they were treated to a slide and tape demonstration of music choices in London radio. Then the discussions began in earnest.

Forty-one groups had applied for the London licences in total. Fifteen of those did not survive the first cull, conducted by David Vick's team along with the 'nominated members' who were Michael Moriarty and John Grant. After further scrutiny of the application documents, and questioning of the applicants, that was reduced further to a short list of eight for FM, and seven for AM. The FM applicants were headed by Capital, which topped the staff scoring, and was effectively an automatic re-appointment. That left two new FM licences to be awarded. The other seven applicants, in descending order of their 'scoring' under the marking scheme, were Crystal, Virgin, Arrow, Metro, London FM, Xfm and Choice.[794] Chrysalis' Crystal application, proposing a soft adult contemporary station akin to the group's Heart FM in the West Midlands, was well ahead of the others, with the next three bunching up closely in terms of scores.

Alun Chalfont began the item by stating simply that since Crystal and Virgin FM had scored highest on the assessment process, they should be awarded the licences.[795] The Authority members were not prepared to move so swiftly to that decision, but embarked on a long discussion. Wide agreement that an adult contemporary format was an obvious gap in the London market, helped to reach a consensus that Crystal should be awarded one of the two FM licences, but from then on matters became more heated. Virgin, Xfm, and Choice all found support, but after much debate the members found themselves evenly divided between Virgin and Xfm. Although Margaret Corrigan was known to favour Xfm, her absence split the voting down the middle, and the chairman was clear that a member who had not been "present and who had therefore been unable to

792 *RA Minutes* 9(94) of a meeting on 6 October 1994.

793 Ibid.

794 *RA Paper* 70(94) 29 September 1994.

795 *RA Minutes* 9(94) op. cit.

participate in the debate should not have their views heard or have a vote". It therefore fell to Chalfont to use his casting vote, which was in favour of Virgin. He said he was concerned in particular that "there was a real danger that Virgin Radio would withdraw completely from the radio scene if it was not successful in its application".[796] The decision left a number of long-standing members of the Authority with feelings of unease: "an opportunity of a bold broadening of choice for listeners in London had, they felt, been foregone, and the licence used instead to broaden choice only in the sense of improving reception of an already available service".[797]

It was not only the members of the Authority who were divided on the issue. The senior staff was split too, and with a depth of feeling which persists to this day. It seemed indefensible to some of them that the Authority had provided Virgin with a licence to simulcast on FM the service it was already offering on AM, when over-much similarity of output between two licences in common ownership had been the primary reason for taking LBC's licences away from it the year before. For the other camp, Branson's threat to walk away from his INR licence, and the consequent damage to INR and to commercial radio as a whole, had persuaded them to become Virgin's supporters on this issue. The usual radio lightning-conductor effect ensured a steady stream of conspiracy theorists with more lurid explanations.

Just as in the year before, the strange day was not yet done. The members now turned to the three AM licences. Here again the re-appointment of Capital Gold was quickly agreed. For the remaining two new AM licences there was an eclectic range of applicants which had survived to the short-list stage. Unusually, there was "a warm staff recommendation in favour of London Christian Radio"[798] which was duly awarded the licence for what became Premier Christian Radio. The final choice lay between what were thought to be not very convincing applications: London FM, which had indicated a willingness to accept an AM licence, Viva, London Business Radio – all of which were extremely close in the staff marking – London Sports Radio Consortium and London Sports Talk Radio. By now the members were tired, and there was a spat about how the questioning of the groups had been carried out, as the meeting fragmented. The eventual decision, to award the licence to Viva, was confirmed only after a vote on whether it should be postponed while doubts about procedures were resolved. At last, the London task was done.

Was it well done? The fate of radio stations is rarely predictable, and the financial climate in which they operated fluctuated greatly through the coming years. The Authority's guiding intention was to offer a wide diversity of services, from stations with plural ownership. That was a respectable aim, and wholly in line with the Authority's statutory brief. Some of those new stations became market leaders, others survived, a few failed; some continued to offer genuine alternative radio, in others the apparent distinctiveness was illusory. Poor AM frequencies were a huge handicap. A couple of the awards followed the passing fashion of the

796 Ibid.
797 Michael Moriarty in a conversation with the author, 18 December 2008.
798 *RA Paper* 71(94) 28 September 1994.

day too closely, and some other tides were missed. In all of that, the Radio Authority would have been forgiven, even praised.

However, the two major award controversies colour everything. First, while the LBC decision was neither a bad one nor unexpected in the industry, it was not managed well and from that huge problems arose, challenging the legitimacy of the Authority. Second, on the one occasion when, arguably, the Authority departed from its strict statutory brief and tried to manipulate the market, in order to prevent Virgin from possibly abandoning its national AM licence, it got into seriously hot water. As a result of those two decisions, plus the inevitable kvetching from unsuccessful applicants, the effect of the two years of Greater London licensing was fundamentally to undermine trust in the licensing process, both within the industry, and among opinion formers. That lack of trust also extended to the Authority itself.

The secrecy over licence awards meant that the Radio Authority was too easily characterised as yet another secretive quango. It was wide open to scurrilous accusation, and had lost all too quickly the freshness it had offered when it floated off from the old IBA on a new voyage. For all that it achieved – and those achievements were legion – the scars of the LBC and Virgin awards never really faded away. They bled repeatedly, and the body was weakened as a consequence. It is credible to argue that, but for its defensiveness on licensing, the Radio Authority would have been better able to protect the local distinctiveness which was the unique feature of ILR. The influence it had lost by ceding the moral high ground, however, meant that the companies were no longer so ready to listen to ideas from the Authority. Once the credibility of licensing was weakened, that diminished the respect for regulation and guidance in all other aspects of commercial radio.

Chapter 18

High summer

1994 – 1996

The mid-nineties was a great time to be in private radio in the UK. It still retained enough of the characteristics of *independent* radio to make it seem a worthwhile undertaking, judged from a broad social perspective, while enjoying the new fruits of *commercial* success. Boosted by national radio, Classic FM, the extensive addition of new ILR stations, the drive of the RAB, and the authenticity of RAJAR, radio was at last breaking away from the 'two per cent' tag, and growing its share of the advertising cake. The national economy was clambering out of recession, and although there were still difficulties, the floating pound – itself a legacy of Black Wednesday – provided a real boost for exports, and helped start a new boom which was to be remarkably long-lived. As the Major government staggered to its long-delayed demise, there was a general sense of renewed optimism. The economy had begun to revive under the outgoing government, and the electorate foresaw a future of market-led prosperity in the pre-destined new administration. The sounds of your life in the middle years of the nineties were Britpop and celebrity, and commercial radio seemed perfectly placed to catch that mood.

It was less fun to be in BBC radio during these years. Under John Birt as director general, from 1992 to 2000, the Corporation became a byword for uncomfortable re-organisation, and the triumph of the management consultant. Radio was treated as being desperately old-fashioned. In terms of share of the UK radio audience, ILR and INR seemed to be drawing irreversibly away from the BBC. In November 1995, BBC director of radio Liz Forgan noted that "the number of commercial stations would more than double by 2005 and the price the BBC would pay for its ambitious editorial plans would be a continuing loss of share".[799] She forecast that "before the end of the 1990s our reach will fall below 50 per cent. By 2005 reach will be around 40 per cent and our share will settle at about 30 per cent." As if determined to prove her right, James Boyle, as controller of Radio Four, began the necessary shake-up of his network in 1996 with a radical approach which lost three quarters of a million listeners. Matthew Bannister had taken over as controller of Radio One in 1993, and through his period of re-vamping the network it had lost its audience in astonishing numbers, falling from 16.5 million listeners in 1992 to 9.3 million in 1998. However, as considered in chapter

799 *Guardian*, 6 November 1995.

21, this was in no sense the end of the story, nor was Forgan's statement necessarily a concession, rather than some careful positioning against expected levels of competition.

As commercial prospects improved, so did the influence of larger radio groups. By the start of 1994, there were seven major quoted companies whose interests lay predominantly in commercial radio. The Capital group, although at that stage it comprised just four stations, was still by far the largest, with a market capitalisation of £280m. Scottish, Transworld, and Metro were all much of a muchness, with capitalisation around £60m and seven or eight stations. The GWR group, with roughly the same capital value, had 14 smaller and more widespread local stations, plus a significant interest in and the running of Classic FM. Southern was half the capitalisation of the others, but also owned eight stations. The growth and coalescing of the quoted groups, the arrival of new majors such as Chrysalis, and the increasing importance of their radio holdings for more diversified groups like Emap, were a dominant theme from this point onwards. This changed the nature of programming content, which became increasingly homogenised within groups and decreasingly local, and it shifted the balance of power within the radio industry from the Authority to the largest ILR owners. The groups represented the impending triumph of *commercial* radio over the constraints of *independent* radio. "People note that audience figures are highest in countries with minimum state interference", claimed a City analyst in 1994. "Some believe that it could be the best thing that could happen to the industry."[800] The goal of full commercial freedom drove forward the major ILR companies through these years, and eventually into free-fall. But for the present, this was the high summer of private radio in the UK, and it was to be savoured.

The sweetest taste for radio investors, and for executives with share options, came from the surging revenue figures. In 1993, advertising income had jumped from £141m to a record £178.3m, and share prices soared accordingly. Radio companies were praised by the City for their new professionalism, and the low cost base which had been established during the recession was seen by them as a chance to drive these revenues straight into profit. Smith New Court forecast advertising revenue for radio in excess of £450m by 2000,[801] which turned out to be an underestimate. These were years of personal and corporate money making, but often at the expense of investment in the programme output, apart from increasing the fees paid to the best-known presenters.

Revenue and audience growth had both been boosted by the impact of the first five regional services, all coming on air in September 1994. If more stations were good news for commercial radio – and by now most, if not all, were persuaded that they were – then there was further encouragement to the burgeoning sector early in 1994. The FM sub-band 105–108 MHz had been in use for sound broadcasting in much of Europe for some time. The UK government had been reluctant to release it for this purpose, since it was used for radio communications by some of the emergency services and there was a cost implication in re-equipping those to use other frequencies. Even when a major international conference on the allocation of FM was imminent in Geneva in 1984, the government seemed

800 Mark Davies Jones of Smith New Court in the *Sunday Times* 13 March 1994.
801 *Sunday Telegraph* 27 February 1994.

reluctant to contemplate making more space available to ILR. Its first submission at Geneva seemed chiefly concerned to super-serve national radio with more filler transmitters, and it was only through the Radio Authority's efforts that this was modified. It took a great deal of arm-twisting before 105–108 MHz was allocated to the Authority for commercial radio. However, once the band space was won, the Authority decided to consult about how it could best be used.

The draft consultation document, considered by Authority members in January 1994, proposed canvassing three options, each thought to be mutually exclusive: a further national network, INR 4; new 'medium coverage' stations, similar to the heritage ILR stations; or small coverage services, "equating with what some might call 'community radio', or even 'neighbourhood radio'".[802] Michael Moriarty felt that this limited "three-decker solution did less than justice to the range of possibilities",[803] and the members came up with a further hybrid option: regional licences in 105–106 MHz; small scale services above 107 MHz; and a blend in between. The consultative document was issued on 8 February 1994, and 21 responses were received.

The response was predictably mixed and self-interested. Each group urged the allocation of band-width for its own use, and few made any wider contribution. Within the Radio Authority, Neil Romain's analysis of the economics of the various options concluded that "overall, INR is likely to enhance the sector's position, although some of the largest local operators may well find that their primary revenue source is restricted, and as a result see a reduction in profitability".[804] However, there was little appetite for advertising another national licence among other Authority staff, while Virgin was irritating some of the Authority members with its campaign to be granted INR4. When the Authority met at a special meeting on 19 May to consider its decision – a chance to get away from the issues of licence re-awards or non-awards – there were complaints of correspondence which "some might regard ... as an attempt to bully the Authority".[805]

The growing evidence of relative success for the new regional licences, and of demand for very local services, produced different priorities. The Authority members therefore chose the hybrid option. This would mean "a combination of additional FM services in large metropolitan areas and, in those and other areas, a number of smaller services of varying sizes";[806] in effect, more regionals and more small scale ILRs. In October, it announced a 'working list' of 32 new ILR areas, almost all to be licensed using space within the new sub-band. The next and final phase of analogue ILR expansion was under way. The frequency planning task, central to licensing and crucial for 'shoehorning' more stations into congested bands, had been brought back in-house at the start of 1994, when frequency planners Nigel Green and Terry Dowland joined from NTL (or in effect re-joined, since they had both been at the IBA). Add Mark Thomas's expert direction, and Martin James' energetic transmitter commissioning, and the unit was well placed to navigate the crowded AM and FM bands. It also meant that the Radio Authority

802 *RA Paper* 102(93) 24 December 1993.
803 *RA Minutes* 1(94) of a meeting on 6 January 1994.
804 *RA Information Paper* 31 (94) 13 May 1994.
805 *RA Minutes* 6(94) of a meeting on 19 May 1994.
806 *Radio Authority Annual Report* December 1994.

had the capacity to do the whole job, even with a staff that still numbered fewer than 40.

The Radio Authority's continuing nervousness over national pop radio was exemplified by its attitude to the perennial chestnut of whether to 'privatise' BBC Radio One. The idea had been around since before the start of ILR, but it had re-surfaced at the start of the nineties in a Tory Reform Group pamphlet, proposing that Radios One and Two should move into the private sector. [807] Predictably, that was opposed by many in the radio industry, who feared that "a strong chart-based national commercial station ... would divert audiences and advertising revenue away from ILR".[808] The Authority, however, was initially attracted by the notion, and contemplated detailed research into the feasibility of transferring Radio One to its own control, retaining universality of coverage, and some initial advertising restrictions. AIRC was displeased. It questioned the Authority's assertion that this would increase total UK radio advertising, and thought the proposed research study overdue. "They have done planning up to now by sticking the finger in the wind", said AIRC Director Brian West. "Economic research should have been done ages ago."[809]

Following the 1992 green paper on the future of the BBC,[810] and the re-opening of the Radio One issue, the Radio Authority commissioned a report from the Henley Centre – in the end, jointly with AIRC – and this was received in October 1993. The report included rather too much relating to the history of the development of the UK radio market, drawing heavily on a RAB-financed market study which the centre had completed earlier in the year, and was much less impressive when considering the impact of a commercial Radio One. Henley anticipated an annual advertising revenue for a privatised Radio One alone of £182m by 1997, further boosted by £36m from sponsorship and other sources. Although it foresaw the significant potential of a privatised Radio One, it concluded that the losses to ILR and INR would outweigh the gains to the sector as a whole. "The largest stations would be most badly hit, reflecting their dependence on national revenues. We estimate that a typical large station with a revenue of £3.3m and a profit of £0.5m would lose over 14 per cent of revenue, wiping out all profit. A medium-sized station would stay in profit but only just, and smallest stations would be relatively unaffected."[811] This was enough to scare off the Authority, and in November it decided to inform the government privately of its view that "the Radio One option had been closed off".[812] It decided against going on the record with AIRC, however, "since this could tie the hands of the Authority in relation to the possible future use of 105–108 MHz", but the option of a second national commercial service on FM was lost at that moment, and its potential commercial and social impact was denied to radio's future.

In other respects, the 1992 green paper had little impact on commercial radio. There were some suggestions that the BBC should leave local radio altogether, but they came to nothing against both the power of the local BBC lobbies, and the

807 *What shall we do about the BBC?* Tory Reform Group March 1991.
808 *Broadcast* 22 March 1991.
809 *Marketing Week* 23 April 1993.
810 *The Future of the BBC — A Consultation Document* Cmnd. 2098.
811 *A Report for the Radio Authority and AIRC* Henley Centre November 1993.
812 *RA Minutes* 14(93) of a meeting on 9 November 1993.

increasingly commercial stance of ILR. Few now expected that ILR could provide the rather quirky, speech-based and older-appeal output which characterised BBC local radio. Almost unnoticed, ILR was starting to be confined in terms of wider expectations within the ghetto of commercial music radio, its potential to make a broader contribution increasingly discounted.

The future growth of commercial radio in regional and small-scale local radio – and later in digital – was to be put into effect by a different Radio Authority. Alun Chalfont retired at the end of 1994, together with John Grant and Ranjit Sondhi. Richard Hooper and Jill McIvor had preceded them, leaving in May 1994. The new chairman, Sir Peter Gibbings, who had been chairman of Anglia Television and a former managing director of the Guardian Media Group, arrived with high expectations from the radio companies of greater commercial sympathy. Gibbings, a kindly and proper man with a keen intellect and high sense of public duty, was surprised to have no influence over the choice of members. One effect of his concern for due process was that the Radio Authority minutes, which had previously been full and detailed, become much more terse and formal. This different approach ends up depriving the historian of a contemporaneous record of the cut and thrust of debates before decisions are taken.

Michael Moriarty, an Authority member from 1991, was a fine deputy chairman from June 1994, when he succeeded Jill McIvor, applying skills and knowledge from his Home Office days, allied with real humanity and a deep concern for the public purposes of ILR. Former Reuters executive Michael Reupke, and public relations campaigner Jennifer Francis, became Authority members on 18 May 1994. Lawyer and racing man Andrew Reid joined as a member on 16 August 1984, as did Lady Sheil as the member for Northern Ireland. This 'second Authority' lacked the camaraderie and popular touch of its predecessor – both internally and towards the wider media world – and also, in some ways, the high commitment which the 'first Authority' had given during the set-up phase. At my first meeting as chief executive, only four members of the Authority were present. The 'second Authority' did not generally win the confidence of the industry, suffering at least in part the fallout from the licensing decisions for LBC and Virgin FM, and the Emap/TWC judicial review, which were made by the previous group.

In June 1995, Peter Baldwin, who had for 16 years been deputy director and then director of radio at the IBA, and then chief executive of the Radio Authority, finally retired. He had stayed on two years beyond the expected retirement age, supported by Alun Chalfont with whom he had an instinctive sympathy. Many, including most of those who worked at Holbrook House, assumed that his deputy, Paul Brown, would naturally succeed him. The recruitment process produced a range of other possible candidates, and by the closest of margins Brown lost out. Several of the Authority staff wept in the office when the news was announced. Paul Brown was hugely popular, with a great capacity for work and a ready ear for those working for and with him. I was appointed to the post, having left radio in 1984 and worked for a decade in retailing, where I had been running a department store. When I was announced as Peter Baldwin's successor in July 1995, one trade magazine asked how someone who had been running supermarkets on the South Coast could be qualified to be the new chief executive of the Radio Authority.[813]

813 *Broadcast,* 10 March 1995.

Paul Brown never offered me any public or private reproach, but I could not persuade him to stay with the Authority. The companies swiftly and wisely recruited him as chief executive of AIRC, when the long-serving Brian West retired, and he continued as the leading servant and public promoter of the industry until the end of 2008. David Vick became Deputy Chief Executive in his place.

Advertising revenues were now rising swiftly, a process fuelled by the Radio Advertising Bureau's skilful optimisation of the industry's opportunities to build national advertising business. Annual revenue had already climbed to £271m in 1995, and £308m in 1996.[814] This revenue rise was driven by convincing mass audience levels, to a high plateau in 1997. In terms of audiences, just as for revenue, these years were the high summer of commercial radio. That was true for profits, as well. The larger the station, the higher its share of national advertising, so the increases accrued disproportionately to the groups with established, large city licences, and their share prices rose accordingly. The major groups become ever more dominant within industry councils, and the interests represented by City analysts (and radio executives' own share options) began to loom over-large.

In December 1995, AIRC proposed setting up a new central copy clearance unit, separate from the ITCA's joint service, the Broadcast Advertising Copy Clearance Centre (BACC), to be known as the Radio Advertising Copy Clearance Centre (RACC).[815] AIRC was looking for a service for advertisers which would be "quicker, more user-friendly and facilitating.[816] The Radio Authority was instinctively sympathetic, but needed to ensure that its responsibilities for the very successful system of consumer protection, which advertising copy clearance represented, would not be put at risk. Having received a range of detailed assurances from AIRC, the Authority gave approval for RACC at its January 1996 meeting, and Yvonne Kintoff moved from the Authority to run it. AIRC's initiative proved very successful. RACC worked faster than its television equivalent, and with some separate rules, although the two bodies stayed close. The separation from television was practically useful, but it also symbolised radio's new sense that it could now stand on its own feet as a commercial medium.

With the 105–108 MHz consultation completed, the Authority's development team embarked on the new phase of licensing from 1995 with almost messianic zeal. It was licensing which most characterised the work of the Radio Authority. Licence advertisement, analysis, award and evaluation was the major single work stream. As the telephone conversations with applicants became, in effect, telephone interviews, that job assumed ever greater importance. The task of finding and planning frequencies as the basis for those advertisements, and then of commissioning the many new transmitters which began broadcasting each year, provided the substantial engineering task. The re-advertisement of all existing licences occupied a great deal of administrative time and thought. When digital radio came along as well, after the passage of the Broadcasting Act 1996, it redoubled this licensing workload.

814 RAB – The Radio Centre.
815 Letter from Paul Brown, AIRC, to Tony Stoller, Radio Authority, 8 December 1995.
816 *RA Paper* 9 (96) 22 December 1995.

In addition to the four Greater London licences,[817] six ILR licences had been awarded in 1994: for St Albans & Watford (Oasis Radio), Fort William (Nevis Radio), North London (London Turkish Radio), Birmingham (Choice FM), Cookstown & Magherafelt (Gold Beat 828) and West Cumbria (CFM). 1995 saw ILR licences awarded for another six new areas: Tunbridge Wells and Sevenoaks (KFM), Shaftesbury (Gold Radio), Darlington (A1 FM), Chichester, Bognor Regis and Littlehampton (Spirit FM), East Lancashire (Asian Sound) and Craigavon (AM Radio). Initially prompted by apparent interest from a local group – the basis on which areas were considered for inclusion on the 'working list' of future licensing – the advertisement of a licence for Southwold in Suffolk attracted no applications at all. The Authority's Annual Report notes with proper pride that "the award of the licence for East Lancashire to Asian Sound brought to 10 the number of 24-hour services specifically for Asian minorities".[818] David Vick also saw this as a useful victory over the BBC, in both real and propaganda terms; a battle he always enjoyed waging. "One of the real triumphs of the incremental licences were the ethnic stations. That also gave us a massive moral high ground over the BBC, which, at that point, was putting out just an hour or two on a Wednesday evening on medium wave for ethnic minorities in places like the West Midlands, and it was doing absolutely nothing for Asian minorities in places like Slough and Reading. The incremental contracts provided a great opportunity for the private sector to prove it could cater for ethnic minorities in a way that the BBC either couldn't or wouldn't try to match."[819]

The pace of licensing accelerated through the decade. In 1996, while completing the task of re-advertising the 125 pre-existing ILR licences, the Authority awarded another round dozen ILR licences: for Stratford-upon-Avon (The Bear), Oban (Oban FM), Great Yarmouth & Lowestoft (The Beach), Heads of the Welsh Valleys (Valleys Radio), Yorkshire regional (Kiss 105), Wigan (Wish 102.4 FM) Yorkshire Dales with Skipton (Yorkshire Dales Radio), Kingston upon Thames (Thames FM), Oxford (Oxygen 107.9 FM), Cambridge (Cambridge Community Radio), East Midlands regional (Radio 106 FM) and Medway Towns (Medway FM).

The remarkable passion generated by local radio, exacerbated by continuing secrecy about the awards process, meant that after awards were made there were frequent accusations of favouritism or prejudice. The excitement of putting together an application, usually in the company of a group of people who together felt themselves both able and representative, built expectation to a fever pitch. Only one group could win each licence, leaving the Authority with one group of uncertain friends and perhaps four or five narked losers, whose natural response was to suggest incompetence or worse by the licensing body. There were many occasions where the award decision was disagreed with, often vehemently, and one instance which went significantly beyond the soured wine of disappointment. That was the award of the East of England regional licence in May 1997, where there was a direct accusation of malpractice.[820] As noted already, the actual award decisions were taken by the Authority members themselves, who guarded that

817 See chapter 17.

818 *Radio Authority Annual Report* December 1995.

819 David Vick in an interview with the author, 24 June 2008.

820 See chapter 19.

privilege insistently. That served to protect the staff, who – even if they disagreed with a decision – valued being distanced as individuals from any accusations of unfairness; not least because they had to continue to engage with the radio industry in all non-licensing functions.

The list of winning groups by now clearly demonstrated the Authority's approach. There were a number of smaller ILR services filling in the gaps on the commercial radio map, some small-scale services – often within major ILR areas and competing with heritage stations – and a handful of new wide-scale regional services. The local licences usually went to new players, or to those companies with neighbouring services where such synergy was seen as necessary. The smallest stations were normally stand-alone, and most often struggled to survive. The regional licences were typically won either by new entrants, or by those who had made their mark with other regional stations. Phil Riley at Chrysalis, and John Myers – variously at Border and then the Guardian Media Group – were both big, tall men, adept at putting together convincing applications (and then running good stations, too).

The existing major groups hardly ever won a licence, other than when re-applying for those they already owned. Was that merely happenstance? Those most closely involved at the time with the analysis and the decision-making are adamant that the major groups put in poorer applications than others. Peter Gibbings recalls how much easier it would have been – politically – to have given licences to the larger groups, but that "we took the hard road" by sticking to the statutory criteria.[821] Michael Moriarty catches the mood of the Authority members. "We did take seriously the phrase [in section 105 of the act] about broadening choice. That was at the heart of [the pattern of licensing decisions], and it wasn't very easy to see how the big conglomerates were going to deliver the authentically local service which we believed these localities deserved, as well as locally-based groups with good local roots. We may have been naïve about that, but it was our mission."[822] Certainly, the staff at the Radio Authority framed all of their analysis and their own thinking in terms of the four statutory criteria for licence awards in section 105. Discussions by Authority members at the awards meetings routinely followed that structure, considering it almost improper to weigh the effect of an award on the competitiveness of the wider industry. Even groups of which the Authority members had the highest opinion for their current operations – such as Scottish Radio Holdings – did no better than those who crossed swords repeatedly with the regulator, such as GWR. The effect of all that was to sour further the relationship between the Authority and its major licensees.

There was still no restriction on the selling on of licences. Many of the licences awarded to newcomers faced early difficulties – often fairly quickly – and there were plenty of people around to offer scared investors a good price if they wanted to sell. Routinely, stations were then bought by the major groups, who had been unsuccessful applicants themselves. As a consequence of the government's repeated refusal to allow any sort of moratorium on such sales, the whole system of licensing came to be discredited to a degree. A lot of money went from the commercial radio groups to buy licences they had failed to win by application, and

821 Sir Peter Gibbings in an interview with the author, 5 November 2008.
822 Michael Moriarty in an interview with the author, 26 September 2008.

therefore out of the industry. Arguably, it would have been better had the Authority members – and the staff providing the analysis – found it within themselves to look behind the quality of individual applications to the bigger picture. But those who had endured the fall-out from the Virgin FM award in 1993 were viscerally conditioned by a determination to avoid a repeat of that. The licensing load on members and staff was also an extenuating factor, explaining why people were reluctant to move beyond the narrow criteria for selection. Many members and senior staff were happier within the comfort of a strict legalistic approach. Those of us who came later to the process both of analysis and decision should perhaps have encouraged a wider view.

These were the years when the larger radio groups took firm shape. The 1990 act had permitted some accumulation of ownership, and as described earlier a number of radio 'majors' had emerged, with Capital, Emap, Scottish Radio Holdings (SRH) and GWR to the fore in these middle years. The award of regional licences to Chrysalis and to Border TV, suggested that one of those might join the major groups in time, and Chrysalis emerged accordingly. An adjunct to Chris Wright's music promotion and publishing enterprise, its radio holdings and earnings came to dominate its parent.

These larger groups – notably TransWorld (TWC), and then especially but not exclusively GWR – found themselves bumping up against ownership limits, and chivvying the government and the Authority to allow them more leeway. In 1994, when the Radio Authority faced its first judicial review concerning its approval of warehousing arrangements over the Emap/TWC merger, the rules established by the 1990 Act limited companies to owning no more than 20 local stations, and of those no more than six could be large licences. The Radio Authority had, for some time, been pressing to increase the number of licences which one owner might hold, not least after its own fright over the Emap/TWC case, where it had allowed 'warehousing' of licences above the statutory limit.[823] At the start of his chairmanship, Peter Gibbings publicly lent support to demands to ease the "understandable frustration" of the ownership limits, believing "that with the greatly increased number of stations now broadcasting, these are unnecessary restrictions".[824] There was some easement in 1995, when the limit on the number of licences in the various categories which any one group could hold was raised by a parliamentary order to 35.[825]

More substantial relaxations in the 1996 Broadcasting Act allowed significantly more concentration of ownership, and a feeding frenzy ensued. There were 14 separate mergers or acquisitions during 1996, and a further 13 in 1997.[826] Scottish Radio Holdings bought West Sound for £1.6m, Downtown for £9.7m and Moray Firth for £333,000. IRG were purchasers of Q96 for £1.2m, Allied Radio for £4.79m, and Central Scotland Radio for £5.231m. The most noticeable acquisitions were made by GWR. Together with Associated and ITN, it bought London News Radio for £7.1m. GWR alone bought East Anglian Radio for £25.2m, the balance of Classic FM for £71.5m, Radio Wyvern for £43.7m. DMG Radio then bought Essex Radio in two chunks, for a total of £15.4m, which was to be operated

823 See chapter 19.
824 *Radio Authority Annual Report* 31 December 1994.
825 Broadcasting (Restrictions on the Holding of Licences)(Amendment) Order 1995.
826 *RA Paper* 56(98), 20 May 1998.

as part of the GWR group. Capital sold £23.8m worth of shares in GWR, placing those with City institutions, and bid unsuccessfully for Virgin Radio.[827] Takeover and merger values for those two years totalled over £300m. This seemed to be the free market operating with a vengeance. Commercial radio had surely arrived.

However, the act also imposed on the whole system two 'public interest tests' (PITs), to be applied by the Radio Authority, as a balance to the greater freedom being allowed to the consolidators. The first public interest test was to be conducted where a radio group was seeking to own two services on the same waveband in any locality; the second was to be conducted when there were proposals for cross-media ownership between radio and newspapers, and as well as plurality and diversity issues it also required analysis of the economic impact and on the proper operation of the market in those industries. These tests were clumsy and unwieldy, bringing the Authority into the arcane worlds of law and market economics. The Authority was well equipped on the first, but poorly on the second, and thus the legal approach tended to dominate.

Where groups were applying for a licence, the award of which might trigger a PIT, the test had to be conducted in advance of the licensing decision, lest the whole process from application to award be even further extended. Despite great thoroughness, the PIT issue was hardly ever material to an award. More generally, it served unintentionally to make hostile takeovers difficult, especially where other, non-radio parties were interested. PITs were far too heavy and complex instruments for the fairly small radio industry to bear, or for its smallish regulator to carry out, hindered companies in planning and action, were costly and slow, and rarely met the public need when major matters were at stake. PITs made the Radio Authority more bureaucratic, well beyond even its own instincts, and did little or nothing to enhance the local service offered to listeners by ILR.

Most of the PITs were dull affairs. The exception was the first cross-media test, concerning the proposed acquisition of Leicester Sound from GWR by its major shareholder, DMG Radio, for the Daily Mail General Trust plc; that is to say Lord Rothermere's Associated Newspapers. GWR was once again up against its ownership limits, and needed to shed a licence. It proposed selling Leicester Sound, which would have brought it within the rules. There was, though, a significant second motivation behind the proposal, which the Authority failed fully to appreciate at the time. Associated Newspapers, under their driving and charismatic chief editor, Sir David English, was more than keen to run radio stations alongside its big regional newspapers. It is now clear that much of the drive to shift Leicester Sound from GWR to Associated Newspapers arose from Associated's own corporate aspirations, firmly driven by David English himself, and that GWR was pretty much obliged to go along with what such a powerful investor required. Ralph Bernard confirms now that "the reason we did that deal ... we needed DMGT's support to buy Classic FM outright. At a meeting they gave me the joyful news that, yes, they would support – but only on condition that we sold them Leicester".[828]

The Authority's initial analysis, considered at its meeting on 11 and 12 December 1996, "expressed a number of concerns in relation to the proposed

827 See chapter 22.
828 Ralph Bernard in an interview with the author, 24 January 2008.

acquisition". There was an exchange of letters with DMG and GWR, ahead of the final decision in January 1997. Some at the Authority suspected that underlying all this was an attempt to 'warehouse' the licence, moving it to friendly ownership only to re-acquire it when the rules were relaxed. The adding of new stations was steadily increasing the universe of ownership points, and thus made future growth possible even for groups near a current maximum holding as new stations were licensed. Given that the proposal was for DMGT to acquire 80 per cent of the station, leaving 20 per cent with GWR and all of the sales management to boot, there were grounds for such suspicions.

The analysis conducted for this unique PIT showed almost all the factors to be negative. There was only the one local English-language ILR station and one dominant regional newspaper, the Leicester Mercury, owned by Associated. The issue facing the regulator was about as extreme as could be imagined in terms of plurality of ownership in a locality. The members' meeting on 16 January at Holbrook House concluded that "it was difficult to conceive of any circumstances in which a merger between local media alone would appear more significant. Since parliament had empowered the Authority to apply the public interest test in cases of purely local radio/newspaper mergers, it was a reasonable assumption that parliament intended that the Authority would, in such cases, pay particular regard to the local media."[829] If the cross-media PIT was ever to apply, it had to be now. And if not now, then why had parliament legislated for the test at all? Extensively advised and counselled by its lawyers, the Authority determined "that the proposed acquisition could be expected to operate against the public interest".[830] In other words, no deal.

The response from GWR was fairly muted, which seemed surprising, but the Authority had not seen where the real drive originated. From Associated and the regional newspaper industry as a whole, the reaction was extremely hostile. Peter Gibbings was virtually summoned to the Newspaper Society (NS), to explain why the Radio Authority was apparently so hostile to the interests of regional newspapers. At a meeting with the full NS Council on 2 July, I recall him setting out the formal proposition, graciously, but not getting much understanding in return. This left the NS and the Authority far apart, and when I was wined and dined by NS luminaries on 2 December we needed to bring the relationship back on to the rails. Fortunately, in David Newell the NS was blessed with a highly pragmatic and agreeable director, and the high levels of tension did not last. The suspicion, however, lingered among newspaper people for the rest of the Authority's life.

It was fundamental changes in the programming philosophy which completed the dominance of the commercial model of private radio. Ahead of the Authority's meeting in July 1995, my first as Chief Executive, we had received a request from the GWR group to be allowed to network programmes across virtually all of its stations. By that time, GWR owned a dozen FM stations: GWR (East), GWR (West), 2CR FM, 2 TEN FM, Leicester Sound FM, Beacon Radio, Hereward FM, KLFM, Mercia FM, Q103, RAM FM and Trent FM. It was already networking between its eight AM stations to a varying degree: Classic Gold stations 1332, 1431, 828, plus Brunel Classic Gold, and Mercia Classic Gold, GEM AM, Isle of Wight

829 *RA Minutes* 1(97) of a meeting on 16 January 1997.
830 *Radio Authority Annual Report* December 1996.

Radio and WABC. Now the group programme head, Steve Orchard, with Australian Dirk Anthony, was to mastermind the transformation of GWR's disparate portfolio of local stations into an American/Australian-style network – as far as the regulator could be persuaded to allow – and introduce trans-Atlantic-style commercial radio networks to the UK.

Orchard approached Paul Brown at the Radio Festival in June. He then wrote to request "the networking of off-peak programming from any group site" for all GWR's FM services. He wanted to extend the precedent of AM networking to all off-peak and weekend output, and some daytime too, although "the stations would be encouraged to 'opt-out' for local-only programming, should local circumstances dictate" – which, of course, they rarely did. "The rationale behind these changes", Orchard wrote, "is our desire to improve the quality of presentation for listeners in local areas, and to help even our smallest sites to compete against the BBC national networks ... We are now actively negotiating with a number of national radio personalities who are ready to 'come over' to the commercial side."[831]

The date of the request is significant, in view of subsequent claims that the radio industry was driven down the commercial road in order to survive. In 1995, advertising revenues were to be £270m, up 22 per cent on the previous year and almost double the take in 1992.[832] Audiences for ILR reached their peak of 24,456,000 at the end of that year, a level they were not to reach again under that research methodology. Orchard's request, therefore, cannot fairly be claimed to have been a defensive reaction. It was perhaps a competitive one; GWR's portfolio was less commercially attractive than its rivals' in the radio industry, who had more coherent and metropolitan groupings of stations. If so, the complicity of the rest of the industry in the positioning of GWR shows, not for the last time, the skill of Ralph Bernard in getting other companies to embrace policies which were to their own medium-term disadvantage, but to GWR's benefit. Arguably, the annexation of the Capital Group by GWR in 2004 was a direct consequence of the larger group's decision in time to dilute its local strengths.

The Authority paper unequivocally recommended agreement to the request. "The recent audience performance of the GWR stations has been impressive, and the group appears to master the computer technology which seamlessly blends local content into networked programming. The changes proposed are either off-peak or to secondary AM services, and are designed to improve the professionalism of the programme offered to the listener. Staff recommend that the group should be allowed to proceed with the changes [but] maintain local input five times each hour."[833] This was the last item on the agenda for the July meeting, and it was approved by the Authority members without significant discussion.[834] The 'national radio personalities' Orchard had promised to lure from the BBC turned out to be Paul Burnett, presenter of BBC Radio One's lunchtime slot from 1972 until 1984. He joined the existing 'network star', Dave Lee Travis, known for his breakfast show as 'the hairy cornflake'.

Here was the pivotal moment in the development of commercial radio,

831 Letter from Steve Orchard GWR to Paul Brown, Radio Authority, 22 June 1995.
832 RAB – The Radio Centre.
833 *RA Paper* 69(95) 27 June 1995.
834 *RA Minutes* 7(95) of a meeting on 6 July 1995.

ushered in by the regulator with scant discussion about its implications. Until this time, the shift from *independent* to *commercial* radio was evident in almost all aspects except the programming output. True, most of what the IBA had regarded as the jewels in its crown – the secondary-rental funded concerts, documentaries and drama – had long-since disappeared, but Classic FM seemed to make up for them more than amply in terms of industry prestige. The social action which was such a feature of independent radio was present to some extent, although not much of it now found its way on to the air, other than in tightly-scripted promotional slots. Charity appeals, usually a day of fun and auctions – copied now by network television – remained the most noticeable of feature of this aspect of ILR.

There had also been a steady rationalisation of station playlists. The arrival from the USA in the eighties of auditorium tests of music tracks was augmented by phone-out research. Such data was then used alongside Selector software (and some other similar but less widespread products) to provide a mechanical tool for music scheduling. This also left less room for local variation. Much of the rationale behind this approach arose from the large number of new stations, which could not readily find skilled popular music programmers. Consolidation gave owners of multiple stations the incentive to use research – and Selector – overseen by a small number of music experts, across all their stations. The option to network took this one huge step further. Before long, even where stations were not networked, many of those within a single group were using identical playlists, even down to the times of day when music tracks were usually played.

With the arrival of networking came the critical shift in the philosophical approach to programming in ILR. The only surprise in this is that it took so long. The radio industry had been agitating since 1984 for the freedom to programme commercially, rather than according to post-Reithian norms of public service. Not until the mid-nineties, sparked by networking, did it realise that it now had the freedom but was yet fully to use it. Since 1973, creating ILR programming had been a 'bottom-up' process. From the mid-nineties, however, it became 'top-down', designed at a group level and implemented by a local management increasingly limited in what it might do on its own initiative. The extent of standardisation varied between groups. It was at its highest within GWR, under the direction of Steve Orchard and Dirk Anthony. Presenters were given playlists to follow without variation, comprising music tracks which had tested successfully. Spoken links were written on liner cards to be read out at the designated time. Local content was weather, news and what's ons. Even the amount of advertising was pre-determined by the 'clock' of each hour, and therefore spots were discounted if necessary to ensure that the commercials were present in the correct number. Community activity was largely off-air. GWR in particular deployed 4x4 vehicles – named Black Thunders – to provide their stations with a local 'presence'. When Jay Oliver had driven his dune buggy around the Black Country for Beacon Radio in the mid-seventies, it had seemed a fairly harmless eccentricity. The Black Thunders, however, were a calculated mechanisation of local profile which deliberately avoided substantial on-air output.

When this approach was introduced into stations which had failed to find success in a bottom-up approach, it frequently produced positive commercial results quickly. Audiences increased for a shortish time, and revenues with them. Many stand-alone companies, and some of the smaller groups, too, had suffered from the rapid expansion of the system, and lacked the on-air talent and management experience in the UK to staff the ever-growing number of stations. A formulaic replacement for lack of competence represented short-term progress towards survival and then increased profitability. Arguably, the scramble for consolidation made this almost inescapable. The rate of licensing contributed too, given that the stations were launching into a commercial environment. To that extent, the Radio Authority must bear a share of responsibility for the changes which more than anything else ceded the heritage of *independent* radio, and undermined the medium. Up to this point, ILR had correctly understood that its localness was the key factor which distinguished it as a medium. Once it saw how much short-term profits could be improved by standardisation and networking (and later automation), and with the admiration of those most fair-weather of friends, the City analysts, it sold its birthright for a souped-up share price. The networking decision of July 1995 underpinned that transaction.

During these years, the regulation of programming was a hybrid of the old and the new approaches. The Radio Authority oversaw the system of 'promises of performance', which were documents of a few pages, forming part of each company's licence, encapsulating the essence of the proposals put forward in its successful application to run the station. This was the promised 'light touch regulation', as understood at that time. If a company wanted to change the promise of performance, that could be requested and was often agreed to. In 1997, the first year when it issued compiled data, the Radio Authority agreed to 118 changes in promises of performance from what were, at that time, 205 licences in issue.[835] It was only if a company moved outside its permitted limits that the Authority would consider any regulatory action. When that happened, both sides had the advantage of being historically close. The half dozen programming and advertising regulatory staff knew, and were known by, virtually all the programme managers, and plenty of the individual presenters too, and much could be resolved without formality. The effectiveness of the radio 'grapevine' usually meant that any major problems were dealt with at an early stage. It is hard to imagine the regime of the nineties getting to the pitch of escalation which led to the £1.1m fine levied by Ofcom on GCap in 2008.[836]

London Greek Radio was fined £1,000 in November 1994 for carrying a 'political' advertisement for a rally to protest against the Turkish invasion of Cyprus.[837]A fine of £5,000, the highest so far, was levied on Virgin in May 1994 for lewd comments made by presenter Nick Abbott on air about LBC's Robbie Vincent.[838] Another fine, a more severe £20,000, followed in December of the same year, when the RA concluded that Virgin had breached taste and decency rules following comments made by callers during a late night phone-in show on sexual fantasies. When imposing sanctions for such breaches of taste and decency, the Authority "felt that the size of the fines should send a clear message that lurid programme content would not be tolerated".[839] The fines were generally lower for breaches of promise of performance. Fortune FM in Greater Manchester was fined

835 *Radio Authority Annual Report* 31 December 1997.
836 See chapter 25.
837 *RA Minutes* 10(94) of a meeting on 6 November 1994.
838 *RA Minutes* 5(94) of a meeting on 11 May 1994.
839 *Radio Authority Annual Report* 1994.

£1,000 in 1996 for failing to deliver its promised music format of standard and show tunes. Choice FM in Birmingham was fined the same amount that year for insufficient reggae music. The licence had been won in an echo of the success of London's Brixton station, but the output had drifted towards the mainstream, to the anger of the local black community – probably fired up by discontented former presenters. Birmingham enjoyed a thriving black music pirate scene. Choice, which had been meant to make illegal broadcasting unnecessary, or at least unjustified, did little to fill that space.

Piccadilly Radio had fallen foul of the regulators for broadcasting offensive material and blasphemy in 1996, leading to a fine of £1,000; and then again broke the Programme Code in 1997, running up a £10,000 fine for the second offence. Unusually, this was one ILR station which failed to learn its lesson, repeating the offences under the Ofcom regime in 2004, when sanctions were more draconian.[840] Two stations in the GWR group were fined in 1996, 2CR for insufficient album tracks and 70s music, and Leicester Sound for insufficient speech and 60/70s music. When Leicester broke its promise of performance again the following year, it was fined £5,000. The station was hurt, the group management all too-easily outraged, and that set the scene for major confrontation between GWR and the Authority's programme regulation in the future. By and large, however, the stations accepted the nature of content regulation, valued the relatively consultative ways in which it was exercised, frequently turned to Authority staff for guidance, and enjoyed working within their fenced space.

Record revenues, record audiences; commercial radio was here for sure, and these were good times. The mood was expansive. When Emap bought the Metro Radio group in August 1995, the two key people in the previous ownership – Neil Robinson and John Josephs – promptly backed a management buy-out of The Pulse in Bradford, to launch an embryonic new group, The Radio Partnership. Emap had paid £98.7m for the 10 stations in the Metro group which had arisen from the single, initially troubled, Newcastle ILR franchise in 1974. Now, with a fortune already earned, Robinson and Josephs joined other potential predators, bidding for the regional licences and making acquisitions if and when the opportunity arose. The radio industry had become acquisitive and competitive; a proper commercial industry, in fact.

That spilled over into programming, too. In Scotland, Jimmy Gordon and Radio Clyde had been steadfastly opposed to using additional frequencies within an area to allow new entrants. Gordon still asserts that it was competition which downgraded ILR output, whereas allowing the heritage stations to acquire more local licences would have ensured complementary rather than destructively competitive programming. Nevertheless, Clyde steadily bought up the other heritage ILR stations, forming Scottish Radio Holdings in 1994, which by 1997 included Northsound, Forth, Tay, Borders, West Sound and Moray Firth. When the Authority awarded a regional licence for Central Scotland to Scot FM in 1994, it nevertheless regarded this as somehow *lèse majesté*.

It then reacted with irritation when the new station bought the rights to live football commentary for the 1995/6 season from the Scottish Football League, in a deal which excluded Clyde's wish to include score-flashes in its *Super Scoreboard*

840 See adjudication of the Ofcom Content and Standards Committee, 24 November 2005.

sports programme on Saturday afternoons. Defending the station in what had become a public spat, programme controller Alex Dickson revealed just how commercial sports broadcasting had become, even in this last bastion of the concept of *independent* radio. "Clyde has invested heavily in Scottish football ... we have parted with more than £1m to Scottish football bosses across the years. It takes £30,000 to kit out just the premier ground with broadcasting equipment ... ."[841] This was big money for radio, and these were figures which would have seemed unthinkable to ILR ten years before. With commercial revenues came commercial expenditures. Such costs were going to continue, even if revenues failed, but in the competitive climate few could hold back on that account.

841 *Herald*, 25 July 1995.

Chapter 19

Challenging the regulator

1994 – 2000

The fashion for mounting formal challenges to the powers-that-be was a notable characteristic of British public life from the early nineties. That arose partly from the disappearance of deference in society, celebrated and promoted by tabloid journalism; partly from the leeching away of authority from successive governments, notably the Major administration through all its Euro-squabbles; and partly from the growth of single-issue politics. Non-governmental organisations (NGOs) liked nothing better than a confrontation with the establishment. It galvanised their support, enhanced their profile and improved fund-raising potential; on occasions, it might even achieve a result in terms of policy changes. The celebration of the 'individual', so potent a colour in the palette of Thatcherism, had the unexpected by-product of encouraging single-person challenges, and the legal system bent over backwards to accommodate litigants in person in public law cases.

The Radio Authority was particularly prone to such set-piece challenges. That reflected their wider prevalence, but it was exacerbated by the 'lightning-conductor' characteristic of radio and its regulation, whereby seemingly minor issues would generate disproportionate passion. The radio companies themselves caught the prevailing mood, and the change in their relationship with the regulator – from an almost forelock-tugging deference to corporate self-importance – encouraged them to stand up to the Authority, even to seek to undermine it. That very much went along with stock market quotes and the shift to *commercial* radio. From 1994 onwards, there were a series of formal challenges to the regulator, by companies and individuals, which illustrate well the changing nature of radio and its regulation.

It is not co-incidental that from 1995 the Radio Authority itself became ever more legalistic in its approach, emphasising that it was both obligated by and limited by statute, and feeling increasingly uncomfortable outside that relatively narrow remit. That was a reaction to challenges – both to itself and in the wider administrative sector – but also reflected the balance of expertise within the Authority, both at Board and staff level, not least in its Chair and Chief Executive. We all felt comfortable with the advice of lawyers, to which great weight was given, and we felt protected by the relative certainties of statute law. That was a proper pre-occupation up to a point, but it produced rather more tunnel vision than was desirable, especially in denying weight to wider commercial considerations.

Public bodies generally feared 'judicial review' of their decisions. That is the process by which courts supervise the exercise of authority by public bodies. A company or individual who feels that the exercise of such power is unreasonable or unlawful can seek leave from the High Court (for England and Wales) or the Court of Session (for Scotland) to challenge a decision or process. If the court finds that the public body has misapplied the law or its own stated rules of procedure, or has acted outside its powers or unfairly, it can set aside the decision, and require the body to go through the whole process again, this time properly. Similarly, if the decision reached is judged to have been perverse or unreasonable, in a strict legal sense, it may be set aside.[842] In certain limited circumstances, the person complaining may be entitled to damages, but it is much more usual for the complainant to be seeking simply to overturn a decision. Not all of the set-piece challenges to the Radio Authority came through applications for judicial review, but the majority did. It is testimony to the legal care with which the regulator operated that it never lost a judicial review.

The first of the nine challenges which this chapter will consider arose in the case of East Midlands Allied Press (Emap) and Trans World Communications (TWC). In June 1994, founder Owen Oyston had agreed to sell his holding of 21.1 per cent in TWC to Emap, for £2.5m in cash and £13.1m in Emap loan notes.[843] Together with the 29.6 per cent it already held, this would be enough to give Emap control of TWC. However, the new company would control eight ILR licences, while the statutory limit at the time was six. Emap had proposed getting around this statutory inconvenience by 'warehousing' the two Radio City licences. This involved Emap putting the licences into a joint venture with its merchant bank, Schroeders, in an arrangement whereby each held 50 per cent, so that neither could be deemed to be in control. The Radio Authority's executive had already, in early May, given its blessing to such an arrangement,[844] although the matter was not considered at that stage by the full Authority, and was only reported to them in an information paper at their June meeting.[845] The permissive decision did not go down too well among the commentariat. One financial columnist observed that "by approving Emap's mooted takeover of Trans World Communications, the Radio Authority is apparently approving a scheme designed to circumvent the very legislation it is charged with upholding".[846]

Worse, the TWC board as whole, led by the Guardian Media Group (GMG) which held 20 per cent, was opposed to the Emap takeover. GMG's Harry Roche wrote to Peter Baldwin on 13 June, challenging whether the Authority had indeed given advance approval to the 'warehousing', so that Emap might with impunity "circumvent the regulations regarding radio station ownership".[847] When no

842 The benchmark decision on this principle was in 1948, in the case of Associated Provincial Picture Houses Ltd v Wednesbury Corporation [1948] 1 KB 223, HL. Lord Greene's seminal judgement in that case was that "if a decision on a competent matter is so unreasonable that no reasonable authority could ever have come to it, then the courts can interfere ... but to prove a case of that kind would require something overwhelming ... ". This threshold is extremely difficult to meet, which is why the *Wednesbury* ground is usually argued alongside other grounds, rather than on its own. The onus is also on the claimant to establish irrationality or perversity, which is hard to do if a public body has shown any sense at all in its decision-making.

843 *Guardian*, 23 June 1994.

844 Letter from Peter Baldwin, RAu, to Terry Smith, Chairman, Emap Radio, 6 May 1994.

845 RA Information Paper 33(94) and RA Minutes 7(94) of a meeting on 2 June 1994.

846 *Daily Telegraph*, 14 June 1994.

847 Letter from Harry Roche, chairman and chief executive Guardian media Group plc to Peter Baldwin, RAu, 13 June 1994.

immediate reply was forthcoming, a follow-up letter two days later confirmed that GMG intended to seek a judicial review of the Authority's decision.[848] The Authority evidently felt the need to secure its defences, retrospectively at least. Its legal advice, sought only on 20 June, was that "a 50 : 50 deadlocked company is *prima facie* not controlled by either party".[849] There was a further issue, over and above the 'deadlock', arising from the Authority's decision in October 1991 not to apply article 11(5) of the Supplementary Order to the Broadcasting Act "in the literal sense", and therefore not to count a 20 per cent holding in a large category A or B licence as "a licence held". This even laid open Emap's pre-existing 29 per cent holding to challenge.

The judicial review was heard in the High Court, by Lord Justice Schiemann, over two days, 27 and 28 July. His judgement, given on 2 August, was that the Radio Authority had not acted outside its powers, and that a 50:50 arrangement did not impute control to either of the parties in the deadlock. On the matter of control, faced with a request from GMG's lawyers to apply a broad, commonsense definition which would rule out 'warehousing' and similar evasions, he judged that "Parliament has proceeded, not with a broad brush, but by using very detailed provisions ... it would be wrong for the court to proceed by using the broad approach urged on behalf of the applicants".[850] Similarly, it was "both possible and right to give article 11(5) the construction which the Respondents [the Radio Authority] contend".

The Radio Authority had won its first judicial review challenge, although in the process it had found itself made most uncomfortable by some previous 'flexing' of the ownership rules. The Authority subsequently acknowledged that the reality was probably that "the Authority could have declined to approve Emap's proposals and that had Emap, rather than GMEN [GMG], sought judicial review ... the Authority would still have won".[851] There was a serious risk that it might appear that the regulator had been taking sides, giving the comfort of prior approval to Emap's proposals, when it could just as easily have either withheld that, or indeed decided against allowing the takeover. It had also virtually sanctioned the prospect of further 'warehousing' schemes, as well as future legal challenges.

Peter Gibbings had been Roche's predecessor as head of the Guardian Group, and when he replaced Alun Chalfont as chair of the Authority from New Year's Day 1995, he arrived with a keen awareness of the Emap/TWC case. The seeming informality of the Authority's processes, and the extent to which it had laid itself open to accusations of favouritism and assisting in the evasion of the law, troubled him. It was a major factor in his insistence that the Authority should henceforward neither approve nor disallow any similar 'warehousing' proposals in advance, leaving the parties to take the commercial risk of having them overturned in the courts. That diminished the certainty which good regulation tries to offer the industry for which it has responsibility, but it avoided compromising the Authority. However, the radio industry, which had warmly supported – and possibly even proposed – Gibbings as the new chair, swiftly came to resent his careful approach.

848 Letter from Roche to Baldwin, 15 June 1994.

849 Letter from Guy Wilson, Allen & Overy, to John Norrington, Secretary RAu, 20 June 1994.

850 R v Radio Authority *ex parte* Guardian Media Group plc 2 August 1994.

851 *RA Paper* 63 (94) 25 August 1994.

The next courtroom drama was provoked not by commercial but by political issues. The British section of Amnesty International (AIBS) had submitted advertising scripts to the joint radio and television copy clearance body, the Broadcast Advertising Control Centre (BACC), in May 1994. These had been rejected on the grounds that AIBS was excluded from advertising by the Authority's Advertising Code, itself echoing section 92 of the Broadcasting Act, since its objects were "mainly of a political nature".[852] As was normal for potentially controversial decisions, it had been referred to and endorsed by the Authority's programming and advertising staff. Paul Brown, then at the Authority, had himself received the comfort of a letter from Frank Willis at the ITC, confirming their supportive legal advice on such issues.[853] This throws a useful sidelight on the warm relationship at a working, regulatory level between the two sundered former parts of the IBA, even though they had gone their separate ways in many respects. Such co-operation was common, and valued by both. The full Radio Authority endorsed these decisions in July, despite AIBS starting to wave around its own legal interpretation that it was not a political body for the purposes of the Act.[854]

In essence, the Radio Authority and its legal advisers followed the conventional interpretation, that the prohibition on advertising by any bodies whose objects were wholly or mainly political covered those which "seek to influence public opinion on issues determined by a government [UK or overseas] or seek to promote a particular political philosophy in relation, for example, to the organisation of society".[855] AIBS wanted to place advertisements highlighting the plight of political prisoners around the world, and the Authority had concluded that this "must, to a considerable extent, involve campaigning in order to influence the policies of governments around the world". On any straightforward narrow reading of the rules, the Authority was self-evidently right. Amnesty hoped to press the argument that it was possible to exclude from these rules "matters which do not fall within the sphere of reasonable political controversy" and that "humanitarian concerns and respect for human dignity come into this category".[856]

The Authority was in a difficult place. It had just narrowly avoided being singed by an over-generous and broad reading of the ownership rules in the Emap/TWC case. That argued for a strict interpretation of the rules. On the other hand, Amnesty and its work on behalf of political prisoners was a 'good cause', popular with the rising political class. Nonetheless, the Authority maintained the exclusion, and judicial review proceedings followed. The High Court's judgement, handed down on 4 July 1995, wholly upheld the Radio Authority's interpretation of the law.[857] Lord Justice Kennedy supported the broad definition of 'political', which extended beyond Amnesty's assertion that it should be limited to one who seeks "to overthrow the United Kingdom government, persuade that government to change its policies or change that government". The restriction on political advertising, he concluded, was proper and included Amnesty. He wisely argued the necessity of "recognising that something which may appear to be an

852 *Radio Authority Advertising Code* rule 8(a).
853 Letter from Frank Willis, Director of Advertising and Sponsorship, ITC, to Paul Brown, 1 July 1994.
854 See *RA Papers* 56(94), 4 July 1994, and 62(94), 25 August 1994.
855 Letter from Radio Authority to Amnesty International, 27 May 1994.
856 Letter from Bindman & Partners on behalf of AIBS to Peter Baldwin, 12 August 1994.
857 R v Radio Authority *ex parte* Amnesty International British Section 4 July 1995.

unnecessary restriction on a good cause could also usefully restrain something manifestly less worthy". Mr Justice McCullough, sitting with him, also noted that since the proposed advertisement was avowedly part of a campaign, it was hard for Amnesty to argue that it was not primarily a campaigning organisation.

The Authority had decided in advance that it would not appeal against the High Court's decision, whatever it was. It wanted to position itself as welcoming the court's clarification of a difficult area of law. Amnesty felt no such constraint. It sought, and was granted leave, to take the matter on to the Court of Appeal. Meanwhile, many were seeking a way out of the dilemma. The Radio Authority conceded that Amnesty's charitable arm would be able to advertise its products, but the NGO was not interested in promoting sales of Christmas cards. It was seeking to break into radio, and therefore also television, and pressed on with its case. As the legal processes ground slowly onwards, the whole issue of political advertising was raised during the Report stage of the Broadcasting Bill in 1996 by Lord Dubs.[858] He was promised by the broadcasting minister, Lord Inglewood, that the government would consider an amendment which would change the Radio Authority and ITC codes to permit advertisements such as that proposed by Amnesty. However, Richard Inglewood eventually concluded that "we are not persuaded that the public interest would be served by the amendment which you propose ... while perhaps allowing advertising which supported worthy political groups [it] would also be bound to allow advertising which benefited others".[859]

What would the Appeal Court determine? This was a powerful court, presided over by Lord Justice Woolf, the Master of the Rolls (and later, between 2000 and 2005, the Lord Chief Justice). John Norrington told the Authority members after the hearing on 28 and 29 October, but before the judgement was known, that "all three of their Lordships appeared to be against the Authority", and that it was "the view of observers, including our own legal team" that the Appeal Court would find for Amnesty.[860] The judgement took a long time to arrive. Just before Christmas, the Authority learnt to its surprise that it had won after all, although it is clear from the detailed judgements that the Court had wished it could have found otherwise. Woolf's judgement was that "with some hesitation I have come to the conclusion it would not be right to allow this appeal".[861] He urged Amnesty to submit a fresh advertisement, with a heavy steer to the Authority that it might want to re-consider just how it applied the 'political' test.

The whole episode reflects considerable credit on the Radio Authority. The regulator had been thoughtful and diligent in considering a very tricky problem (and was praised by the Appeal judges accordingly). It had set aside its own prejudices, and concentrated on applying the law as best it could, even though that risked unpopularity with an important national constituency. The Authority

858 The delightful Alf Dubs was born in Prague, and was one of the Jewish 'kindertransport' children rescued from the Holocaust. As MP first for Battersea South and then Battersea, and subsequently as a life peer, he has campaigned tirelessly for human rights. When he was chair of the Broadcasting Standards Council, he was part of the winding up of the heritage regulators for their replacement by Ofcom. He was, and is, a steadfast friend of the *independent* principle in broadcasting. For him to be prepared to press for a measure of political advertising, even on the emotive subject of political prisoners, shows how far the liberal establishment was won over by Amnesty's arguments.

859 Letter from Lord Inglewood, DCMS, to Lord Dubs, 17 June 1996.

860 *RA Paper* 133(96) 31 October 1996.

861 R v Radio Authority *ex parte* David Neill Bull and Nigel Wright 17 December 1996.

seemed to have learnt much from the Emap/TWC episode, and to have got its house back in order. There is more than a suspicion, on the other hand, that Amnesty had chosen to go for the Radio Authority, thinking it a softer target than the ITC. Amnesty re-applied for permission to advertise on 22 October 1997[862] and a rather reluctant Authority decided that it could accept the heavy guidance given by Woolf, and change its mind. It cautiously approved the principle when it met the following month.[863]

Late in 1995, the Authority faced its first judicial challenge to a licence award. The Sheffield/Rotherham and Doncaster licences were both operated by Emap, from their Sheffield station, with common programming. The simultaneous re-advertisement of these licences opened up the possibility of a separate application for Doncaster, which duly came from Trax FM, a subsidiary of Lincs FM, where the tenaciously determined Michael Betton was keen to build a network of local stations reaching out north and west from Lincoln. Emap's application proposed the Doncaster operation continuing as an extension of its Sheffield output. Lincs wanted much more of a stand-alone service for Doncaster. In November 1995, the licence was awarded to Hallam. Betton, and his chairman, Ednyfed Hudson-Davies, were furious, feeling that the regulator had departed from its statutory duty to give priority to localness of output, and challenged the decision though the process of judicial review.

The case was heard in the High Court on 11 January 1996. The atmosphere on those days was strange. Both sides employed eminent barristers – the Authority as usual retaining the brilliant, much sought-after David Pannick QC, as it did whenever possible – but the close contact over many years, and the working friendships between the principals, made it almost chummy, outside the formal proceeding. I remember both Betton and Hudson-Davies showing a cheerful courtesy during those informal contacts around the court room. The presumption against a court overturning an administrative decision such as a licence award is quite strong. Trax needed to show that the Authority had misdirected itself in how it applied the rules, as the test of unreasonableness would never work in their favour. Mr Justice MacPherson found for the Authority, saying that he saw "no strength in the points raised by the applicants".[864] None of the established radio companies ever again chose to challenge a licence award through judicial review, although there were many occasions when they felt at least as aggrieved as Betton had done over Doncaster.

Subsequently, the Authority advertised a separate new stand-alone licence for Doncaster in June 1998. There were three applicants, and it was awarded to Trax FM in February 1999. The published appraisal of the winning application was notable. "The successful applicant group has a long and active history of commitment to the Doncaster area, starting with its first application for a licence there in 1995 and continuing more recently with two short-term RSL broadcasts. This has not only served to generate strong local support from local decision-makers and the general public, but should put the group in a good position to ensure their service appeals to local tastes and interests … members noted that there would be

862 Letter from Bindman & Partners on behalf of Amnesty to David Lloyd, director of programming, RAu 22 October 1997.

863 *RA Minutes* 9(97) of a meeting on 3 October 1997.

864 Regina v The Radio Authority ex-parte Trax FM Ltd, QBD no. CO-4198-95.

some programming shared with Trax FM in Bassetlaw (another Lincs FM licensee) but did not consider that this would unacceptably detract from the service's local relevance and appeal."[865] Michael Betton could be forgiven for not appreciating the unintended irony.

The other licence award to attract a judicial review was a challenge mounted by a litigant in person, Francis Wildman. The Authority had considered early in 1997 the award of small-scale alternative location licences (known as 'sallies'[866]) for East Kent. Three licences were awarded, to CTFM (Canterbury), Neptune Radio (Dover) and Thanet Local Radio. These awards caused no end of trouble. There were allegations of bias and procedural impropriety in the award to Neptune, chaired by former Authority member John Grant, requiring an internal investigation before they could be refuted by the Authority at their April meeting.[867] The winning group for Thanet fell out among themselves, requiring the Authority to check whether the award should still stand.

And then along came Francis Wildman. He had submitted an application, A-TEN FM, with proposals to cover Ashford, but neither the locality nor his application was chosen. After complaining without success, Wildman sought leave to seek judicial review in June. A litigant in person (although allegedly supported by the law faculty of a local university), he received extended courtesy and help from the courts throughout the long saga, but his initial application was refused. He persisted, and in September 1997 was eventually granted leave. The substantive hearing to review the Authority's decision to grant the East Kent licences took place in the High Court on 14-16 June 1998.

Judgement was delivered by Mr Justice Jackson on 17 June 1998, who decided that the Radio Authority had acted lawfully in its decision not to grant Wildman a local radio licence for Ashford.[868] He agreed that the Authority was entitled to invite applicants to specify the precise localities for which they intended to provide a service, under the 'sally' procedure. Wildman had additionally argued that the decision made by members not to award a licence to the applicant had been irrational, but this argument was rejected too. There followed a wrangle over a further appeal, and then over the costs which had been awarded to the Authority, which it eventually decided not to pursue.

Meanwhile, in the summer of 1998, Wildman had another go at judicially reviewing the Authority. On this occasion, Mr. Justice Owen rejected an application for a judicial review of the decision of the Radio Authority to allow part-networking on Capital Radio's AM services through amendments to the promises of performance of five of the Capital licences. Leave to appeal the decision was refused.

While the Wildman case followed its long drawn-out course, there was a challenge to the Radio Authority from a much more threatening quarter, BBC2's *Newsnight* programme, and a rash of accompanying accusations. These events fundamentally changed the Authority's attitude to the radio industry. Aside from the gravity of the challenge itself, they mark the moment when the Authority was brought up hard against the implications of the shift from the friendship of

865 Local licence award Doncaster. Appraisal of successful applicant. Radio Authority, 19 February 1999 .

866 See chapter 23.

867 *RA Paper (129(97)* 25 September 1997.

868 Radio Authority News Release 96/99 17 June 1997.

independent radio to the harsh calculations of *commercial* radio, and it did not like what it now understood.

Newsnight alleged improper practice in the award to Vibe FM of the East of England regional licence in April and May 1997. That was the key accusation in a long list of points which they put to me over the summer of 1998. This culminated in the BBC recording an interview with me on 21 September, part of which was used in the eventual report.

The essence of the East of England allegation was that the Authority members had made a decision to award the licence to Esprit at their April meeting, but had been improperly asked to reconsider under some sort of unspecified pressure. The Authority had indeed made rather a meal of this award, unusually postponing a final decision from April to May. The supposed 'smoking gun' was that the eventual winner was a group which included a participant in which a charitable trust of Peter Gibbings' family had held an interest at the time of the April meeting. That interest was on the point of being disposed of at the start of April, and had been sold by the May meeting. There were rules about how to handle any possible conflicts of interest set out in the legislation, so because of the trust involvement deputy chairman Michael Moriarty rather than Peter Gibbings had chaired the April meeting. By May, when the shares were no longer owned by the trust, Gibbings had resumed the chair, and it was in May that the final decision was made. The Radio Authority maintained robustly that no decision had been reached in April.

What gave the programme apparent substance was that *Newsnight* claimed that the allegations were supported by John Norrington, who had retired at the end of January 1998 as secretary to the Authority. The programme-makers seemed determined to nail their story using whatever means, to the extent even of having a secretary phone the Authority, pretending to be a friend of an absent member of staff who they hoped might give them information. That was excused by the BBC as merely "a sloppy piece of work".[869] To an increasingly embattled Radio Authority, it felt like further evidence of an attempt to do it down.

A separate, highly serious matter had arisen at around the same time, and the BBC decided to meld the two issues. In November 1988 the Authority's deputy head of programming, Janet Lee, was arrested by police in connection with her visit to India under a Commonwealth Relations Trust bursary. She was suspended from work while police investigations continued and returned to work in April 1999, after the Authority was told that no charges were to be brought against her.

The Authority was the subject of much other tittle-tattle. It may be that some who by now judged it to be in their interest to undermine the regulator were not averse to using this opportunity. A national magazine editor suggested to me at the time that this was a deliberate and concerted campaign. Certainly, a slew of extraordinary stories began to do the rounds. Punch magazine reported "outlandish rumours ... bags stuffed with cash ... handed over in Claridges in exchange for a licence ... a station using blackmail to secure a licence".[870]

Newsnight broadcast its report as part of its edition on 1 December 1998, under

869 Letter from Stephen Mitchell, BBC deputy head of news programmes to Tony Stoller, RAu, 26 August 1998.

870 *Punch* 13 March 1999.

the title "Radio skulduggery". The programme set out its allegations about how the East of England award had been made, including that minutes had been fabricated. The broadcast material included an interview with a disappointed applicant, Ian Davies, and film of John Norrington playing golf, although without any broadcast interview with him. The programme also featured my own recorded interview, input from Susannah Simons on behalf of GWR criticising "bureaucrats sitting in an office in London", Robert Stiby complaining that "some sort of decision seems to have been covered up", and Guy Hornsby explaining how much money he had made by selling on a radio licence. By way of introduction, *Newsnight* ran a report on the arrest of Janet Lee.

The Radio Authority had complained vigorously to the BBC at the most senior levels that it was being treated unfairly, but to no avail, leaving a scar on those high level relationships which took a long time to heal. After the broadcast, Paul Brown for CRCA, not many years departed from a senior participation in the process, wrote to secretary of state Chris Smith, "expressing concern about the way radio licences are awarded".[871] Capital chief executive, David Mansfield, spoke of "a number of decisions and incidents [which] have come to light which cause us to further question our regulator". No wonder paranoia ruled at Holbrook House for a while.

I had decided that the only appropriate action would be to commission an independent inquiry into the licence award allegations (the police investigation regarding Janet Lee was a wholly separate matter), and duly asked the forensic department of the Authority's auditor, Grant Thornton, to conduct that at once. The choice of investigator was a presentational mistake on my part. I thought of the auditors as both independent and stern, but there were others who felt that the relationship between corporations and their auditors was not always so arms-length. Events such as the Enron scandal have since borne that out, but it was certainly not the case in this inquiry, which was a thoroughly uncomfortable episode for everyone at the Authority. The investigators interviewed at length everyone involved with the decision, with both electronic recordings and short-hand to provide transcripts of those interviews; some people were interviewed two or three times, for hours at a stretch. The investigators spoke to or were in letter contact with all the interested parties, and noted receiving letters from John Norrington between 2 December 1998 and 3 February 1999, "in which he detailed his observations and recollections on matters pertinent to the Inquiry".[872]

Grant Thornton published its report on 3 March. The report made a number of recommendations where it thought that the Authority's processes, as revealed by its investigations of this incident, should be improved, and which it said "can be characterised as a 'tightening-up' of process". All of these had either already been adopted by the Authority or were then put into place. The report's central finding was unambiguous. "The RA cannot operate effectively as the licensing and regulatory body of the independent radio industry if the integrity of its processes, members or staff is suspect. Mr Norrington's accusations raised fundamental questions about the fairness of the East of England award. However, our Inquiry

871 *Financial Times*, 3 December 1998.

872 Grant Thornton, *Report of the Inquiry into allegations concerning the award of the East of England Licence*, March 1999.

leads us to conclude that the East of England award process was fair ... In our view, none of the issues which were the substance of Mr Norrington's allegations had an impact on the outcome of the licence award."[873] The Authority was exonerated, and *Newsnight* rather grudgingly reported as much in a broadcast statement within its edition of 3 March 1999.[874]

The whole episode caused great distress to many of those involved. They felt they were in the trenches while bombs rained down. One director has spoken to me of "a sense of being very much under the cosh, with everyone climbing onto the bash-the-Radio-Authority bandwagon". I recall that senior Radio Authority staff also lost much of the trust and some of the affection that they had developed over the years for key industry players, most of whom had been at best silent, and at worst had seemed keen to endorse the now dismissed allegations.[875] Afterwards, it was more difficult to give companies or individuals the benefit of the doubt, on matters which previously would have been settled with friendly informality.

Later, it became clear that for many of the Authority staff, not just those directly involved in the events and the investigation, it was a seminal time in a much more positive way. On 25 February 2003, as part of a training course preparing for the shift to Ofcom, they were asked to recall good times and achievements during their years, as a technique for helping sustain a positive approach through the impending changes. Many identified the exoneration of the Authority from the *Newsnight* charges as a highlight. It had, after all, been a shared burden, and became eventually a shared affirmation too.

The next set-piece challenge to the regulator came – not unexpectedly – from Ralph Bernard, but at an unexpected moment. As part of its investigation and promotion of third tier, not-for-profit radio, the Authority had convened a conference at the Royal Society of Arts on Monday 12 February 2001, to discuss its plans for Access radio experiments. Bernard, a long-time supporter of this type of radio – partly, but not entirely, to free ILR stations from their remaining public service obligations – had been invited to speak. However, on the previous Friday he telephoned me to warn me that he intended to talk about other, wider matters. Neither I, nor the assembled chair, deputy chair and members of the Authority, were prepared for what came next. On the Monday morning, Bernard, who had been growing increasingly frustrated by what he saw as the obstructionism of the Authority to his commercial moves, launched into a vitriolic attack on the Authority. The immediate occasion was a complaint that GWR's Vibe FM was departing from its promise of performance, evidenced in monitoring by playing a Kylie Minogue song in a supposedly dance format.

The Radio Authority's Martin Campbell recalls the issue more soberly than it was reported at the time. "GWR had a stake in the dance station Vibe (East of England) and the spotlight had been put on that because it had started to become not very dancey. We had complaints that it was 'Kylie FM', and in my conversations with Ralph I said ... that we certainly wouldn't want to say he couldn't play Kylie – but why not play the dance tracks or dance mixes rather than the out-and-out

873 Ibid.

874 BBC News Press Publicity, 3 March 1999.

875 A notable exception was Kelvin MacKenzie, who generously offered me quiet re-assurance that such clouds pass on to rain on someone else. As the great tabloid editor of his era, he had reason to know how the game was played.

pop? Ralph immediately runs to the press saying we told him he couldn't play the Pert Pop Princess."[876]

Bernard told me that "the Kylie Minogue incident was the last straw".[877] Given a platform at the RSA, he held forth at length on the evils of regulation, and this regulator in particular. "There's a problem in the radio industry", Bernard opined. "It goes like this. The BBC has the freedom to manoeuvre, to change its formats and services ... the commercial radio regulator thinks that the people who run the large groups of stations don't know or understand their audience as well as the BBC ... that *it* always knows best ... the definition of formats is dictated to highly experienced operators by people who have absolutely no experience in the programming of radio stations in the modern broadcasting environment ... sometimes by reference to pieces of paper which could be seven years old ... like all dictators, it feels threatened by democracy ... and thrives in a system of utterly unchangeable decision-making".[878] There was more in this vein. Following some tales of his own radio experiences, Bernard concluded by hoping for a new regulator which would be "fair, open and a supporter of radio creativity".

Authority members were outraged to be attacked at this occasion. Bernard himself recognised that what he said might seem to be simply "the sour grapes outpourings of a known critic of the regulator". Whether it was thought to be that or not, it was also quite clearly a *cri de coeur* from commercial radio to be allowed to be just that, to put aside the last vestiges of the old public service obligations of independent radio. When in September GWR revealed a 70 per cent fall in its profits, the frustration and the need to find someone to blame became manifest. The reality was that a 'beauty parade' system for awarding licences seemed out of tune with the notions of the proper operation of a market. Licences were awarded competitively on the basis of programme proposals, which then had to form an enforceable part of the eventual licence. Since the larger groups failed to win new licences, they set out to buy stations from the winners. They were then saddled with a set of obligations which they spent the succeeding years trying to change or evade. Bernard was pointing out in the starkest terms the fundamental contradiction in the statutory arrangements which seemed to permit commercial radio, but allocated licences in such a way that the stations could not freely change their formats.

The last two legal challenges came from the Christians. Premier Christian Radio went to the brink of judicial review, in a dispute over whether it could benefit from the automatic renewal offered to other ILR licensees in exchange for a commitment to DAB; and United Christian Broadcasters took the UK government to the European Court of Human Rights (ECHR), over the continued exclusion of religious groups from holding national radio licences.

Premier Radio was placed in an uncomfortable and disadvantageous position by an oversight in the drafting of the 1996 Broadcasting Act. That provided automatic renewal for licences committing to digital, but the wording of the statute meant that religious groups were disqualified from holding digital licences. The government had accepted the Authority's representations that this needed to be

876 Email from Martin Campbell to the author, 1 October 2008.

877 Ralph Bernard in an interview with the author, 19 November 2007.

878 Speech by Ralph Bernard to Radio Authority Access Radio Seminar, 12 February 2001.

corrected, but it was not going to be done until the upcoming communications bill. The advertisement of the final digital multiplex for Great London in 2001 was just too soon for Premier, which found itself denied both automatic renewal and the digital outlet it craved. The company pressed the matter through lawyers, by encouraging its listeners and supporters to lobby, and through MPs. However, as I wrote to one MP, the Authority's "legal advice is quite clear that we cannot anticipate the likely legislative change", acknowledging that "we have a good deal of sympathy for the predicament Premier Radio finds itself in".[879]

Lawyers were retained, cases were prepared, and the hearing was fixed for 5 and 6 November 2001. However, on 12 September – while the world was reeling from the attack on New York's Twin Towers[880] – Premier Radio withdrew, citing the high potential cost and the lack of a backer to support them in this challenge. The Radio Authority completed its life with the record of never having lost a judicial review. In such a challenging industry, through confrontational times, it was a vindication of due process. Premier Radio re-applied for its analogue licence, which was duly re-awarded on 14 March 2002, and it is still the standard-bearer for home-grown Christian radio in the UK.

United Christian Broadcasters (UCB), founded by New Zealand evangelist Richard Berry, was an altogether different proposition. Its British organisation describes itself as "a Christian media ministry within the UK and Ireland, formed to promote the good news of the Kingdom of God".[881] Whereas the Radio Authority had nurtured Premier Radio since it began in June 1995, was proud of the diversity it offered, and was on good and friendly terms even when there was the threat of legal challenge, UCB was a global mission, much more remote from the UK regulator. It held a licence for its radio services on satellite, and was starting to eye the possibilities on the national digital multiplex. That was prohibited by the 1996 act, and the Authority had proposed to the government removing only the prohibition on local digital ownership by religious bodies. This issue provoked more public representations than anything else in the entire bill, a testimony to radio's lightning-conductor effect, and the ability of UCB to mobilise its supporters. DCMS officials told me at the time that they had received upwards of 100,000 letters, pressing that the disqualification of religious bodies from owning national radio licences should be lifted.

UCB took the UK government to the ECHR. It lost, but used that judgement to pursue the matter further in parliament. The UK case was explained in a letter from secretary of state Chris Smith to the shadow minister, John Whittingdale. "The government's view is that the retention of the disqualification for religious bodies to the extent proposed by the bill is compatible with the ECHR. The aim of the ban (as summarised by the European Court of Human Rights in ruling inadmissible the challenge ...) is to promote the efficient use of scarce resources, i.e. radio spectrum in order to safeguard pluralism in the media, cater for a variety of tastes and interests, and avoid discrimination between the many different religions practised in the United Kingdom."[882] In the report stage debate, Whit-

879 Letter from Tony Stoller, Radio Authority, to Keith Hill MP, 19 June 2001.

880 Three of the Radio Authority directors were in the eastern United States on 9/11; Martin Campbell and Neil Romain in Washington, DC, myself in Charleston, SC.

881 www.ucbmedia.co

882 Quoted in *Hansard,* House of Commons, 30 January 2003, col 829.

tingdale put the opposite case, which had by now attracted wide support from Christian groups in the UK. "Church groups across the board have strong concerns about the maintenance of the disqualifications, even though they are of a more limited type. Religious groups find it especially offensive that they are disqualified."[883] In the end, the government stood firm against the intense lobbying, and religious groups remain prohibited from holding national licences, or taking on the 'gatekeeper' role as owners of digital sound multiplex licences.[884]

The last set-piece challenge from the radio industry to the Authority came over the question of 'automation', which can mean all sorts of technical support for the programming output. At its simplest, pre-recorded programmes can be played by the station's computer systems, just as they were once played from disc or tape by technical operators in the studios. This has long been part of radio's history. The point at issue in 2000 was the use of computers to play out music, advertisements and 'voice tracks', so that it appeared to the listener that there was a live presenter running the radio programme. It was an essential part of running small-scale radio stations. While the computer did the broadcasting, the presenter could be preparing, gathering the news, and doing all the multi-tasks which a station with perhaps fewer than a dozen staff needed to accomplish. However, looking across the Atlantic to the trend in the US, the Authority feared that automation in this manner would be used as mere cost-cutting by stations which could afford more, and thus jeopardise programme quality. For these purposes, 'voice tracking' is all recorded in advance, usually by a very cheap freelancer, perhaps doing a whole evening's and overnight programming for a small – often derisory – fee. Where it was linked in with the growing amount of networking and the standardisation of group output, then localness would be further put at risk.

Radio Authority director of programming, Martin Campbell, took the view that a shift from promises of performance to formats (discussed in chapter 22) meant that the regulator would have no way of controlling this practice if it threatened to get out of hand. He therefore proposed to the Authority, at its meeting on 5 October 2000, that there should be limits set on the amount of daytime automation. With the members' agreement, he wrote to all stations on 11 October, "to impose constraints on the amount of daytime automated programming. FM stations are to be allowed a maximum of two hours of automation during daytime. AM stations ... will be limited to four hours of automated daytime programming each day. Networked AM stations, with an obligation to carry only four hours of local programming, cannot automate those at all."[885] Although the letter flagged possible "exceptional circumstances", it hardly hinted at any flexibility. In the Authority's mind, these limits were set to allow all that was currently done via automation, but to prevent its increase.

The stations went potty. Small companies felt that their lifeline was being cut. Chris Carnegy, who often spoke for the wider constituency, wrote that "the idea is repugnant in principle. In practice, the imposition of any fixed overhead cost is also disproportionately damaging to smaller licensees. I hope the Authority will come to its senses very fast."[886] Groups of smaller stations were affronted by the

883 *Ibid*, col 830.

884 Communications Act 2003, section 348.

885 Letter from Martin Campbell, Radio Authority, to all managing directors and programming directors of AM and FM stations, 11 October 2000.

principle and troubled by its practical implications. Michael Betton wrote that "I cannot understand why the Authority should wish to become involved in the way a service is delivered rather than what is delivered. Successful radio broadcasting has always been a trick, where all the senses can be stimulated just by the audio we broadcast. Listeners cannot measure the quality of what we broadcast except in audio terms."[887] The large stations (in suspicion of whose intention these rules had been crafted) saw an opportunity potentially thwarted. Steve Orchard argued that "GWR has experience in increasing local content via automation. For example, GWR's Bath output carries automation for 50 hours a week in order for us to improve localness *(sic)* ... Would members prefer live but less local, the only viable alternative?"[888]

A few brave souls wrote in support of what former programme-maker Campbell had intended to be protective of ILR programming, and rather more expressed private but deniable support. But none could accept with comfort that this new rule was to be imposed without prior warning or consultation. Paul Brown at CRCA wrote on behalf of the industry that the new limits "seem to us to comprise a revision of the Authority's rules that should be included in a revised programme code".[889] This would oblige the Authority, under the Broadcasting Act, "to consult every person who is a holder of a licence" before issuing any final revision.[890] Emap's boss, Tim Schoonmaker, was more succinct. "Dear Martin. Thank you for your letter of 11 October on Automation. Where's the consultation? Regards, Tim."[891]

At their meeting on 2 November 2000, the Authority members considered the responses received from the industry and licensees to the new limits on daytime automation.[892] Both they and the senior staff still failed to see how iconic this issue was becoming for the radio industry. Although noting "that the letter could perhaps have better reassured the smaller stations that the Authority would be most unlikely to insist on changes to their existing levels of automation ... [members] confirmed the decision taken at the October meeting to set limits on daytime automation. Members considered the policy to be central to the Authority's duties to maintain quality and that it did not constitute micro-management ... In making their decision on automation limits, members were aware of industry views. Moreover, members were of the view that, because the change represented a change to licences through formats and was not a programme code matter, there was no duty to consult."[893]

The Authority had got it wrong. A letter from CRCA's solicitors threatening judicial review caused everyone to stop and think again. Sky Television ran a special programme on the topic. The companies would not let the matter rest, seeing in it both a practical issue – should the regulator have rules for 'inputs' to programming rather than merely for 'outputs' – but also sensing a winnable trial of strength

886 Letter Chris Carnegy, 107.8 South City FM, to Martin Campbell, 16 October 2000.
887 Letter from Michael Betton, Lincs FM, to Martin Campbell, 23 October 2000.
888 Letter from Steve Orchard, GWR, to Martin Campbell, 27 October 2000.
889 Letter from Paul Brown, CRCA, to Martin Campbell, 12 October 2000.
890 Broadcasting Act 1990, section 91(3).
891 Letter from Tim Schoonmaker, Emap, to Martin Campbell, 12 October 2000.
892 *RA paper* 104(00) 22 October 2000.
893 Published note of the meeting of the Radio Authority, 2 November 2000.

with the Authority, which had been growing in confidence since putting the *Newsnight* episode behind it. They were right; the regulator backed down. On 22 May 2001, I wrote to all ILR licensees that "while the Authority holds firm to the view that this is an important and necessary area in which it needs to ensure that it has the necessary minimum of powers ... it may now be possible to move these rules and guidance from being individual licence conditions to forming part of the programming code instead ... To make such a change will require further, formal consultation."

The spread of what former Capital MD Richard Eyre had called "the McDonaldisation of ILR programming"[894] suggests that the Authority was right to be concerned. However, it had set about the matter without understanding either that consultation was now the order of the day, or that the companies would oppose the practical limits as well as the principle. In a parallel with the years around the Heathrow Conference in 1984,[895] the radio industry was coming to resent the very existence of a regulator with notions from its *independent* radio past. When I read Richard Hooper's speech[896] (he was unwell) to the Manchester Broadcasting Symposium on 1 May 2001, urging that the companies should pay more attention to quality in ILR programming, Paul Brown's reaction was vituperative. "I can understand why the Radio Authority wants to ensure that its value is understood and its relevance is maintained in the new order. Fine, but I wish it would make a greater effort not to devalue its current licensees in the process."[897] The battle lines for the communications bill debates had been drawn, and work was starting on digging the trenches.

There had, therefore, been nine set-piece challenges to the Radio Authority. Six had been judicial reviews or similar – Emap/TWC, Amnesty International, Doncaster, East Kent, Premier and UCB – and in each case the Authority had done well. One, around the *Newsnight* allegations, was trial by media, and damaged the regulator for quite a time, although it emerged stronger and more robust. Two other challenges saw direct confrontation between industry figures and the regulator. The second of these, the 'automation' controversy – marked the end of creative contact between the industry and its old regulator. From that point onwards, the arm-wrestling would be over the terms of the forthcoming Communications Act, and the structural arrangements it would introduce. The radio companies' attention turned towards keeping Ofcom as harmless as possible, the Radio Authority's to preserving in the new dispensation whatever it could of value from the old.

894 Speaking at the RSA Seminar on Access radio, 21 February 2001.

895 See chapter 11.

896 *Regulating for high quality*, Richard Hooper speech to the Manchester Broadcasting Symposium,1 May 2001.

897 *Radio Magazine*, 4 May 2001.

Chapter 20

Radio by numbers

Digital radio

To that fine radio actor Peter Jones, as the voice of *The Book* in *The Hitchhikers Guide to the Galaxy*, three questions were central. "Why are people born? Why do they die? Why do they want to spend so much of the intervening time wearing digital watches?"[898] Digital transmission technology is to post-modern broadcasting what the arrival of the antipodean red-top tabloid culture had been to newspapers, at once liberating and trivialising. For digital radio enthusiasts, the new technologies promised riches; a huge increase in the number of available services, highest quality transmission and reception, text and pictures to accompany the sound, a whole new shining future. For radio folk as a whole, there was a vague sense that transmission methods would have to go digital, since everything else was, but also an Augustinian wish for that to happen later: 'give me digital, but not yet'.

To the bright new Radio Authority in 1990, DAB represented the chance to get even. The BBC had retained the best AM frequencies and dominated the FM bands. The Radio Authority did not want to be left behind by the BBC on digital; better than that, DAB would even the score. The Radio Authority's director of engineering, Mark Thomas, oversaw and promoted the technical planning and introduction of DAB to commercial radio in the UK. He recalls that two technical limitations seemed to be holding back independent radio; having enough good analogue frequencies to get full and wide coverage, and the subjective quality of reception. "DAB offered the possibility to overcome those constraints."[899]

DAB as a whole was also the product of a hunger among broadcasters generally for more spectrum. BBC engineers at their Kingswood Warren headquarters had been working on digital radio and digital television systems for some time. In 1987, the Eureka 147 digital audio broadcasting project was established, with an initial consortium of a few European broadcasters, many equipment manufacturers, European car makers and transmission companies. The only UK member was the BBC. In 1994, the industry watchdog in Europe, the European Telecommunications Standards Institute (ETSI), formally accepted the DAB standards, confirmed at a world level in 1995. The BBC was now getting behind DAB in a big way. Radio MD Liz Forgan obtained a budget of some £10 million to bring DAB to a target

898 BBC Radio Four, 1978.
899 Mark Thomas in an interview with the author, 1 July 2008.

of 60 per cent of the UK in five years. She persuaded David Witherow, later to be deputy chair of the Radio Authority, to come back from retirement to run the project. He was to be a leading figure in most of what was good about the planning of DAB at the BBC, its international promotion, and later by encouragement also at the Radio Authority. At the BBC's instigation, a European DAB forum was established in 1995. Based initially at the EBU headquarters in Geneva, it soon became the World DAB forum with its offices in London.

The BBC launched its first DAB services, experimentally, in September 1995. At this stage there was no support from the commercial radio companies in the UK, although GWR's Quentin Howard – who started as station engineer for Wiltshire Sound in Wootton Bassett, and went on to be the leading proponent of DAB in the UK – was already interested. In Europe, the state broadcasters in France and Germany were also experimenting with DAB transmissions, and there was consequently a risk that the public service broadcasters might monopolise the new frequency band. That encouraged the Radio Authority to become involved with DAB, and Peter Baldwin to chair sub-group D1 of the European Broadcasting Union (EBU), which was responsible for the essential European dimension.

The Radio Authority issued its first policy statement in October 1992. Its approach was unequivocal. "The Radio Authority believes that DAB is the single most significant advance in sound radio transmission technology since the development of FM broadcasting; its impact is likely to be even greater than that of FM."[900] It asserted that the technology was already well-developed and that radio sets could be available by 1995 "and not necessarily at high cost". It noted the key decisions which would be needed: the allocation of at least 12.5 MHz in the region of the 230 MHz FM band; and a solution to the transmission arrangements. Since digital radio works by weaving together or 'multiplexing' a number of different programme channels on a single frequency – as different strands of line are woven together to make a rope – the analogue licensing structure, whereby programme origination and transmission existed as a single licensed entity, also needed rethinking.

From this point onwards, the Authority was effectively committed to advancing the cause of DAB, and in particular the Eureka system which it asserted was "the only one likely to be adopted on a worldwide basis". That was neither strictly true then, nor particularly prescient. Most digital technologies are highly fluid in their early stages, before the decision of the market; that is, in effect, which system the large manufacturers choose to back, and whether their choice is then endorsed by potential consumers. Digital radio was no exception. Even at that time, the National Association of Broadcasters (NAB), representing the whole swathe of US commercial radio, was getting behind the 'in band on channel' system (IBOC). This worked by squirting the digital signal within existing analogue channels. European engineers doubted whether it would work, and claimed that this was part of the American 'not invented here' syndrome. American radio station owners, though, were far more concerned to ensure that there were no new challengers to their decades-old hegemony over radio broadcasts. IBOC meant that the existing licence-holders kept the freehold, whereas DAB would mean the possibility of upstart new competitors. Whether the IBOC technology worked or not hardly

900 *Initial Policy Statement on Digital Audio Broadcasting* Radio Authority October 1992.

mattered.[901] Other digital radio technologies were already extant or envisaged. Satellite digital radio transmissions, which were better suited to the large territories of Africa and Asia, as well as the Americas, were developed and deployed by World Space. Digital Radio Mondiale (DRM), similar in many ways to DAB but using AM rather than FM, arrived later than DAB but had many promoters.

The Radio Authority and the BBC, however, had already nailed their colours to the DAB mast. For the Corporation, their substantial part in the development of DAB, and leading role within the Eureka 147 consortium, kept their field of vision pretty narrow. The Authority was clearly influenced by the BBC, as was the European Broadcasting Union (EBU), where the BBC always has disproportionate influence. In turn, the EBU's preference further confirmed the Radio Authority's Peter Baldwin and Mark Thomas in their commitment to Eureka, and they promoted it to the UK government accordingly. Through the joint efforts of the Authority and the BBC, a UK national digital platform was convened at the start of 1988, giving some formal shape to the meetings of regulators, equipment manufacturers, transmission providers and programme providers. The commercial companies were still not enthusiastic. Apart from Quentin Howard, and subsequently GWR creator Ralph Bernard, few were more than lukewarm at best.

The Authority, however, was pressing ahead. Relying on its statutory obligation to "secure the provision within the UK of a diversity of national services and a range and diversity of local services"[902] it commissioned from GEC-Marconi in 1994 a study of how quite limited spectrum might be used. This concluded that local DAB networks should consist of "dense low-powered transmission networks in order to deliver a reliable high-quality service yet, at the same time, using a limited spectral allocation as efficiently as possible".[903] That in turn provoked argument within commercial radio, as the more transmitters the higher the cost of establishing DAB. The Authority had determined that the local DAB areas should replicate – so far as possible – the pre-existing analogue 'total survey areas'.[904] The choice of dense transmitter networks flowed inescapably from that, and consequently the relatively high costs of building the transmitter infrastructure, and in effect gave the existing ILR UK operators the opportunity to dominate digital radio.

One area in which the Radio Authority has been criticised is for its choice of the coding standard mpeg2. This is a continuing topic of debate, as modern approaches to digital broadcasting try to eschew formal standardisation. Mark Thomas, though, is clear. "At the time it was the only coding standard we could use. As a technical standard for DAB, we chose the only one then available. There

901 In the event, the dominance of the American analogue networks was challenged from space. Two rival digital audio services launched in July 2002. Sirius and XM both offered up to 100 channels for around $10 per month, relayed by satellite to a new generation of radio receivers. To the furious opposition of the NAB, they also acquired terrestrial repeaters to fill in some of the signal gaps in major cities. The market sorted out the competing satellite audio technologies, with Sirius buying out XM, a deal approved on 25 July 2008 only after an 18 month battle with regulators and Congress. According to Associated Press that day, the agreement came thanks to the casting vote of Republican FCC nominee Deborah Taylor, once the companies agreed to pay $19.7 million to the U.S. Treasury for violations related to radio receivers and ground-based signal repeaters. At that date the new company was claiming over 18 million subscribers.

902 *Broadcasting Act* 1990 s.85 (2).

903 *Radio Authority Annual Report* December 1994.

904 See chapter 6.

are now more efficient coding standards, and if you were starting from scratch you probably wouldn't choose mpeg2." The French have chosen a different approach, based on a continuing sense of capacity scarcity, but that may be more apparent than real. And the current DAB standard has at least prompted fairly extensive manufacture of radio sets able to receive it. David Witherow stresses that it is still "a very sound system. People like using DAB to listen to the radio."[905]

An international conference at Weisbaden in July 1995, under the auspices of the International Telecommunications Union (ITU), offered the solution to the frequency needs. The Radio Authority was very active in the UK delegation in Germany, led by the Radiocommunications Agency, and in the pre-planning. Mark Thomas and senior frequency planner Nigel Green spent weeks in the negotiations, to good effect. The eventual conclusion was that television in the UK would vacate the part of the FM band which had been used for the old 405-line black-and-white transmissions. The dividend for radio in the UK was sufficient frequency space to allow for seven DAB frequencies: one for national BBC services; one for national commercial services; and five to allow the necessary intricate pattern to bring local digital services – BBC and ILR – to most of the UK. Further negotiation was needed between European administrations, where coordination was essential. The story going the rounds was that if the UK used certain frequencies, they might open all the remote-controlled garage doors in France. Perhaps sadly for the headline-writers, the experiment was never tried.

Once the frequency question was largely settled, attention turned to the statutory and administrative arrangements. These were strange times for UK governance, with the Major administration fading and a technology-obsessed New Labour waiting for its cue just off-stage. In the general wreckage of policy burn-out, and amid the frustrations of waiting for the economic wheel to turn, the government was looking to its legacy. After the forced resignation of David Mellor, the Department of National Heritage had seemed to be in limbo, not least under the passive oversight of Stephen Dorrell. It was digital broadcasting's good fortune that he was swapped in July 1995 with Virginia Bottomley, who was looking to re-build her reputation as secretary of state at the Department of National Heritage (DNH) after a torrid time in the Department of Health. With a thorough and sympathetic junior minister in Richard Inglewood in the House of Lords, DNH introduced legislation for the UK's digital broadcasting future.

That mostly meant digital television. Ever since the takeover of British Satellite Broadcasting by Sky, finalised in November 1990, the satellite broadcaster had been anticipating a transition from analogue to digital satellite television (DST). Since 1992, DNH had been working in earnest on a way to introduce digital terrestrial television (DTT). That was far from a certainty. Mathew Horsman, in his compelling history of BSkyB, notes that "the main attraction lay in freeing up analogue capacity, for use by other commercial operators … with BSkyB such a strong competitor, and with its digital satellite plans, if delayed, already well advanced, there were doubts that DTT could even get off the ground".[906] Nevertheless, enthusiasm for multi-platform television encouraged the government to proceed. BSkyB's Sam Chisholm saw programme delivery as "a horizontally

905 David Witherow in an interview with the author, 1 October 2008.
906 Mathew Horsman, *Sky High*, Orion Publishing, 1997, p 199.

integrated business, that's why [we're going to be] in digital terrestrial, cable, satellite, smoke signals … what ever platforms there are".[907]

Introducing the second reading of the broadcasting bill in the House of Commons in April 1996, Virginia Bottomley spoke of standing "on the verge of a new broadcasting revolution even more significant than the change from black-and-white to colour television. Digital technology has the potential to bring new and different services to viewers. It can bring more income to the broadcasting industry and more jobs across the country. Digital television offers improvements in picture quality, increased potential for wide-screen broadcasts, interactive television, more subscription services and, above all, many more channels and greater choice."[908] She also pointed out "that introducing digital television and turning off the analogue signals offered a great potential 'digital dividend' when the spectrum could be sold off to the burgeoning mobile telecommunications industry".[909]

Radio offered no such dividend, at least according to the technical wisdom at that time. The band space occupied by radio services was much less useful for re-deployment, and any credible prospect of turning off analogue radio much more distant. The potential gain lay in the better quality which DAB promised, replacing the uncertainties of AM reception, and improving on FM. 'CD quality sound' was the mantra. In the event, compromises on transmission standards, and the Radio Authority's 'light touch' decision to impose only broad and permissive rules regarding minimum bit rates for individual programme services, partly thwarted that aspiration. DAB proved to be as good as properly installed fixed FM, but not that much better. It had the advantage of maintaining that quality when mobile, unlike FM, but the power demands of the earlier generations of DAB microchips made this difficult to exploit. Battery use remained a problem into 2009. Ease of tuning, and the ability to provide text alongside audio data, were useful add-ons, but it was pretty obvious that DAB lacked a 'killer application' to drive it into the market.

The UK was now in position to be at the forefront of DAB development worldwide, another attraction to both the beleaguered outgoing government and its bright-eyed successors. All the necessary conditions were soon to be in place. David Witherow recalls that "what Britain had right, and other countries did not, was having all the different ducks in a row. You had the technology; the interest of a key broadcaster, determined to see it through; appropriate regulation; spectrum; and incentives to get the commercial radio people on board."[910] If DAB was going to be the new future for radio, then Britain could lead it. However, there needed to be enough other nations prepared to follow, in order to produce the critical mass which any modern technology would need to get the full benefits of scale and the mass-production of receivers, and that never materialised.

During that second reading debate of the broadcasting bill in April 1996, Virginia Bottomley made clear that the government's goals for digital radio were extra channels and extra choice, better quality and greater diversity of output. To

907 Quoted in Horsman, *op cit*, p 198.
908 *Hansard* House of Commons 16 April 1996 col.537.
909 Ibid. col 556.
910 David Witherow interview, op. cit.

achieve those, they were to guarantee the INR stations a slot on the national commercial digital multiplex, and to give local stations who took up a place on the relevant local multiplex an automatic eight-year extension to their analogue licences.

This was the key inducement. The major radio groups, GWR apart, had concluded that paying twice to transmit the same signal to the same audiences was not a good commercial deal. I clashed with David Mansfield, by then CEO of the Capital Group, at a seminar arranged by the corporate lawyers Ashurst, Morris and Crisp on 4 July 1996. He was pretty dismissive of the prospects for DAB at what seemed a crucial time in its early fortunes. The idea of effectively incentivising the radio companies to join DAB, with an automatic licence renewal, changed the calculations. Nowhere was that felt more keenly than at Classic FM, whose unexpected success would surely have attracted rival bidders for the INR1 licence, ready to offer hefty cash bids. The same was true at Capital, where what was still the flagship London FM station would also have attracted intense competition. In taking this step, the government was responding to lobbying most of all from Classic FM, which pointed out the electoral risks of alienating its listeners in the heart of Middle England. There were also strong representations from Radio Authority chair, Peter Gibbings, that unless some inducement were offered, the commercial companies would walk away from digital.

The government's offer to INR and ILR of automatic renewal of their analogue licences in exchange for signing up to DAB, was the perfect tactical response. It effectively locked in the major radio companies to apply for DAB licences, and it got Major's troubled electoral efforts off the hook of the Classic FM vote-loser as well. However, it also forced the entire industry to become supporters of DAB, when some commercial scepticism would have been useful. DAB proved infinitely stubborn to launch, at that time and in that way. The clever political 'fix' denied the broader issue the policy consideration and debate which perhaps it needed, however necessary the fix seemed in the short term. By 2008, the price to the ILR companies – keeping going with unprofitable DAB licences or otherwise surrendering those renewed analogue licences – now seemed alarmingly high, but no one had foreseen the scale of the eventual collapse of commercial radio.[911]

David Vick recalls that it was Jean Goose and her team of DNH officials who had cut the Gordian knot over digital radio licensing.[912] The Authority had assumed that it would be licensing individual programme services, and was fretting over how it could then assemble them on to a single multiplex. Someone at DNH realised that the solution was to advertise competitively each of the local multiplexes, and then to allow the multiplex operators to assemble their own cohorts of programme service providers. There were therefore two types of digital licences: one, which was competitively advertised, for the multiplex itself; the other, available on demand, to provide a programme stream, the digital sound programme service licence (DSPS). As Vick notes, "this turned the whole thing on its head and was quite a revolutionary idea, because the multiplex operator was to become a quasi regulator, choosing the range of services ... In terms of loosening regulation,

911 See chapter 25.
912 David Vick in an interview with the author 24 June 2008.

this was a very significant development. For the first time it wasn't simply the regulator choosing what the programme services would be ... either the programme formats or the providers of those formats." Stations seeking digital licences in order to win analogue renewal needed to have some guarantee that they would be included, and the BBC local stations enjoyed a statutory right to be carried on the relevant local commercial multiplex. Otherwise, the mix was to be whatever the market could provide, subject to some light constraints on diversity of services.

For the Radio Authority, the practical implications of the promised analogue renewals were huge. In order to ensure that each ILR station had the opportunity to benefit equally, it had to design an advertisement process to dovetail into the pre-existing schedule for the re-advertisement of existing licences. That needed to give every existing operator the chance to obtain licence renewal by going digital, rather than facing a competitive application process, and to do that within a narrow two-year span for every regional and every first generation local station. To David Vick this was "one of the most satisfying things I have done ... I spent a Sunday morning in my study where I sat, jotting ideas down, and everything fell magically into place".[913] It was this process of re-advertisement which drove the Radio Authority forward, and more than anything explained the seemingly headlong rush into digital radio.

For several years, while the BBC under Christopher Bland and John Birt looked coolly upon digital radio – preferring the sexier pastures of digital television and the internet – it was the steady stream of new digital multiplex licences which kept alive the impetus for DAB. It probably moved at a pace which was more than the market could bear, but this was a time of good profits for commercial radio. Since the stock market was uncritically obsessed with all things digital, winning a new multiplex licence seemed only to strengthen a radio company's share price, whatever the objective reality of the short and medium-term commercial prospects digital offered. Senior radio executives were unlikely to try and set the record straight for credulous City analysts.

The Broadcasting Act received the royal assent on 24 July 1996. For the next 18 months, the Radio Authority's senior team – with director of programming David Lloyd prominent alongside David Vick and Mark Thomas – worked out how digital radio could be brought into being, while the radio industry and some interested outsiders laid their corporate plans. By November 1997, the Authority had prepared and consulted on draft Notes of Guidance, setting out how the licensing and advertisement processes would work. Alongside the idea of separate licensing of multiplex and programme providers, these presented minimum bit-rate standards for different types of programming services, but generally permitted "commercially-led flexibility in how the multiplex capacity is divided in the interest of listener choice". This represented a real step change in the extent to which the regulator was passing over what had previously been its own responsibilities for selecting and overseeing programming to the new digital companies. The Radio Authority was aware, too, that in time this would reverse back into analogue, reducing the levels of regulation, as and when digital radio truly established itself as the platform of choice. It said in June 2000 that "it is important that

913 Ibid.

format regulation shall not be seen to be inappropriately different between digital and analogue", and suggested that any new legislation should include provisions which could be "activated once digital patronage had risen sufficiently to require the same approach for both platforms".[914]

The national commercial digital multiplex was advertised on 24 March 1998. Despite attracting only one application, it was a highly eventful award process. The application from Digital One was heavily dominated by the GWR Group, and reflected what had become a passion for digital radio in Ralph Bernard and Quentin Howard. The three INR stations were guaranteed slots on the multiplex, and for a while it seemed that they would apply jointly. In the event, Virgin withdrew just before the application was submitted on 23 June. Talk Radio, by now owned and run by Kelvin MacKenzie, was part of the applicant group, but withdrew from the consortium in January 1999. That left just GWR and the transmission provider NTL, the latter well-placed to earn fees from building and operating the transmitters. NTL also had more than half an eye on the converged future, and was keen to get experience in programme-supply as well as transmission.[915] It seemed inconceivable to Ralph Bernard that the Radio Authority would do anything other than gratefully award Digital One the national licence. At Holbrook House, however, brows were furrowed, calculators were red-hot and there were frequent calls to the Authority's financial and legal advisers, Pricewaterhouse Coopers and Allen & Overy.

The Authority's finance staff, led by Neil Romain, could not satisfy themselves that the business plan in the Digital One application was good enough to meet the key statutory criterion of sustainability – "the ability of the applicant to establish the proposed service and to maintain it throughout the period for which the licence will be in force".[916] When the Authority met on 3 September 1998, it was to consider four possible options: to award the licence to Digital One unconditionally; to award it conditionally; to defer a decision; or not to award at all, and consider re-advertisement. At the heart of the staff's concern was that Digital One was undercapitalised, and that its revenue forecasts were unrealistic. "The suggestion that the project is undercapitalised is reinforced by the level of return that the financial forecasts anticipate from a relatively small initial investment, particularly for GWR, who retain control of the operation and yet appear to have persuaded NTL to assume a greater degree of risk, by getting NTL to commit the initial financing of the transmitter network."[917]

There was great unease, too, over the extent to which GWR might dominate the programme services to be carried on the multiplex. The application envisaged nine programme streams. Apart from the three INR simulcasts (one of those largely GWR-owned), only two other non-GWR services had been identified. NTL would provide 'classic gold rock' – although the production of this would be contracted out to GWR – and ITN would supply rolling news. Anticipating discussions with the Office of Fair Trading, the paper before the Authority

914 *Radio Regulation for the 21st Century*, Radio Authority, June 2000, pp 26, 27.

915 Stephen Carter, at that time NTL's chief executive, in an interview later with the author, 11 September 2008.

916 *Broadcasting Act* 1996 s 47 (2) (c).

917 *RA Paper* 83(98).

members warned that "on the face of it, competition concerns could arise if GWR were in the position of both controlling the multiplex (and therefore access to it) and providing a disproportionate amount of content services on it".[918]

These and other reservations had been rehearsed with GWR over the summer. On 20 August, Digital One's chairman (and GWR chief executive) Ralph Bernard wrote to me, to request a delay of the Authority's decision on the licence award until October. He wished a number of issues to be clarified before the Authority committed itself to making and announcing a decision, which might otherwise, he feared, "include conditions that would make it difficult for Digital One then to maintain the integrity of its business plan".[919] By this time, I recall the situation to have been very fraught, with Digital One doubting the Authority's *bona fides*, and those of us at the Authority contemplating what an enormous set-back it would be if the national digital licence went unawarded. At the September meeting, therefore, the Authority members were readily persuaded to defer a decision while Digital One – effectively GWR – did what they could to convince us.

When the Authority met next, on 9 October, most of the issues had been resolved. The Authority was reminded that "the biggest concern expressed by members at their September meeting was the lack of evidence offered by Digital One to support the assumptions in their business plan ... Members were [therefore] unable to satisfy themselves as to the ability of Digital One to maintain the service through the licence period",[920] one of the key statutory requirements. However, "as a result of further questioning, Digital One have now been able to satisfy the chairman, deputy chairman and staff that commitments to provide 57 per cent of the total anticipated fee income are in place, with substantial interest expressed to supply the balance". The Authority therefore determined to award what it believed to be the 'first and only' national commercial digital licence to Digital One. "Although no additional conditions were to be imposed on the award, Digital One would be told that the Authority would be reviewing arrangements with DSPS providers during their licence term, and act in accordance with the Authority's general competition duties if the need arose. Staff were instructed to investigate NTL's agreement with GWR for content provision to ensure that it is not anti-competitive."[921]

It was a bad way to start a new medium. Although there was remarkably little public speculation about the delay, the new operator was resentful of how the licence process had gone, and the regulator was fearful of the decision it had felt forced to make. It is a mark of the strength of enthusiasm at the top of Digital One that the company proceeded to build the new network and get the programme services on air, despite the falling away of the other INR operators and a pretty lukewarm radio industry. Digital One launched with five of its proposed ten programme services on 15 November 1999, with Planet Rock and Core being digital-only programme streams. Other services, including the innovative One-word speech station, arrived over the following months. However, the phrase "the

918 Ibid.
919 Letter from Ralph Bernard, Digital One to Tony Stoller, Radio Authority 20 August 1998.
920 RA Paper 113(98) 1 October 1998.
921 *RA Minutes* 9(98) of a meeting on 9 October 1998.

first and only national commercial digital multiplex",[922] used by the Authority in its press statements and elsewhere, remained as an unexploded bomb buried in the ground under the feet of digital radio.

There was an expectation among DAB's proponents that 1997 would be the year when DAB took off. The major European equipment fair, the Internationale Funkausstellung in Berlin, was expected to launch a whole range of DAB sets, ready for the consumer retail and car market. I recall flying to Germany, and the acute sense of disappointment at the small number of prototypes on display, all big, expensive and power-hungry. At that stage also, the recurring hopes that European car manufacturers would factory-fit DAB sets into new cars began to seem a constantly receding chimera. The exhibition two years later had more sets on display, but still not enough to convince the BBC's Chairman, Christopher Bland, that DAB was going to run. From that point onwards, the BBC's new radio MD Jenny Abramsky – herself a strong supporter of DAB – had greater difficulty in getting new BBC funding and support for the new radio system. In Germany and other nations, without the impetus from the private radio sector, the poor response of the radio set manufacturers was to slow DAB almost to a stop. The paradox is that enthusiasm among the public broadcasters was disappearing just as commercial radio in the UK was starting to commit investment to DAB.

The first advertisement of a local digital multiplex was for the Birmingham area on 20 November 1998. The Manchester advertisement followed on 17 December, and then the first of the three multiplex licences for the Greater London area on 19 January 1999. Local multiplex licensing was in the main much less eventful than the Digital One saga, but was still a very intense workload for the Authority, and for the industry, in putting together applications. From then on, the Radio Authority set itself to advertise one multiplex a month for the next two years, in order to provide every established ILR company with the opportunity to renew its analogue licence through digital participation rather than in open competition. In the event, few of those multiplex licences were competed for. Other than the London multiplex, all the early areas saw just one consortium applying, led by the owners of the dominant ILR station. In the second phase of advertisements for wider coverage licences, some rivalries between heritage ILR stations and successful new regional services produced some competitive applications. Those newer companies also tended to see digital as an opportunity, where the older stations took a much more defensive approach.

The award of local radio multiplex licences was again by a 'beauty parade' process. There were seven statutory criteria. In broad terms, those were the four classic requirements for analogue ILR from the 1990 act, plus coverage, speed of roll-out and fair competition.[923] The relative lack of competition for digital multiplexes prevented these from becoming as sensitive a topic as they had been for analogue. Even so, had the legislation happened a year later, the incoming government would presumably have put market mechanisms into place for the brave new digital future, rather than the old public service requirements which governed the award of digital radio multiplexes.

922 *Radio Authority News Release* 12 October 1998 "The Radio Authority has decided that, after considering the proposals submitted by the sole applicant, it is offering the first and only national commercial digital multiplex licence to Digital One Limited".

923 *Broadcasting Act 1996*, section 51(2).

Eight local multiplexes were awarded in 1999. CE Digital, which was owned on a 50:50 basis by Capital and Emap, won three of them: for Birmingham, Manchester, and Greater London I. Emap Digital was awarded South Yorkshire and Tyne & Wear. SCORE Digital, underpinned by SRH, was awarded the multiplex licence for Glasgow, Capital Radio Digital for Cardiff & Newport, and Now Digital (GWR) for Wolverhampton, Shrewsbury & Telford. The consortia formed and re-formed for these early applications, usually around the owner of the dominant ILR analogue station in the locality.

The following year, 2000, saw 12 local multiplex awards, spread between Now Digital, SCORE, Emap, Switch and MXR: Bristol and Bath; Coventry; Edinburgh; Greater London II (to Switch Digital); Humberside; Central Lancashire; Leeds; Liverpool; North-East England; Northern Ireland; Teesside; and Central Scotland. Generally, the pattern of DAB licensing looked likely to consolidate the carving up of the localities and regions in favour of the heritage companies, plus those who had gained a foothold through regional licences. New entrants were not going to come into commercial radio in any numbers through digital multiplex ownership.

A further 13 local multiplex licences were awarded in 2001. The one for Aberdeen caused a kerfuffle. Scottish Radio Holdings regarded it as pretty much their birthright to own digital radio multiplexes in Scotland. They were outraged on 6 April when it was announced that the Switch Digital consortium, fronted by Kelvin MacKenzie's Wireless Group, had beaten SRH's SCORE consortium to this licence. The Authority's decision was based on the much more extensive coverage proposed by Switch Digital, and "a considerably more rapid timetable for the roll-out of coverage, with all its transmitters established from the commencement of the licence period".[924] However, it further confirmed SRH in its belief that the Radio Authority was failing to give the group due credit for all its achievements when it came to the prize of licence awards.

The other 12 were less contentious. They included three wider-coverage 'second multiplexes' in localities where regional ILR stations were strongly represented – North West England, South Wales/Severn Estuary and the West Midlands – and were all awarded to MXR. This consortium included Chrysalis, Capital, Guardian Media Group and Jazz FM, plus a 5 per cent holding from the Ford Motor Company, probably keeping its options open over factory-fitting DAB radio sets into its cars, if the platform ever became sufficiently important. The more local multiplexes also went to the main local analogue incumbents, straightforward decisions since there was no competition, to SCORE, Now and TWG: Ayr; Bournemouth; Bradford & Huddersfield; Dundee & Perth; Exeter & Torbay; Inverness; Peterborough; and Southend & Chelmsford. These outcomes were not optimal for the medium-term health of the radio companies. Mirroring the analogue ownership pattern, GWR through Now Digital was acquiring the largest number of multiplex licences but for generally smaller localities. The surprising award of the Greater London III multiplex licence to the Digital Radio Group – led by GWR and the Wireless Group – left the major regional digital operator out in the cold. The Authority had awarded all the regional multiplexes in England to

924 Radio Authority, Aberdeen: Assessment of Local Licence Award, 24 April 2001.

MXR, Chrysalis' digital vehicle. Denying it Greater London undermined its potential viability.

Six new multiplex licences were awarded in 2002: Leicester; Norwich; South Hampshire; Swansea; Swindon/West Wiltshire; and Yorkshire. Six more followed in 2003, as the Radio Authority moved to complete its task of providing nationwide coverage for local and regional DAB, before it was replaced by Ofcom: Sussex Coast and Kent; Stoke on Trent; Nottingham to Now Digital; Cornwall; and Cambridge. Half of the licences in these two years went to GWR, leaving it ever more exposed

In just four years, between late 1998 and the end of its life in December 2003, the Radio Authority had awarded one national digital multiplex carrying 10 programme services, and 45 local digital multiplexes offering over 270 sound programme services. Every analogue licence renewal which was permitted by the Broadcasting Act was accommodated within the digital licensing timetable. As an exercise in effective and sympathetic bureaucracy, it was of the highest quality. As a commercial strategy for the commercial radio companies, and notably GWR, it was an exercise fraught with the greatest risk. GWR was far and away the largest and most enthusiastic owner of digital radio multiplexes, mostly through Now Digital, as well as of Digital One, and was hugely exposed should the new platform fail.

Among the new programme services, there were a few genuine innovators. Digital One carried two music channels which were wholly new, and available only on digital: Planet Rock, and Core. In London, the three multiplexes had room for all sorts of services, ranging from the new but predictable – Capital Disney and the classic soul of The Groove – through the fairly orthodox attempts to attract early adopters of the new technology – Saga Radio, Arrow Radio with adult rock, Young Asian Radio – on to the downright experimental – Gaydar Radio, Passion for the Planet and Abracadabra. Almost none of them made it. On local and regional multiplexes, all the promised offerings of local and community-linked output lasted for only a brief time, if they got started at all. A combination of competition from the new BBC digital services, the slow take-up of DAB, the high rentals charged by the multiplex owners, the hostility towards digital felt by the new owners of the major groups in 2007 and 2008, and the failure of the channels themselves mostly to catch the imagination of listeners, meant closure sooner or later for almost every new DAB-only channel, with the exceptions (as DAB channels late in 2008) of Planet Rock, Passion for the Planet and Gaydar. More successful were the brands of streamed radio which were offered on other platforms, such as Kerrang! and Smash Hits Radio. A place on the digital television satellite or terrestrial bundle, or dissemination via the internet, meant that this type of new service could survive and even perhaps prosper, but not one has yet changed the face of UK radio. Commercial DAB attempted too much, too soon.

The small-scale stations faced a particular dilemma. It was the nature of DAB coverage, as designed by the Authority, that it fairly closely matched the larger coverage areas of the heritage or regional ILRs. The cost charged by the multiplex operator to those holding digital programme service licences was much the same whatever the service. An annual cost of, say, £80,000 was wholly beyond the means of small-scale stations, which were often surviving on monthly turnovers of around £30,000–40,000. Yet if they declined to join the local digital multiplex, they would face eventual competition for their analogue licences. It seemed unfair, and so it was, but it reflected the *realpolitik* of the mid-nineties. They were best advised simply to bide their time; there was the prospect of digital radio in the medium term on another frequency band, L band, which would give much smaller coverage areas. Meanwhile, the price for them of automatic renewal seemed higher than its value. More than 10 years later, the issue of accommodating small stations on DAB was still not solved.

Transmitting services was one thing; winning listeners to digital quite another. To begin with, the availability of digital radio sets remained very restricted. In contrast with television, where set-top boxes and integrated digital TVs were widely available, the lack of receivers to buy in the shops, and the high prices of those which were available (particularly compared to analogue receivers), greatly restricted DAB when it launched. Ofcom reported that "the first DAB radios were in the hi-fi, kitchen–portable and personal portable product segments, which addressed less than one quarter of the total radio receiver market. This narrow offering, combined with the relatively small range of models available within each category and the large price premium for DAB radio receivers over analogue radio receivers, constrained take-up."[925] The first portable 'kitchen' DAB radio for under £100 was not available in the UK until 2002. The best sales period in these early years was around Christmas 2003, when 176,000 DAB sets were sold. Even so, as at the end of September 2004, five years after the launch of full-scale DAB nationwide, only 800,000 DAB digital radio sets had been sold in the UK.[926]

Far Eastern radio set manufacturers such as Sony and JVC showed little interest in the Eureka DAB specification, and by that time no mass market radio sets were being produced anywhere in the western hemisphere. Some of the mass-producers tried fairly short runs of sets, and a number of UK-based manufacturers saw DAB as a real opportunity to develop products for this specialised market. The *Pure Evoke* set has become a modern media fashion statement. When Radio Authority staff were given one each as a Christmas bonus in 2002, the whoops of delight as the sets were unwrapped at the Christmas lunch high up above Leicester Square could probably have been heard throughout the West End. Some car radios were available from 1999 onwards, but the holy grail of factory-fitting of DAB sets into new cars on the production line remained seemingly forever out of reach. There were frequent announcements, especially from the Digital Radio Development Bureau (DRDB), that an agreement had been reached with Ford, or General Motors, but nothing substantial had happened up to the end of 2009.

The launching of the DRDB October 2000, jointly by the BBC and CRCA, seemed to give DAB its best chance. Broadcasting academic and commentator, Steven Barnett, by his own admission a digital radio sceptic, was converted to the cause: "There is real industry co-operation. [Jenny] Abramsky and Paul Brown, chief executive of the Commercial Radio Companies Association, share conference platforms and information in a collaborative spirit which digital TV operators can only dream about. There are no damaging rifts or confusing arrays of subscription deals, and the big commercial battalions within a notoriously competitive

925 *Radio: Preparing for the Future* Ofcom 15 December 2004.
926 DRDB/Gfk quoted in *Radio: Preparing for the Future*.

industry are prepared to bury hatchets (and even sleep with the BBC) to get digital radio off the ground."[927] Yet still, as he noted, "its biggest challenge will be to persuade manufacturers to make affordable, portable hardware". The DRDB achieved something of a masterstroke towards this aim. It provided seed-corn funding to help develop a later-generation microchip for use in DAB sets, and this lifted the market from virtual stagnation to modest niche availability.

The great step forward in terms of digital programming came from the BBC. Commercial radio had begun to offer some new services, but none had remotely enough appeal to drive the consumer take-up of digital radio sets in the type of mass numbers needed. However, late in 2001, Jenny Abramsky at the BBC at last persuaded the Corporation to take a step forward with digital radio content provision, prompted not a little by the extensive provision of digital channels in the commercial sector. The BBC proposed to the government, whose approval was needed, the launching of six brand new digital channels. The response from the commercial sector was mixed, and CRCA lobbied DCMS to restrict the BBC's new services. This was partly the companies' usual Luddism in the face of potential growth in radio outlets, but some of the more innovative digital providers were rightly fearful that their prospects would be blown away by the BBC's market dominance. Oneword, the national plays, books, comedy and reviews channel launched by Unique on the Digital One multiplex in May 2000, was wholly stuffed by the new Radio 7 channel, backed by the Corporation's huge archive. Given that the BBC also acquired some of the commercially available spoken word catalogue, Oneword faced a struggle for material which it could not survive. It was replaced on its multiplex channel early in 2008 by a recording of birdsong.

Nevertheless, the renewed presence of an ambitious BBC in the market changed the equation, and prevented the failure of DAB at that time. The BBC launched Five Live Sports Extra, 6 Music, 1Extra, the Asian Network and eventually Radio 7 in the course of 2002. There was extensive cross-promotion between BBC analogue and digital services, and BBC TV also promoted digital radio. Sales of sets began to rise, yielding a total of seven million owned across the UK by May 2008. Even so, this was still little more than a drop in the ocean of electronic kit sales in this country, and was not matched even closely anywhere else in the western economies. With estimates for the number of analogue radio sets in use in the UK varying between 150 million and 200 million, the total of DAB sets was still comparatively tiny. The highly disputable claim by a government working group in 2008 – nearly ten years after the launch of Digital One – that a total of seven million DAB sets was "impressive", seemed to be whistling in the wind.[928]

If DAB as a stand-alone platform seemed likely to atrophy, what rescued digital radio as a whole was digital television. There was irony in this, since the overhasty launch of digital radio, and the UK's consequent locking into the DAB technology, came about as a result of the timing of the legislation to bring about digital television. BSkyB's daring exploitation of DSTV, and the rescuing of the failed ITV Digital platform by BBC DG Greg Dyke's brilliant Freeview initiative, made switch-over by 2012 from analogue to digital television unexpectedly practical. That was helped by the market reality that people are accustomed to buying new

927 *Observer*, 7 January 2001.
928 *Interim Report of the Digital Radio Working Group*, 23 June 2008.

television sets and usually dispose of their old ones. Who ever throws away a radio set?

Yet people will certainly listen to radio on new hardware. The spaces on the terrestrial and satellite digital television multiplexes were ideally suited to carry audio channels as well, using very little bandwidth in the process. The DAB programme providers had the product and an anxious need to expose it. As soon as some previously DAB-only channels became available on BSkyB and Freeview, digital radio at last began to show some real growth. Allied to internet radio – the anorak's dream – there was the historically improbable phenomenon of people going back to listen to the radio on fixed, wired appliances. It was as if the 'wireless set' and 'steam radio' never really went away when the transistor arrived; as if they had simply bided their time, and then re-invented themselves as the internet and digital radio.

When Ofcom had succeeded the Radio Authority at the end of 2003, it was drawn naturally to digital radio, as a converged regulator of converged technologies. Stephen Carter recalls that, not unlike the telecoms issue of local loop unbundling, it seemed necessary to take bold steps to drive digital radio forward. It had been around for too long without making a break-through, while the bureaucracy had "allowed people to make a meal of the process".[929] In December 2004, Ofcom published its provisional thoughts and plans for consultation, in *Radio – Preparing for the Future.*[930] It was confident that "digital radio offers significant benefits to both broadcasters and listeners, including greater choice, enhanced services such as on-screen programme information, ease of use and reduced audio interference".[931]

Ticking away in this document, and the subject of some impassioned exchanges between Ofcom and Digital One, was the unexploded bomb of 'first and only', the soubriquet which the Radio Authority had consistently applied to the DAB multiplex awarded to Digital One. In 2004, new spectrum was released in VHF Band III which could be used to increase the number of digital radio services offered through DAB. Ofcom saw it as space which might be used in a platform-neutral way for many sorts of digital provision, and, critically, as a chance to offer a second national commercial digital multiplex. It floated this idea ahead of the formal consultation, and Digital One/GWR went ballistic.

The legal arguments continued for almost three years. GWR asserted that it was entitled to rely upon the statements of the Radio Authority that there would not be any other competing national commercial multiplex. Ofcom argued that no such commitment could be binding when circumstances changed, a decade afterwards. In July 2005, Digital One's Chairman, Ralph Bernard, alleged that his company had been "seduced" into investing in digital radio, having been assured it would be the sole national multiplex operator in that format. "I'm sick of fighting regulators", he was reported as saying, "but I'm now about to be pitched into battle with Ofcom".[932] He believed that there was "anecdotal evidence" that the regulator had already decided to license either one or two more national digital multiplexes. "Less than a quarter of the way through [Digital One's] licence period Ofcom is

929 Stephen Carter interview, op. cit.
930 *Radio: Preparing for the Future* op. cit.
931 *Ofcom News Release*, 15 December 2004.
932 *Guardian*, 5 July 2005.

about to flood the market with as many as 20 more national stations. Ofcom has no legal right to change the basis of the investment that was seduced from my company, and neither has it a moral right." When Ofcom announced on 21 December that it would indeed advertise a further national digital multiplex, Digital One threatened judicial review. It dropped that only late in March 2006, when it claimed that it had got "got certain assurances from Ofcom that the second multiplex would only air stations that don't compete with Digital One, and [GCap] is happy with that".[933] The claim was hastily denied by Ofcom the following day. Since those files remain closed, it is not possible yet to verify who understood it best. Digital One told me later that year that it had only backed off from taking the matter to court because of the unacceptable costs involved.

Ofcom advertised the second national digital multiplex on 1 December 2006. It received two applications. The 4 Digital Group was a consortium with Channel 4 in the lead, including Emap Radio, UTV, UBC Media Group and Virgin Radio owner SMG. Their rivals were wholly owned by National Grid Wireless, but offered programme content provided by GCap, the BBC, Digital One, RTL, Premier Christian Radio, Somethin' Else, FUN Radio and Colourful. According to Channel 4 chief executive Andy Duncan, previously the BBC's director of marketing, 4 Digital intended "to shake up the moribund commercial radio sector and fight back against a dominant BBC".[934] Ofcom's licensing committee was duly impressed, and on 5 July 2007 awarded 4 Digital the multiplex licence. 4 Digital proposed to launch in July 2008. It planned speech services to compete head-on with the BBC, recruiting BBC Five Live controller Bob Shennan to design the programming.

Meanwhile, the wheels were coming off DAB. Ralph Bernard departed as chief executive of GCap on 23 November 2007. He had carried the banner for DAB within the commercial radio sector ever since putting together the Digital One application back in 1997. His short-lived successor, Fru Hazlitt, announced on 19 March 2008 that GCap now saw its future on broadband and FM. "The majority of people who are listening through DAB receivers are listening to stations that are simulcasting on FM. The majority of DAB receivers out there are FM-enabled too ... If you put that against a background of the cost structure of DAB, it cannot be an economically viable platform."[935] GCap sold its 63 per cent stake in Digital One (effectively over Bernard's dead body) to communications infrastructure company Arqiva, and shut down national digital stations The Jazz and Planet Rock. The cuts failed to save GCap from takeover by Global Radio, and Hazlitt soon moved on. She left in June 2008, but her brief reign seemed to have pretty well scuppered DAB among the heritage commercial groups.

Then in July 2008, the "long-term viability" of digital radio was questioned by the BBC Trust, which said that all digital-only stations, including those offered by the Corporation, had "yet to make a breakthrough".[936] In the BBC's annual report for the year to the end of March, the trust said it "remains to be seen" whether the launch of the second national commercial digital multiplex backed by Channel 4 "will boost the market sufficiently to ensure its long-term viability".

933 *Guardian*, 29 March 2006.
934 *Guardian* 19 March 2007.
935 *Guardian* 11 February 2008.
936 *Guardian* 8 July 2008.

They were not to have long to wait to discover that such a boost was not to be forthcoming. Channel Four announced in October 2008 that it was pulling out of DAB entirely.

The future of DAB is at best immensely uncertain. DAB has had several of what David Witherow calls "windows of opportunity", notably "in the second half of the 1990s and the very early years of the 2000s. Now, the technology is moving so quickly; internet radio and iPods are coming along at a mad rush. Ideally, DAB would have got itself pretty well established Europe-wide and in other countries as well."[937] The launch of the DRDB, co-incident with the BBC's return to the field with new national services and the availability of cheaper DAB sets, all in 2001/2, was a new window. Even as late as July 2007, the arrival of Channel Four Radio on to the second national digital multiplex seemed to be another. All had been slammed shut, and the blinds drawn down. As the end of the decade approached, it remained to be seen whether the *Digital Britain* approach would allow some light back in.[938]

No one seriously doubts that digital transmissions will somehow become more and more important for radio. The dissemination of services over the internet alone would ensure that. The hunger for content of the new wireless technologies – 3G, 4G and the rest – makes radio programming a desirable asset. Yet it remains improbable that people will stop listening to free-to-air wireless transmissions on their transistor radios in order to pay for the privilege of less appealing services over their converged mobile phone, at least in sufficient numbers to make sound radio redundant. Digital will eventually change the face of all radio. In the history of independent radio, though, nobody has yet conclusively demonstrated just how or when.

937 David Witherow interview, op. cit.
938 See chapter 25.

Chapter 21

Things can only get better

1997 and all that

1997 was a pivotal year in so many ways, both nationally and for commercial radio. Two themes ran through it – buoyant hope for an improbably bright future; and a rather emotional nostalgia for an imagined past – which produced a level of expectation beyond anything that might reasonably be sustained. That had two consequences: first, a failure to make timely provision for a number of things that really needed to be fixed; and second, an almost sullen reaction to the inevitable disappointments. And, once again, what happened to private radio illustrates what was happening in the wider British polity.

National hope centred on what ought to have been the iconic day of 1997. On 1 May, Tony Blair's New Labour was swept to power on a tide of public optimism unprecedented since 1945. The Labour Party won the general election in a landslide victory, leaving the Conservatives in tatters after 18 years in power, with Scotland and Wales devoid of Tory representation. Labour had a formidable 419 seats, the Conservatives just 165. It seemed that Blair, at 43 the youngest British prime minister of the century, had been given the power to make everyone's dreams come true. The spin masters at the heart of the New Labour project understood the potency of cheap music, just as did the radio programmers. The campaign's theme tune, D:Ream's *Things Can Only Get Better*, hit exactly the right note of wide public expectation.

However, the emotional nostalgia found an even more public focus which, for a while, overwhelmed the events of early May. On 31 August, Diana, Princess of Wales, was killed in a Paris car crash, triggering an outpouring of collective and manufactured grief not seen since the death of Victoria, almost a hundred years before. Elton John had written *Candle in the Wind* the year that ILR began, in 1973, in a memorial celebration of Marilyn Monroe. That September, he re-wrote it for Diana. The huge popularity of what became a maudlin anthem caught that other theme of 1997, a general sense of a new, touchy-feely Britain. D:Ream was expressing the naïve optimism prevalent at the time. Elton John, in both his original and revised song, was yearning back towards an imagined past. Briefly, the popular music scene of 'Cool Britannia' seemed to have merged with a re-invented nation. Noel Gallagher of Oasis had a union flag guitar, Geri Halliwell of the Spice Girls wore the shortest of union flag dresses. These were the feel-good months, the months of mass-produced emotion, the Pangloss months.

For commercial radio in mid-1997, the year seemed ripe with the promise of moving forward from commercial success to real wealth, and from mass audiences to a settled place at the media industry top table. Blair was the natural successor to Thatcher and Major in terms of broadcasting policy (as elsewhere), running with the notion of the primacy of market forces, the desirability of reducing regulation, and the bright future of the digital communications age. That seemed to bode well for the aspirations of commercial radio. The RAB's Douglas MacArthur, close to the Scottish Labour leaders, coined the phrase "the commercial radio generation",[939] for those then aged 15-44 who had grown up since the launch of ILR. This included not only the industry's audiences but their new young political leaders, too, who knew their Blur from their Suede, and even took the risk of inviting the Gallaghers to parties at Number Ten.

The departmental deck-chairs were moved again, although not significantly for broadcasting. The old Department of National Heritage became the Department for Culture, Media and Sport (DCMS), and the caring and cultured Chris Smith was appointed its first secretary of state. Effectively succeeding Virginia Bottomley, he displayed the same approachability and genuine interest in radio; but whereas she had a natural empathy with the local commercial stations, he had an instinctive affinity with the national cultural channels. Radio was well served by both of them. The year, which included devolution for Wales and Scotland (but not of responsibility for broadcasting, which remained a 'reserved' matter), also saw further recognition for Jimmy Gordon, who was created a life peer in October.[940] It was well-deserved honour for one who brought intellectual distinction and political nous to radio across four decades.

The total adult audience for commercial radio had risen from 27,745,000 in the first quarter of 1995 to 28,408.000 at the end of 1997.[941] Commercial radio's share of all radio listening in the UK stood at 50 per cent or more for four out of five quarters of RAJAR nationwide research, between spring 1997 and summer 1998. Audiences for commercial radio reached a peak of 28,739,000 adults each week at the start of 1998,[942] and then showed up as high as 32,180,000 when the research specification changed. Advertising and sponsorship revenues were glorious, reaching £354m in 1997. National revenue of £202m in 1997 was more than three times as high as in 1992. Radio at last shed the tag of the 'two per cent medium'; its share of all UK display advertising rose from 2.8 per cent in 1992 to 4.9 per cent at the end of 1997.

The licensing of new stations was at its fastest ever rate in 1997. Remarkably, the Radio Authority awarded as many as 22 new ILR licences in the year. ILR passed the milestone of 200 licences with the award of the stand-alone Basingstoke licence to Kestrel FM on 4 September.[943] Using 105–108 MHz even allowed for one further Greater London licence, won at last by Xfm. For its proponents, though, this was too late, as the indie music boom was ending, or at least indie music had become one of the several mainstreams of commercial popular music. The station struggled to keep solvent, eventually being bought out by – of all companies – 'mainstream' Capital Radio.

939 See, *inter alia*, CRCA's evidence to the Commons Select Committee for Culture, Media and Sport, contained in its Fourth Report, 6 May 1998, para 102.

940 as Baron Gordon of Strathblane.

941 Hallett Arendt, *RAJAR* analysis, q4 1997.

942 *RAJAR/RSL*/Hallett Arendt Q 1 1998.

943 *Radio Authority News Release* 79/97, 4 September 1997.

There were new regional licences in the East of England (Vibe FM), which became a saga of accusation and challenge, and the Solent (Wave 106). Local licences were awarded in Canterbury (CTfm), Dover & Folkestone (Neptune FM), Thanet (Thanet Local Radio), the Western Isles (Isles FM), Wolverhampton (The Wolf), Ullapool (Lochbroom FM), Liverpool (Crash FM), Southport (Dune FM), Brighton (Surf FM), Eastbourne (Sovereign Radio) Hastings (Arrow FM), Huddersfield (Huddersfield FM), Peterhead (Peterhead Waves Radio), Basingstoke (Kestrel FM), Macclesfield (Silk FM), Warrington (Wire FM), Havering (Active FM), South-East Staffordshire (Centre FM) and West Somerset (Quay West Radio). Sixteen of those licence awards were for areas with adult populations below 300,000, and seven were for areas with fewer than 100,000 adults. Quay West was to serve only 20,000, and Lochbroom in Ullapool just 1,500, the smallest coverage area by population in the UK. This was deliberate, reflecting "the Authority's commitment to providing services for smaller coverage areas".[944]

The natural levels of civic aspiration within these new groups might have provided a last flowering of the *independent* radio notion within ILR, but all too often they hit the reality of the commercial world in short order. Here and there, some applicants had from the beginning more of an eye to the profit to be gained from selling on the licence than on the romance of running a station for their home town, but those were a smallish minority. Yet there were significantly more who were soon persuaded to sell on by the difficulty of making a go of it in the early months, the blandishments offered by potential buyers, and the natural anti-climax which came after the winning of the licence itself.

This licensing workload was not just for permanent stations.[945] There was a continued and significant growth in the number of Restricted Service Licences (RSLs). 350 were issued in 1997. The Authority tested freely-radiating low powered AM transmissions for hospital and student radio in the first half of 1997. These low-powered services, dubbed Long Term RSLs (LRSLs) were made widely available for non-commercial purposes covering just a single site from July 1998. They proved understandably attractive, and represented a genuine aspect of *independent* radio in the *commercial* radio years. There were even satellite licences to manage; an audio service up-linked from the UK needed a special licence, which imposed no content obligation beyond the statutory requirements, but was another process. Thirty-one satellite licences were in issue in 1996, 39 in 1997. The Authority's small staff – still fewer than 40 – had also to manage the first stirrings of digital radio licensing.

High prices were paid for licences. Even the smallest stations attracted buyers. GWR sold KLFM to Dawe Media for over £1m in January, and St Albans FM to Essex Radio for £725,000.[946] When The Radio Partnership bought Radio Wave in August, it paid £2.1m for a company with a net asset value of just £216,000. In his interview for BBC2's *Newsnight* attack on the Authority in 1998,[947] Guy Hornsby confirmed that he had sold his two Kiss FM licences in Manchester and Leeds "in less than three years for seventeen and a half million … £17.6m to be precise …

944 *Radio Authority Annual Report* December 1997.
945 See chapter 23.
946 *RA Paper* 56(98), 20 May 1998.
947 See chapter 19.

The return I got from it was absolutely amazing ... I still can't get over it now."[948] No wonder investors loved commercial radio, and that a hint of derision can be heard in the BBC reporting: "The most astonishing deal came this summer when just months after going on air the London rock station Xfm was sold to Capital Radio for £14m. Chris Parry, one of its founders, was worth £4m overnight. Radio stations are changing hands like second-hand cars."[949] The free market was rampant, precisely in line with the vaunted philosophy of the times. Those who disapproved – be they BBC journalists or Radio Authority regulators – were rowing against the stream. The government would still not contemplate any restrictions on the selling on of licences, so until the market hit its inevitable downturn, radio licences were worth an improbable fortune.

The limits to expansion set by the ownership rules laid Capital open to accusations that it was not using its revenues to the best benefit of its shareholders. As Richard Eyre recalls, Capital was under great pressure from City analysts to make a "strategic investment".[950] The company had looked around the radio world, even at India – where "things take longer than you think", and Capital was about 14 years too early. The Mayor of Paris turned down Capital's interest in a failing radio station there with the comment: "This is a French problem. I cannot be seen to be employing a British solution". For a while, Capital explored a record label tie-up with Telstar, under the Wildstar brand. With the station's move from Euston Road to Leicester Square, there was a problem about what to do with the ground floor of the new premises. Branded catering ventures were all the rage just then, so Eyre and David Mansfield devised the idea of the Capital Radio Café during a car journey.

After an initially successful joint venture agreement with the My Kinda Town restaurant group, Capital paid £57m in November 1996 to acquire the company and its theme-based restaurants, including the Chicago Pizza Pie Factory, Tacos, Beach Blanket Babylon, and Harry J. Beans. The station persuaded itself that it was in the entertainment business, and that cafés could be part of that. When the deal was done, Eyre "felt a sense of fulfilment and achievement which was to last for about 6 hours until the first analyst meeting, where the purchase went down like a lead balloon".[951] Capital Radio's share price plummeted. By 1999, My Kinda Town continued to provide the company with a bad case of indigestion. The company sold off the Havana restaurant chain and closed down its restaurant division, keeping only its four Radio Cafés, which were then regrouped directly under the company.

Commercial radio may have thought it was riding high in 1997, but BBC radio was at the start of the process of turning itself around. Liz Forgan, a founding commissioning editor and then director of programmes at Channel 4, had joined the BBC in 1993 to become managing director, BBC Network Radio. She laid the groundwork for the revival with a series of key appointments.

First, Matthew Bannister became controller of Radio One in 1993, succeeding the long-serving Johnny Beerling, whose bitterness is still evident. "Matthew Bannister did what John Birt wanted, and brutally changed the image of Radio

948 *Newsnight*, BBC Two, 1 December 1998, Guy Hornsby.
949 Ibid., Richard Watson.
950 Richard Eyre in an interview with the author, 20 November 2007.
951 Ibid.

One back to a 'youth music station'. In doing so he took no cognisance of the existing audience, or the need ... to manage change so that those disaffected listeners could be moved to Radio Two."[952] The station had been hugely popular, with 16.5 million listeners,[953] but seemed tired. Its large audiences and lack of ambition made it vulnerable to the perpetual speculation about privatisation, especially in the run-up to Charter renewal in 1994. Bannister addressed both concerns. He got rid of most of the older DJs, replacing them with a new wave. He also got rid of almost half the audience, reducing it to 9.5 million in five years.[954] However, a new style was set and the risk of privatisation averted. As Britpop arrived, it seemed retrospectively to justify Bannister's approach. He has been quoted as saying that his proudest achievement was "changing Radio One 1 from a middle-aged radio station to a champion of new music, particularly the major role it played in the success of Britpop and the UK dance scene in the 1990s". In 1997, however, the jury was out, and commercial radio felt even more confident that it had seen off a major competitor.

Second, Forgan managed to get the BBC to commit to digital radio with DAB, persuading David Witherow back from retirement to run the BBC's digital project, which gave it both a skilled manager and a persistent promoter. Third, she got Jenny Abramsky to launch Five Live, recovering the ground lost by the BBCs first efforts to provide a demotic news, talk and sport station alongside Radio Four. Forgan departed from the BBC in February 1996, reportedly over a disagreement with John Birt, then BBC director general, over the decision to move BBC Radio News from Broadcasting House to Television Centre. Her successor was Matthew Bannister, who also remained as controller of Radio One until March 1998.

However, before she left, Forgan's fourth masterstroke had been to appoint Jim Moir as controller of Radio Two at the end of 1995. Having been head of Light Entertainment and Variety on BBC television, many assumed this would be a quiet end to Moir's career. Nothing about the amusing and ebullient Moir is quiet, however. He turned Radio Two around, switching to an aggressive CHR format in daytime – effectively stealing commercial radio's clothes – and bringing in star broadcasters like Johnnie Walker, Janice Long, Paul Gambaccini, Lynn Parsons, Bob Harris and Alan Freeman, all of whom joined Radio Two during Moir's controllership. For historians looking for a key date, it is probably the RAJAR research for the final quarter of 1997, which showed Radio Two's audiences at last turning upwards after years of decline. They were to rise from eight and a half million in mid-1997 to thirteen and a quarter million by the end of 2006.[955] Steve Wright – sacked by Bannister from Radio One – arrived to present Saturday and Sunday morning shows. He was soon to fill the afternoon slot, Johnny Walker was to do drivetime, and the new Radio Two began to have a huge impact. Wright had begun his career on radio at little Radio 210 in Reading, in 1976. By the end of the century, his *Steve Wright in the Afternoon* show was to characterise the return to dominance of BBC radio, and commercial radio's decline.

952 Johnny Beerling, *Radio 1, the inside scene*, Trafford publishing, 2008 p 284.
953 *RAJAR*, Q2 1992.
954 *RAJAR*. Q3 1997.
955 *RAJAR*, Q4 1997 and Q4 2006.

This was a seminal moment for commercial radio, just as had been Richard Park's invention of ILR's CHR format for Capital Radio in September 1987. The two events serve as monuments at either end of for the high plateau of fortune for the commercial sector. From 1988, ILR had stood largely unchallenged in the centre of adult pop music radio. In those middle years, Radio One's reinvention was not an immediate threat to commercial radio's core audience, and Bannister's shedding of millions of listeners had been a major factor in boosting the commercial radio audience to then unprecedented levels. ILR had the CHR format largely to itself, and had flourished accordingly. However, after 1997 it faced formidable competition in Radio Two; hugely resourced, able to attract genuine star presenters, well-managed with an instinct for popular appeal, and enjoying the cross-promotion afforded by BBC television.

Henceforward, everything would change, exacerbated by commercial radio's falling investment in its programming. The demands of those ruthless paymasters – the City analysts – for ever increasing profits and continued growth by acquisition, imposed different priorities on those who ran quoted commercial radio groups. Consequently, when challenged in the very heart of its product by Radio Two, commercial radio withdrew into the shell of its existing formats, rather than striking out as boldly as the BBC had done. It cried 'foul' endlessly, claiming – with some justification – that the BBC was free to change format while it was constrained by its promises of performance. However, since the 1984 Heathrow Conference commercial radio had been pressing the virtue of free market competition on successive governments, so it received scant sympathy and no effective redress. While bemoaning the impact of this new competition, ILR failed to take any effective action on its own account. The groups relied on cost-cutting to go on delivering profits and keep up share prices in the short term. After 1997, that was never going to be enough.

Making headlines in the business pages in the summer was the bid made by Capital to take over Virgin Radio. That required the Authority to consider the public interest test aspects of such a deal, while ceding a decision on competition issues to the Monopolies and Mergers Commission. The Authority concluded in July that it would approve the takeover, subject to "satisfactory assurances by Capital in respect of the independence of news provision on Virgin, if also supplied by IRN".[956] The Commission took a good deal longer to reach its conclusion, not least since the matter only reached it formally on 31 July, and it was not until December that it announced that the merger should only be allowed to proceed if Capital divested itself of Capital Gold in a manner approved by the OFT and the Radio Authority.[957] While David Mansfield and his board pondered over Christmas, Chris Evans' Ginger Productions nipped in and snatched Virgin away from under their noses.

Things felt good for the Radio Authority in 1997, although as other chapters report it was not short of major issues to resolve. It was mostly comfortable with where it and the industry had got to, and with what it saw as the likely future. In October, the members reassured themselves that the small-scale stations they had

956 *RA Minutes* 7(97) of a meeting on 3 July 1997.
957 *Capital Radio plc and Virgin Radio Holdings Limited, a report on the proposed acquisition,* Monopolies and Mergers Commission, announced 3 December 1997, published January 1998.

been licensing were, in the main, viable – even though "it was not until the 3rd or 4th year that stations generally started to make money".[958] The structures for commercial digital radio were also put in place during 1997, and the mood was very much one of satisfaction with the current approach.

With so much moving forward, such good commercial results from the companies, and commercial radio audiences stretching ever further ahead of those for the BBC, Peter Gibbings felt justified in being expansive. In his statement written for the Radio Authority's 1997 Annual Report, he spoke about an industry which was "young, new, vibrant and energetic" and which "had now firmly established itself as the dominant radio service, overwhelmingly so for those aged fifty or below ... We at the Radio Authority retain great enthusiasm for the achievements and prospects of commercial radio, both new and not so new, whether analogue or digital."[959]

However, for the Authority, just as for the industry, the year was incubating a monster. For the regulator, it was the events surrounding the award of the East of England regional licence in April and May which were to reverberate through 1998, challenging the legitimacy of the regulator, and permanently changing its relationship with the industry.[960] All of the Authority's remaining hopes that the commercial radio companies might work with it to rescue some of its aspirations for *independent* radio were lost in the wreckage of that fall-out. For the Authority's dealings with its industry, things could only get worse; just as they were to do for the industry itself, with a resurgent BBC and commercial radio's inadequate response over the coming years.

958 *RA Conference 1997*, minutes.
959 *Radio Authority Annual Report*, 31 December 1997.
960 See chapter 19.

Chapter 22

Weddings and wind-ups

1998 – 2000

To the outside observer, and to most within the industry, commercial radio seemed to carry the bloom of its best year into the end of the decade. The figures still pointed upwards. Researched audiences for commercial radio were to be as high as 32,180,000 when the research specification changed in 1999. Advertising and sponsorship revenues reached £594m in 2000, when radio took 6.4 per cent of all display advertising in the UK.[961] New radio stations were springing up across the UK. As well as the three INR services, there were 226 ILR licences in issue by the end of 1998, up from 130 in 1991, and destined to reach 272 by the end of the Radio Authority's stewardship in 2003. Each year heard between 350 and 450 restricted service licences, bringing accessible, alternative radio to communities nationwide. DAB multiplex licences began to be advertised from 1998, pointing to another of the periodic re-inventions which had sustained radio since its inception. The relaxation of ownership limits in the 1996 Broadcasting Act was allowing a considerable measure of consolidation, which the rapid introduction of new stations served to enhance.

Radio was also fun. On 21 January 1998, Capital presenter Steve Penk arranged for the impressionist Jon Culshaw to put through a hoax call to prime minister Tony Blair live (or 'as live') on his mid-morning show, pretending to be the opposition leader William Hague. Culshaw duped the Downing Street switchboard, but not it seems Blair, who nevertheless enjoyed the encounter – and the resulting publicity – as much as Penk, Culshaw and Capital. The exchange was typical of the good-natured chutzpah of the times.

> "*Blair:* I'm just trying to work out who it is really. Well, of course, it's ... You've done very well to get through the network, you must have taken them in very well on the switchboard.
>
> *Culshaw/Hague*: Well no, all it was I was talking to John Prescott the other day and he said that you wanted that Cher exercise video that you were interested in ... We went to the car boot sale (*Blair laughing in the background*) me and Ffion and managed to find it, and so perhaps I can let you have it in the Commons today.
>
> *Blair*: I think it would be very helpful, just hand it over at Prime Minister's Question Time. It will be a better exchange than usual."[962]

Commercial radio was self-confident, fun, newsworthy, and – even where it stepped over the line – still not entirely irresponsible. BRMR in Birmingham, ran

961 RAB/Ofcom, www.rab.co.uk/rab2006/showContent.aspx?id=243
962 http://news.bbc.co.uk/1/hi/uk/49473.stm

a 'two strangers and a wedding' promotion in at the end of 1998, an idea taken from Australian radio. In order to win the competition, two contestants who had not met or made contact with each other were chosen by the station's listeners, and then matched by a panel of relationship counsellors and astrologers. They then had to get married legitimately to win prizes. The groom was 28 year-old Greg Cordell, a sales manager, and his chosen bride was Carla Germaine, a 23 year-old former model. The couple married, on air nationally and online, but they soon split after the wedding. Once the divorce was complete, the bride married Jeremy Kyle, who at the time was one of the radio station's DJs.

Yet these last years of the century were to be institutionally ill-tempered, a period of confrontation between the radio industry and its regulator. The major groups wanted more acquisitions than the Authority's policing of the ownership rules would permit, and more freedom to network and to seek economies in programming costs. They increasingly objected to the level of regulatory interference with their programming freedoms, even though in practice this was steadily reducing throughout the period. The industry wished to be wholly *commercial* radio; the increasingly legalistic regulator – plus, it has to be said, the government and the legislation – still hoped to have the best of both worlds, retaining the best features of the *independent* heritage. It was not going to make for a comfortable relationship or a cheerful industry.

Licensing new services, and completing the re-advertisement of existing licences, dominated the Radio Authority's workload, and provided a constant dynamic for the industry as a whole. 1998 saw 23 ILR licence awards, the highest-ever total and one which strained the Authority's resources to the limit, and sometimes beyond. There were high piles of applications along the walls of offices, which could not be accessed because of the back-log. Awards were taking more than six months, and, not surprisingly, there were complaints. Nevertheless, in that one year new regional services were authorised in North West England (Century 105) and North East England (Galaxy FM). The first of those services was to provide rather more than the usual quota of speech, the second for the niche format of dance music, thereby sustaining the Authority's broad approach. That was also to continue with the local licences: Lewisham (FLR), Chelmsford (Chelmer FM), Caernarfon (Champion FM), Fife (Kingdom FM), Cheltenham (CAT FM), Stroud (The Falcon), Bassetlaw (Trax FM), Chesterfield (Peak FM), Mansfield (Radio Mansfield), Hinckley (Fosseway Radio), Loughborough (Oak FM), Rutland (Rutland Radio), Arbroath (RNA FM), Dundee (Discovery FM), Bolton & Bury (Tower FM), Oldham (Oldham FM), Telford (Telford FM), Bournemouth (the NRG), Portsmouth (Victory FM), Southampton (Southampton City Radio) and Winchester (Win FM). Coleraine had been advertised but not awarded to either of the two applicants, and was re-advertised in 1999.

In many wider localities, it was possible to offer one or more licences for low-powered transmission, but not to meet all the possible local demands. In such instances, the Authority invited applicants to propose not only what programming they would provide, but which part of the wider area they would provide it for. Looking for a catchy label, David Vick had decided to call these Small-scale Alternative Location Licences, or 'sallies'. The term 'sally' came by default within in the industry to be used for all the smaller coverage services, whether they were alternative location or not. What had been intended as fun became all too easily used to be used as a derogatory term, making running such a service profitably that bit more difficult.

Generally, 'sallies' caused more problems than they solved. For example, in March 1998, an advertisement invited applications "within the area of the mainland around the Solent, extending between Poole to the west and Havant to the east and northwards to include Winchester [in which] the Authority anticipates making available a maximum of four frequencies".[963] The Authority had received 14 applications by 2 June, and awarded licences on 10 December to four of them: The NRG in Bournemouth; Southampton City Radio; Win FM in Winchester; and, in a piquant echo of earlier ILR history, to Radio Victory in Portsmouth. The published appraisal of the winning Radio Victory application showed that the Authority had been underwhelmed by the quality of some of the applications. "Members were, however, disappointed to note that [the board] consists predominantly of non-local directors and investors, as a larger representation would have reassured members that the local knowledge and connections would not be dissipated. The financial forecasts appeared achievable, although based on relatively low staff and marketing costs."[964] The smallest licences often attracted the fiercest competition, and the most virulent disappointment. It was clear that so long as there was a discretionary basis for the award of radio licences, the licensing body – whichever one it was – would lose friends with ease. Peter Gibbings felt the need to stress that "in the midst of lively debates about our licensing decisions, it is easy to forget that commercial radio is in good shape, growing and prospering, and to underrate the Radio Authority's role".[965]

As the Authority neared the bottom of the frequency barrel, the use of low power – typically 100 watts – for small-scale services gave coverage which many thought scarcely economic. It allowed the frequency planners to maximise the number of services which could be fitted in, but was a mistake for stations which had to be commercial rather for community services. Many operators of small stations argued that they were constantly hindered by their low power. That, in turn, made them vulnerable to takeover, and offered poor prospects for providing much more than basic programming fare. That many did more than the minimum, mindful of the legacy of independent radio, is a tribute to their commitment and operational ingenuity.

The potential use of 106–107 MHz had been left open in 1994, when it was decided to offer new regional and small-scale local services on the rest of this sub-band. On 8 April 1998 the Radio Authority consulted formally on the use of this last remaining block of analogue frequencies. It announced on 27 October that it had decided to use these frequencies also for a mix of small-scale and regional services, and that "the Authority recognises that both these forms of local radio have demonstrated that they perform a valuable role in satisfying listeners' requirements and serving the public".[966] It signalled, however, that "the use of the alternative location (sally) approach will not continue to any substantial degree". The band was "no longer 'virgin' territory; the flexibility inherent in the 'sally'

963 Radio Authority News Release, 2 June 1998.

964 Local Licence Awards; Solent/Mid Hants Area' assessment of winning applicants, issued with Radio Authority News Release 135/98, 24 December 1998.

965 *Radio Authority Annual Report*, 31 December 1998.

966 Radio Authority News Release, 27 October 1998.

approach would be neither practical nor efficient if superimposed on the existing frequency plan". This was the last of the long series of major strategic decisions about using spectrum for analogue radio.

As well as new ILR advertisements, and DAB licensing, in 1998 the Authority began the process of re-advertising those existing ILR licences which did not qualify for automatic renewal through a digital commitment. As the workload gathered pace from 1999 onwards, a new 'fast-track' procedure was introduced, also derived from the 1996 act.[967] This involved 'pre-advertising' existing licences, and where no serious letters of intent were received from potential competitors – backed by a bond for anything between £20,000 and £100,000,[968] depending on the size of the licence area – the incumbent would not have to undergo the costly and unsettling process of full re-application. There was a catch, however. This required a prior commitment from the operator not to seek any major changes in their format, to prevent them using the lack of competition to dilute their output. It streamlined the award process, as it was intended to do; but it had the unintended consequence of locking stations into their pre-existing programme formats, and of increasing their frustration over such limits.

In 1999, in addition to nine re-advertisements, the Radio Authority awarded 17 new ILR licences. These were for Wakefield (Ridings FM); Central Scotland (Beat FM); Doncaster (Trax FM); Peterborough (Lite FM); Fenland (X-Cel FM); Bath (Bath FM); Bristol (107.3 The Eagle); Weston-super-Mare (107.7 WFM); North Lanarkshire (Clan FM); Kintyre, Isley and Jura (Argyll FM); South Hams (South Hams Radio); Bridlington (YCR BRID); Coleraine (Q97.2 Causeway Coast radio); North London (Choice FM); Bridgend (Bridge FM); Newbury (Kick FM); and Burnley (Two Boroughs Radio). They ranged from large regionals to the smallest local stations. Sometimes the awards went to neighbouring stations, which could provide support and gain from synergies, but all too often they were for a small, stand-alone service. That was a matter of conscious policy. The weighting in scoring, derived from the staff analysis, often left competing applications close together. The judgement of the Authority members, under those circumstances, most usually favoured the independent new entrant, especially where it could demonstrate genuine local roots.

The high workload continued into 2000, albeit at a slightly slower rate. Seven new ILR licences were awarded. These were for Bridgewater (BCR FM); Burgess Hill & Haywards Heath (Bright 106.4); Dumbarton (Castle Rock FM); Grimsby (Compass FM); Hertford (HertBeat FM); Knowsley (KCR FM); and South Wales (Real Radio). This brought the total number of ILR licences in issue to 248. 18 existing ILR licences were re-awarded. Five attracted competition, but all went to the incumbent. With one digital multiplex licence awarded a month, there was little respite for either the administrative or the technical teams at Holbrook House, nor for the Authority members who continued to take the decisions.

By then however, the membership had greatly changed, and the 'third Authority' was in place. Richard Hooper succeeded Peter Gibbings in January 2000. He took to the post with aplomb, his natural bonhomie a contrast to Peter Gibbings' careful formality, and with the memory of his time on the 'first Authority' – and

967 Broadcasting Act 1996, section 94 , creating a new section 104B for the 1990 act.
968 Radio Authority News Release, 20 May 1998.

around the LBC licence non-award – consigned to history. Just as Gibbings had needed to bring a disciplined approach when he arrived in 1994, so Hooper's easy informality was just what the Authority as a whole needed as it emerged from its time of siege.

Hooper also had the advantage of a new group of members, whose enthusiasm and bonding was an object lesson for such a small public body. Andrew Reid, Michael Reupke and Brenda Sheil had all retired during 1999, Jennifer Francis before the end of her term on 31 March 1998. Unlike the first two Authorities, the third group were all appointed only after they had applied for the posts, and this brought a new type of member. Feargal Sharkey, David Witherow and Sheila Hewitt had all arrived on 1 December 1998 through this process. They were joined by Sara Nathan on 26 July 1999, and Mark Adair on 10 July 2000; and then by Kate O'Rourke, Thomas Prag and Geraint Talfan-Davies at the start of 2001. Together, this was a group who respected each others' views and expertise, although never afraid to challenge them robustly, and who enjoyed the *ésprit de corps* which Richard Hooper assiduously cultivated. The Authority had been notably well served by Michael Moriarty in his challenging times as deputy chairman, and it was fortunate to find a replacement for him in David Witherow, from 18 January 2000, who brought the same quiet, intelligent, focussed and sympathetic qualities. Changes at senior staff level had seen Martin Campbell replace David Lloyd as director of programming and advertising, who headed off into the radio industry where he continues to thrive; and Eve Salomon arrived from the ITC to succeed John Norrington as secretary. The management team, even more than the Authority members, were cheerfully collegiate and comfortable with each other's views and company.

Strong revenues, and a host of new ILR licences, fuelled the appetite for consolidation. Rising share prices for the quoted larger radio groups – and the share option riches that offered their senior executives – made expansion a virtual necessity. Within the rules derived from the 1996 act, the groups were able to increase their ownership of licences, although not to the extent they would have liked. It seemed there could be few new entrants at the national level, for all the Radio Authority's encouragement of them for smaller licences. However, Kelvin MacKenzie, once a highly controversial and successful editor of the *Sun*, had for a long time wanted to be a radio owner. In November 1998, he bought the ailing INR licence, Talk Radio, for a reported £15.5m. Among his backers was his former employer Rupert Murdoch, who was thought to be only too pleased to have pinched MacKenzie back from the *Daily Mirror*. The take-over led to a mass clear-out of presenters and a more sports-orientated programming schedule. 21 November was coined 'Black Thursday' by hardcore Talk Radio listeners, but the *Guardian* was pretty up-beat. "The disgraced – Mandy Allwood, the Hamiltons – have always been welcome at Talk Radio, and this week new owner Kelvin MacKenzie continued the tradition by signing Geoff Boycott and Derek Draper. He has had a busy fortnight hiring and firing."[969] MacKenzie had an eye for a headline, not least to make up for a non-existent marketing budget. The report went on to note that "MacKenzie, the creator of Live TV's News Bunny has now introduced into James Whale's programme a Weather Fairy – a camp young man

969 Anne Karpf, *Guardian*, 21 November 1998.

reading the weather forecast. The station's weather sponsor, British Gas, is not amused and has pulled out."

The conservative radio industry was equally unamused, especially once MacKenzie started rocking the industry boat over audience research, and falling out with the BBC over sports rights. His company – the Wireless Group – became the fifth largest UK radio group when it purchased The Radio Partnership in 1999, gaining control of nine local commercial stations, but it frequently fell out with the other majors. Over two years from 2002, MacKenzie withdrew the stations his group owned from CRCA, partly over his dispute with RAJAR and partly to save the cost of the subscription.

MacKenzie had success with the regulator. After discussions through the second half of 1999, members approved a switch in the format of INR3, though still fully within the terms of the original speech licence advertisement. At midnight on Monday 17 January 2000, MacKenzie relaunched Talk Radio as 'talkSPORT', the UK's first national commercial sports radio station. The programming lineup was drastically altered, beginning with the Sports Breakfast show, a mid-morning motoring show called The Car Guys, with further sports programming in the afternoon and evening. Almost all the station's talk show presenters were axed at the time, with only James Whale and Ian Collins surviving. To complement its new format, talkSPORT purchased the rights to a range of sports including Manchester United, Arsenal and Newcastle matches in the UEFA Champions League, the FA Cup, England Internationals, UEFA Cup, England's winter cricket tours to South Africa, Zimbabwe, Pakistan and India, British Lions Tours to South Africa and New Zealand and rights to the Super League, Rugby League World Cup, and World Title Boxing Fights. The Murdoch style is evident here, and it scared the BBC rotten. After talkSPORT picked up overseas commentary rights for England cricket teams for a couple of years, and resurrected the career of Geoffrey Boycott as a commentator/summariser, the BBC was determined never to be beaten again. The effect was to push up the costs of radio sports rights, to the point where talkSPORT was priced out of the market. It retaliated in June 2000 with unauthorised coverage of the Euro 2000 football tournament, based on television pictures. Its commentaries came from the Jolly Hotel in Amsterdam and not live from the grounds, although later the station was obliged to acknowledge on air that its commentators were watching a relay and not the actual game.

One impact of the 1996 Act and its 'digital incentive' was that the three INR licences were to enjoy automatic renewal, once they were committed to the national digital multiplex. The Authority already had experience of four different types of licensing: according to specific statutory criteria, for ILR services and digital multiplexes (the so-called 'beauty parade'); by single sealed highest cash bid for INR; largely on demand, for cable, satellite and digital sound programme services (the so-called 'dog licences'); and on a discretionary basis, for RSLs. To that was now to be added administrative pricing. The measure used was broadly to work out what the licences would now fetch if they were bid for again on the open market, with similar formats, and then set three prices as required by the act.[970] The first was the rental payable to the Authority, to cover the cost of

970 Broadcasting Act 1996, section 92, creating a new section 103A for the Broadcasting Act 1990.

regulation, which was by far the smallest and least contentious of the figures; the second was annual 'cash bid'; the third was the percentage of qualifying revenue (PQR) which would be paid over to the Treasury's Consolidated Fund, in addition to the cash bid, effectively a tax on advertising revenues.

Classic FM, the first INR to go on air, was the first up for renewal. Both the Authority and GWR had their financial advisers and lawyers, with long-standing lawyers Allen & Overy being joined by Price Waterhouse Coopers in advising the regulatory team headed by Neil Romain. There followed what was effectively a negotiation, but leading to a final decision by the Authority members. At its meeting in September 1999 the Authority set the annual cash bid for the continued INR1 licence at £1m, to be adjusted annually for inflation; and raised Classic FM's PQR from 4 per cent to 14 per cent.[971] This was on the high side of the company's expectations, but it had the misfortune – in this respect – of seeking licence renewal at a high point in commercial radio revenues. The decision to load the payments on to PQR, rather than a fixed cash bid, at least made things easier when times got tougher.

The following year, using the principles established for the INR1 re-award, the Authority re-awarded the INR2 and INR3 licences. In April 2000 Virgin's licence was re-awarded, requiring an annual cash bid of £1m, and with a PQR up from 4 per cent to 12 per cent.[972] In December 2000, what was now TalkSPORT enjoyed much easier renewal terms, an annual cash bid of £500,000, and with its PQR raised from 4 per cent to just 6 per cent. The difference is explained by the predicted revenues and value of the different licences over the new licence period, but the technical accounting arguments did not play too well between the competing companies. In addition, the Authority had concluded that "the costs to the INR licensees in launching a digital simulcast service cannot be taken into account in the calculations, although allowance will be made for the expected migration of the radio audience from analogue to digital". The companies, and especially GWR, were to press the government during the following years to put the moneys it received from INR to support the roll-out of digital radio across the UK. As usual, there was little sympathy and scant financial help forthcoming for radio.

In programming, the system of promises of performance had worked tolerably well towards the legislative aim of keeping companies broadly to the programme output which they had proposed in order to win their licence. For some of the heritage stations, they served as protection against their new commercial competitors. Generally, the older stations had more broadly-drawn promises of performance; the newer companies had needed to demonstrate that they would bring something distinct to the locality, and consequently theirs were typically tighter-drawn documents. Martin Campbell, aware of external pressures towards less onerous regulation, but also with his practical programme manager's experience of what worked and what didn't, took the initiative to streamline this licence element. At the Authority meeting January 1999 he argued successfully for replacing the typically five-page promises of performance with a one-page 'Format'.[973] That would keep the essential elements of the original licence, but would get away

971 RA Minutes 8(99) of a meeting on 1 and 2 September 1999.
972 RA Minutes 4(00) of a meeting on 6 April 2000.
973 RA Minutes 1(99) of a meeting on 7 January 1999.

from detailed micro-management, seeking to ensure that "each station ... has more freedom to make changes within its format, but the character of the station must stay the same".[974] The Authority's programming staff worked with the ILR companies from February to September 1999 to agree the new format documents for every station. It was done region by region, to avoid having stations within competing areas possibly operating under different regulatory regimes. New stations were still given a promise of performance, but that was usually converted to a Format after about six months.

The other change in programming regulation at around the same time was the introduction of a warning system for breaches of the rules, to apply in between an informal word and a formal sanction. In line with what happened on football fields, the Authority introduced what were called 'yellow cards'; a warning, but not a sending off. When, through complaint or its occasional monitoring, the Authority became aware of breaches of the spirit or the letter of the new Format, it would issue "a 'yellow card' letter which seeks assurances that content will clearly be brought within the Format guidelines".[975] The list of yellow cards, typically two or three in a quarter, was then published in the Authority's Quarterly Bulletin. It was a good initiative, broadly popular among the companies (except where they received a card, and sometimes even then, as it was preferable to a formal sanction). It was also necessary, since during 1999 and 2000 there was a run of serious compliance breaches, which could thus properly be distinguished from the more routine cut-and-thrust of format regulation.

In a foretaste of the scandals which were to hit the BBC and GCap in 2007 and 2008, the Radio Authority was tipped off by a complaint about a fixed competition run by 107.5 Cat FM in Cheltenham. It alleged that during May 1999 the station had run a competition to win 40 CDs for which the prizes did not exist but a fictitious name was used as a prize winner; that in the following week a similar competition to win 20 video tapes was running, but the prize did not exist; and that a recent competition for a cash prize had been won by a person connected to a senior manager of the station to ensure that the station would not have to pay out an accumulation of money. On investigation, Cat FM confirmed that the first two prizes had not been available when the competitions were run, and that the management had terminated the competitions by inventing fictitious winners. The station argued that the cash prize competition had proved impossible for listeners to solve, and that the station management had arranged for a person connected to a senior member of staff to give the right answer on-air. It said that no prize money had actually been paid.[976] The fine of £20,000 was huge by the standards of that time, given that Cat FM was a very small station, of marginal profitability.

Next came an attempted fix of a different order. Oxygen FM in Oxford began broadcasting in February 1997 as a student station, making extensive use of volunteers. It had a highly ambitious, self-imposed format requiring debate, discussion, and science and arts programmes. To follow up a complaint in March 1999, the Radio Authority made a number of what seemed to be routine requests for tapes of the output during the period in question, and in particular for 1 March.

974 *Radio Authority Annual Report*, 31 December 1999.
975 Ibid.
976 Radio Authority News Release 107/99, 6 July 1999.

However, when listening the Authority's programme staff noticed an item in the news bulletin, supposedly broadcast on 1 March, which had, in reality, only happened the week after. It transpired that, since they couldn't find the logging tapes for 1 March, Oxygen had doctored their tapes for 8 March, and sent them to the Authority hoping that no one would notice. When the Authority requested the 8 March output, it was sent the output of 15 March labelled as 8 March. On my instruction, we actually went to Oxford to collect tapes for an entire week, only to find that none of 21 tapes contained broadcasts of the week in question. The Authority was therefore unable to check whether the programmes which Oxygen was required to put on-air were being broadcast.

The rule was simple, and well-established. Absence of tapes was a major offence in itself, to be treated at the same level as the alleged programme breach. When it was compounded by such duplicity, Oxygen going to get the book thrown at it. That was harsh on the new owners, who had only recently taken over the impoverished station with the intention of making it less a student service and more a commercial radio station, but this was going attract a fierce sanction. Meeting in September, the Authority not only fined Oxygen £20,000, but also shortened its licence by two years, meaning that it would come up for potentially competitive re-advertisement in 2003. Given the value that ILR licences still attracted, this was playing hardball.

Then, in December 1999, an increasingly uncomfortable Authority found itself required to adjudicate on the two worst instances of offensive programming which had to date come its way. As a consequence of the first case, quite unmentionable descriptions of 'your worst ever job', Xfm was fined £50,000 for two editions of its breakfast show, the maximum then permitted under the Broadcasting Act for an ILR station. Peter Gibbings' statement reflects the members' sense of outrage. "These broadcasts, which have been listened to and fully considered by all members of the Authority, heedlessly ignored the requirement that the portrayal or description of bestiality must not be broadcast. Language and sexual references made throughout both programmes also breached the Authority's Programme Code. The fact that this material was broadcast at breakfast time, when a significant number of young people might be expected to be listening, makes the offences all the more serious."[977] It was, though, not just the Authority's sensitivity. There were specific rules in the Programming Code, intended to protect children and designed to give guidance to broadcasters, but Xfm had ignored them in its enthusiasm to be 'cutting edge'.

At the same meeting, Hallam FM was also fined a maximum £50,000. This was for "two broadcasts which condoned rape and contained a gratuitous description of paedophilia, and for the failure to supply a tape for a further complaint to be investigated".[978] Peter Gibbings again spoke for the regulator. "The whole Authority condemns this kind of broadcasting which demeans the radio medium. We are determined that similar material shall not be broadcast again."[979] In this instance however, prompt action by the station to prevent a re-occurrence was welcomed.

977 Radio Authority News Release 180/99, 14 December 1999.
978 *Radio Authority Annual Report*, 1999 op. cit.
979 Radio Authority News Release 181/99, 14 December 1999.

Not so at Xfm, where "the members of the Authority were not left feeling fully confident that the owners of Xfm, Capital Radio Plc, had taken steps which should ensure that these offences cannot be repeated. The Authority has notified Capital Radio Plc that if Xfm does not put in place effective compliance measures, then in the event of any further serious breaches it will consider a further range of sanctions which may call into question the continuation of Xfm's licence."[980] This was tough stuff indeed. David Mansfield had written, in response to a formal invitation to make representations before sanctions were considered, that "Xfm has a format that requires us to break new ground ... the station is not being programmed to challenge the Radio Authority and we are not pursuing 'shock tactics' to grow ratings ... the mitigating circumstances are that in developing a new format we will make errors and regrettably this was such an occasion".[981] Given that there had been a whole series of such 'occasions', and that Xfm had given Authority staff assurances at a meeting on 20 October that they had got a grip on the continuing problem, the Authority members had been unimpressed. I met David Mansfield to try and get the station, and Capital's relationship with the Radio Authority, back on track. Once the key axis in the development of independent radio, it was fracturing within the new commercial radio atmosphere to a point almost beyond repair.

Other programming issues continued to arise elsewhere. There were fines in 2000 for Manchester's Asian Sound Radio, twice, and for the luckless Oxygen, now under its new management, for enhancing its transmission power without authority, known as 'signal over-deviation'. The most eye-catching sanction, though, was on Virgin Radio. Chris Evans' company, Ginger Productions, had bought Virgin for £85m in 1997, snatching it from the grasp of Capital Radio. Ginger sold on to Scottish Media Group in January 2000 for £225 million, netting its proprietor a reported "£75 million divvy",[982] and the opportunity to continue with his regular broadcasting on Virgin. In the run-up to the London mayoral election in 2000, Evans had invited into his live show on 21 March the then independent candidate, Ken Livingstone, and given him a resounding endorsement of his candidature, in flagrant breach of the impartiality rules of the Broadcasting Act. It was no momentary aberration, either. The Authority's adjudication, following its May meeting, noted that "Chris Evans first aired his support of Ken Livingstone at length after the 8.00 am news, and again an hour later in an interview with a news reporter. The station had not only failed to ensure that its presenters were fully aware of the rules on due impartiality, but had allowed a second breach to be aired." The Authority itself avoided being manipulated into providing material for the pre-election bunfights, by not considering sanctions until 11 May, seven days after the election.

Most of the 'yellow cards' issued during these years arose from either over-enthusiasm or naivety among presenters or producers. There were very few rogues in the radio industry. The regulation of content against the norms of avoiding offence, ensuring taste and decency, political impartiality and the like – to be known for a while as 'negative content regulation' – worked pretty well. It enjoyed

980 News Release 180/99 op. cit.

981 Letter from David Mansfield, Capital Radio, to Eve Salomon, Radio Authority, 19 November 1999.

982 *Guardian*, 14 January 2000.

a large measure of consent from the programmers themselves, and was helped by a close relationship between them and the individual regulatory staff at the Authority. With 248 ILR licences in issue by the end of 2000, plus INR, long and short-term RSLs and now DAB, the level of complaint and the number of sanctions was comfortably small. It could all be managed by a total content regulation staff of eight or nine.

An issue of a wholly different magnitude rumbled through the years from 1996 onwards. Owen Oyston, after selling TransWorld to Emap for £71.1m in June 1994, had since returned to radio as the owner of The Bay (Morecambe Bay), Heart Beat 1521 (Craigavon), Gold Beat (Cookstown), and City Beat 96.7 (Belfast). In 1996, he was jailed for six years at Liverpool Crown Court for raping and indecently assaulting a 16 year-old model. As the case proceeded, the Radio Authority had repeatedly to consider whether Oyston's conviction made him not a 'fit and proper person' to hold an ILR licence, under the terms of the Broadcasting Act.[983] Relying on detailed legal advice, the Authority decided to take no steps until the appeal process had run its course. Oyston's appeal was turned down on 9 December 1997, by which time the Authority had considered nine separate papers on the matter, and received three counsel's opinions. Now it concluded that "in the light of Oyston's conviction for rape and indecent assault now confirmed by the Court of Appeal ... Oyston was not a fit and proper person to hold a licence".[984]

That was by no means the end of the matter. In the face of pressure from minority shareholders, the Authority had to consider at its meeting on 2 April 1998 whether the licences should simply be revoked, or transferred to Oyston family trusts. With the assurance that "during such time as significant radio company shares are held by the trusts Owen Oyston will not be entitled to be a beneficiary".[985] and that the transfer should be completed within six weeks, it agreed to the latter. The licences to operate all four stations were consequently at that point transferred to a new company, Classworld, which was at pains to meet all of the Authority's conditions. However, it struggled to make the two smaller Northern Ireland stations viable, and eventually at the company's own request the Gold Beat and Heart Beat licences were revoked in May 1999. Oyston was released from prison on parole on 7 December 1999,[986] still maintaining that he had been the victim of a conspiracy by "former Conservative ministers and a businessman".[987]

These were the high years for New Labour, with its huge parliamentary majority and its chosen mission to transform the UK into a market-friendly, cutting-edge yet compassionate modern society. The makers of this new political philosophy were considering changes for broadcasting structures and regulation well before the 1997 election. Geoff Mulgan, in the *Demos Quarterly* in 1994, had urged a recasting of regulation in the communications sector: "Regulation has been a policy success story over the last decade. But on its own it solves few problems. Moreover, its mindset has been so shaped by the regulatory theories developed by

983 Broadcasting Act 1990, S 86(4)(b).

984 *RA Paper* 150(97) 9 December 1997; *RA Minutes* 11(97) of a meeting on 11 December 1997.

985 Radio Authority News Release 28/98, 9 April 1998.

986 Oyston maintained his innocence steadfastly, and does so to this day. He became a test case when he challenged the principle which denied him early release, since by refusing to admit the offences he was excluded from a course for sex-offenders which was then a pre-requisite for consideration of parole. On 14 October 1999, he succeeded in overturning that principle in the High Court.

987 *Guardian*, 8 December 1999.

economists in the 1960s, 1970s and 1980s that it may not be well suited to coping with the more complex issues of network openness and social uses. The task now is to build on the successes of independent regulation but to adapt it to a wider set of goals. To do this a range of reforms is needed."[988]

Top of Mulgan's shopping list was the creation of an integrated regulatory agency for all communications which he dubbed 'Ofcom'. "Our regulatory structures are now far too complex, with a mess of overlapping roles. The regulation of networks needs to be consolidated into an Office of Communications, covering all broadcasting and telecommunications networks ... Competition should be seen as a means, not as an end. For example, some regulatory interventions might be justified in terms of increasing telework (and parenting time) or cutting travel to work times. In each case competitive principles (such as negative tendering) may be useful as means." Since Mulgan went on to be the head of Tony Blair's cabinet office strategy unit, his views had more than a fair chance of getting a good hearing. Meanwhile, 'competition' shifted from a means to become an end in itself.

The Institute of Public Policy Research (IPPR), the other think-tank much beloved of New Labour, followed up with two publications in 1996. Christina Murroni, Richard Collins and Anna Coote published *Converging Communications, Policies for the 21st Century* which argued for "a media specific regulator, Ofcom, to undertake both carriage and content regulation",[989] the order of duties signalling the fundamental shift in received wisdom about media regulation. Following an IPPR seminar, Richard Collins edited a series of short papers on the same theme. Don Cruickshank, then the director of telecommunciations (the post which would later become DG of Oftel) saw the telecommunciations and media industries "converging, colliding, coming together ... let's drop the 'tele', in due course let's drop 'broadcasting', [and call it all] the new communications industry."[990]

It became the government's orthodoxy that technological convergence would change broadcasting, telecommunciations and the use of the public resource which was the electro-magnetic spectrum generally, and that this would therefore require converged regulation; a super-regulator. Sure enough, a green paper, *Regulating Communications*, was published in July 1998, after barely a year of the Blair administration.[991] It was concerned with "the likely implications of digital convergence for the legal and regulatory frameworks covering broadcasting and telecommunications". Public policy objectives were to be "serving the consumer interest, supporting universal access to services at affordable cost, securing effective competition and ... competitiveness, promoting quality, plurality, diversity and choice in services, encouraging investment in services and infrastructure [and] providing economically efficient management of scarce resources". This was a break with the previous mainsprings of public policy intervention in broadcasting beyond anything that even Margaret Thatcher had contemplated.

The Radio Authority did its best to trim its sails to the prevailing wind. Just like the IBA in the late eighties, it clung to an unrealistic belief that it had a chance of surviving to regulate radio, if only it could adjust its practices a bit. A rather

988 Geoff Mulgan, *Demos Quarterly*, Issue 4 1994.

989 *Converging Communications, Policies for the 21st Century*, Institute for Public Policy Research, 1996.

990 Don Cruickshank, *Regulation or Convergence*, in *Converging Media? Converging Regulation?*, ed. Richard Collins, IPPR 1996.

991 *Regulating communications; approaching convergence in the digital age*, Cm 4022, July 1998.

reluctant 'second Authority' was persuaded to allow a little light to shine on how it went about its business. I wrote a paper in 1998 arguing that two new pieces of legislation – the Freedom of Information Act, imposing disclosure requirements, and the Utilities Regulation Act, enforcing greater transparency – would soon require change, and it would be better to be "pro-active in advance of legislative coercion".[992] From July 1988, it was agreed the Radio Authority would give reasons for its licence awards, in the form of an assessment of the winning application; release all relevant technical data about transmission parameters in the context of frequency planning; and make public how it would assess the statutory expectation of 'quality' for assessing applications for the re-advertised ILR licences.[993]

The 'third Authority' under Richard Hooper was much keener. It moved quickly to open up most of what it did. This was, at least in part, Hooper's way of finally lifting the shadow cast by the *Newsnight* events. After just a few months in post in 2000, he announced six new initiatives to increase the Authority's openness and transparency. These covered the publication of the agenda and contents of Radio Authority meetings; the publication of guides which fully explained a number of the Authority's processes; the publication of a Register of Interests; public access to the Register of Gifts and Hospitality; the publication of the Code of Best Practice for members; and changes to the Authority's 'purdah' rules which governed contact between members/staff and applicants. Hooper is particularly proud of the extent to which these were pioneering for Ofcom's later practices, and produced changes at the BBC and elsewhere. "The key thing for me in that early period was agreeing on disclosure of reasons for awards, of members' interests and a public register of gifts and hospitality. That, and publishing notes of meetings (not minutes). We can claim that we were the first to do that. Simon Milner told me that the BBC governors went to notes on the website as a direct result of the Radio Authority doing it first."[994]

The radio companies rejoiced, suspending briefly their now-usual curmudgeonly attitude. One chief executive told the *Radio Magazine* – at that time a samizdat weekly which uncritically spread news, gossip, and sundry tittle-tattle through the industry – that "at last we have someone in charge who understands the situations we can find ourselves in. Making the Authority more open will benefit everyone and hopefully will end long held suspicions on decision making ...".[995] The new processes represented very much best practice. The Radio Authority's systems and internal codes were imported almost unchanged into Ofcom, and remain current there. The cheerful mood among the major companies, however, was as usual only transitory. What they hoped for were not so much open decisions, as decisions favourable to them. Within eight months, CRCA was telling the same magazine that it was threatening legal action against the Authority over its admittedly clumsy handling of the regulation of automation.[996]

The next and key stage in the Authority's preparation for the debate about the future regulation of radio was developed during the Authority's annual strategy conference at Lainston House, on 6 and 7 April 2000. The resulting document,

992 *RA Paper* 54(98) 27 May 1998.
993 Radio Authority News Release 70/98, 14 July 1998.
994 Richard Hooper in an interview with the author, 28 May 2008.
995 *Radio Magazine*, 11 April 2000.
996 *Radio Magazine*, 14 December 2000.

Radio Regulation for the 21st Century,[997] was the quintessential production of the later Radio Authority. I wrote the draft, the Authority's directors and members debated it long and intensely. The outcome was a vision for preserving the separateness of radio, as a protection from being swamped by the big battalions – and huge financial figures – of spectrum, telecoms and television. Sent to the two departments, DCMS and DTI, who were jointly to introduce the new structures, it was widely admired; and, apart from the proposals regarding a third tier of not-for-profit radio, entirely ignored.

Once the draft had been signed off by the Authority members on 11 May, all the companies were invited to a series of presentations – in London (24 May), Glasgow (25 May), and Birmingham (30 May), at which they were given a preview of what the Authority would be saying to the government. This was unprecedented for the Authority, and while in the modern context it seems dreadfully belated for a piece of 'consultation', at the time it felt like a significant step. The companies themselves had mostly yet to get their act together. GWR's Ralph Bernard had addressed the All Party Media Group on 2 February, setting out his company's wish for less restrictive ownership rules, but that had been about it so far. The sight of a newly self-confident and articulate regulator, putting forward arguments for a modern version of *independent* radio, was not very palatable.

The industry and the regulator had spent much of these years winding each other up, with less charm than Jon Culshaw and Steve Penk had deployed with Tony Blair. By this stage, neither really enjoyed the marriage. The companies looked to the appointment of Lord Eatwell as the new chair of CRCA, from January 2000, to produce dividends quickly, to redress the balance of lobbying weight and to move them towards a new relationship elsewhere.

997 *Radio Regulation for the 21st Century*; submission to DCMS/DTI, Radio Authority, June 2000.

Chapter 23

RSLs and Access Radio

The strange triumph of the social engineers: 1991 – 2006

For those who hoped for a separate third tier of not-for-profit radio in the UK, pretty much all had seemed lost with the passage of the 1990 Broadcasting Act. The drive to establish community radio as a separate sector had been undermined by the extremism of its proponents, which marginalised it from the seventies onwards, and then let down by its fair weather supporters in the government when they abandoned the community radio experiment in 1986. By the start of the nineties, it was received wisdom that ILR, especially the smaller ILR stations and the incrementals, were to all intents and purposes 'community' stations, with the added benefit that they made no demands on the public purse. There seemed little need for anything else.

The Radio Authority had a mixed inheritance in terms of alternative radio services. Hospital and student radio services, broadcast by technologies which kept their signals within confined areas – usually within individual buildings or groups of buildings – had been licensed directly by the Home Office. That now passed to the new Authority. It also acquired the remnants of some experiments with radio on mostly new town cable networks: Basildon, Milton Keynes, Telford and Thamesmead. The 'special event licences' which had also been issued directly by the government became the Radio Authority's responsibility, and from them the Authority fashioned an unlikely but hugely successful alternative radio sector, Restricted Services Licences (RSLs).

The only real challenge to the consensus that community radio was an idea whose time had passed was the continued and widespread existence of land-based pirate radio stations. Although there were significant exceptions, many of these could be stigmatised as offering an extreme form of music, revelling more in their illegality than in any sense of service, or linked to the growing scourge of the drugs trade and other criminality. Some argued that the existence of so many illicit stations meant that they should somehow be legalised, but increased public access to the radio airwaves was provided by a remarkable new way of making radio even more available and accessible to the smallest of localities: the Restricted Service Licence (RSL).

Restricted Service Licences

RSLs were one of the most striking of all the innovations by the Radio Authority in the early part of the nineties, once it was split from the more hidebound thinking of the IBA. They were and are beyond question a form of community radio, and one unforeseen by either its early advocates or by the shapers of broadcasting policy. They were independent, and made extensive use of volunteers and all available short-cuts to make highly local broadcasting affordable. RSLs were often commercial, either in their own right or by being trials for later applications for ILR licences. They expanded what had been merely special event coverage – though that remained significant – to serving communities of interest such faith groups at the time of religious festivals, and all sorts of imaginative trials in the use of radio. The high level of demand for such licences, their success in operation and the variety of forms which they took, revived the notion that community radio might still somehow thrive in the UK, and offer something different from the increasingly commercial nature of ILR.

By the mid-nineties this new sector had become an established and widespread feature of independent radio, and its effect was quietly revolutionary. Susan (Soo) Williams, the leading figure in taking forward the licensing of RSLs, both at the Radio Authority and at Ofcom, points out that "RSLs remain the only mass-audience broadcasting licences which can be issued to individuals".[998] All others, including the access stations and the post-2004 community licences, could be held only by corporate bodies. RSLs could be wholly commercial, or not-for-profit, just as their operators chose; self-styled 'community' applications received no priority.

The RSL concept was worked out between David Vick, and the Home Office's Bob Macey in 1990. The licences were to be 'restricted' in terms of highly localised coverage areas, low-powered transmissions, and a duration almost always of no more than 28 days. Macey and Vick devised how the Home Office's responsibilities might come across to the Authority, and how greater freedom might be given to the new body to take this form forwards. Neither quite realised that they were creating a new institution. By supreme good sense, the 1990 act is strikingly brief on the power given to the Authority. "The Authority may ... grant such licences to provide independent radio services as they may determine".[999] It has specific obligations to provide national and local services, but otherwise shall "discharge their functions as respects the licensing of independent radio services in the manner which they consider is best ... ."[1000]

RSLs had a steady start, impressive but less dramatic than the top line figures suggest. The first RSL to be licensed was atypical, Ski FM, which broadcast for the winter months of 1990/1 in the Scottish Highlands. There were 178 licences issued in 1991, and 241 in 1992. 90 of those in the first year and 83 in the second were for Radio Cracker, an interdenominational Christian youth project. That swelled the numbers, but also demonstrated the appeal of RSLs to religious groups. Muslim stations for the period of Ramadan were soon to become a feature of the sector, beginning with the cleverly-named Fast FM in Bradford in March 1991. After an apparent dip to 175 in 1993 – but without many Cracker licences – the number rose to 262 in 1994 and continued to grow steadily to 464 in 2000, the latter figure including 41 RSLs to celebrate the new Millenium. For the last years of the Radio Authority, licensing ran at 450 in 2002 and 492 in 2003. Even after

998 Susan Williams in an interview with the author, 21 May 2008.
999 Broadcasting Act 1990 section 85 (1).
1000 Ibid. section 85 (3).

the arrival of licensed community radio in 2004, RSLs continued as a strong part of the radio mix, staying at almost 500 each year.

The mix fluctuated, partly through happenstance, but also the changing interests in this third sector and the extent to which more mainstream concerns intruded. Illustrating this last aspect were the 'trial' RSLs, run by groups wanting to apply for ILR licences. In 1994, as ILR was entering into a new period of expansion, around a third of RSLs were trial services. This rose to 42 per cent the following year, "reflecting the large number of groups who ran RSLs in preparation for applying for a local licence in certain popular areas (e.g. Greater London, East Kent and Merseyside)"[1001] before settling back down again to 36 per cent in 1996. It is arguable whether such trials really taught potential applicants anything about running an ILR service in an area, but they were seen as very valuable statements of intent, enhancing applications and providing a good opportunity to drum up expressions of the 'local support' required.[1002]

Digital (DAB) RSLs were used to trial the new platform in 1996. They were promotional to a degree, but also valuable in running genuine technical tests of the new transmission medium. There were two such experimental ventures in London that year, while NTL linked with Virgin to run a DAB service at the Radio Festival in Birmingham in July, and then on its own for the Radio Academy Technical Conference in November. The last of these was an early sales pitch by the transmission provider, but it was of note since the Authority agreed to license RSLs to provide exactly the same output on FM, AM and DAB, so that NTL could demonstrate the difference in reception to delegates.

What caught the imagination more than the trials, however, were those services which had grown from the expanded interpretation of 'special events'. Take 1996 as a typical year.[1003] Radio stations for local festivals, including arts or music events, carnivals or air fairs, were the largest single group; 61 in 1996, compared with 52 the year before. Licences were issued to cover the Edinburgh International Science Festival, a dance music event in Sheffield, the Manchester Irish Festival, Knutsford May Day celebrations, an Irish language festival in West Belfast, RAF Cosford's open day, the Merseyside Caribbean Carnival, and for county shows across the UK. Sporting events came next in popularity, with 35 licences in 1996 – including nine 'matchday' football services, nine motor racing circuits, Wimbledon, the National Hunt Festival, motor racing in Ulster, and yachting at Cowes Week. Then there were student services, for freshers' weeks, student elections, school projects and media training. Nineteen RSLs that year were run for religious purposes, during Ramadan, Diwali, Easter and Christmas.

The attractions of running services for the month of Ramadan in locations with growing Muslim populations were considerable. There was great community prestige to be garnered, and good commercial prospects too. Given the limited

1001 *Short Term Restricted Licences; Annual Report 1996*, Radio Authority.

1002 The system was even used by the Radio Authority to show how regulation could be rationalised without extensive legislation. In an early example of operational convergence, the frequency planning function for RSLs was moved from the DTI's Radiocommunciations Agency to the Radio Authority in September 1997, with the ready agreement of both organisations. As the contemporary jargon had it, Holbrook House was now a 'one stop shop' for RSL operators. However, it failed to head off the absorption of radio regulation into Ofcom.

1003 *STRL Annual Report 1996*, op. cit.

availability of frequencies for RSLs, there was soon great competing demand for quite scarce licences. RSLs were usually offered on a first come, first served, basis, but for Ramadan licences this began to be impracticable. The opening date for the receipt of applications, one year ahead, risked a stampede. On 29 November 1999, the Authority received 16 applications for Ramadan which commenced on 29 November 2000; six for Bradford alone.[1004] The Authority moved to a ballot for Ramadan RSLs, but even that attracted complaints of favouritism. 23 licences were issued in 2001, to cover Ramadan and Eid between 16 November and 16 December 2001, the highest number so far. To ensure even-handedness between Muslim and Sikh groups that year, when Muslim Ramadan and Sikh celebration of the birthday of Guru Nanak Dev Ji were both celebrated during November, it was necessary to set aside the rule of one licence per locality.[1005] RSLs had become, and remain, a major attraction for religious groups in the UK. Christian organisations had been prominent at the start of RSLs with Radio Cracker, but then faded from the scene for quite a while. However, they began to apply for RSLs too, in growing numbers from 2000, and in 2001 there were 21 such licences issued for Easter, Pentecost and Advent, and for coverage of Bible Week gatherings and conferences.

A further category of restricted service licence had been introduced from the early nineties, providing for the long-term operation of free-to-air broadcasting services at defined locations such as universities, hospitals, prisons and military bases. The Radio Authority in its later years greatly developed the use and scope of long-term RSL services (LRSLs). A number of pre-existing student and hospital services took up the free-to-air licences, which the Authority increasingly made available. This allowed them to switch off old and expensive 'induction-loop' systems, and to serve more effectively their clearly-defined, site-specific coverage range of typically less than a maximum of 1 kilometre from the transmitter. These LRSLs were prohibited from serving listeners outside the bounds of the site concerned even if longer distance reception was technically possible. The scope of LRSL licensing was expanded by Ofcom in 2007, to include availability to commercial sites such as shopping centres and holiday camps, although Ofcom tried to maintain a "clear distinction between such licences and the community radio licensing scheme".[1006]

The British Forces Broadcasting Network (BFBS) was a prominent entrant into the LRSL arena in 2000. It had seven such licences, three for Northern Ireland army barracks – Thiepval in Lisburn, Lisanelly in Omagh, and Shackleton in Ballykelly – and for RAF Aldergrove, near Belfast. In England, there were BFBS stations for Catterick Garrison in North Yorkshire, Bulford and Tidworth garrisons in Wiltshire, and for the Sir John Moore Barracks in Folkestone. This last was home to a regiment of the British Brigade of Gurkhas. The Authority reported that programming output from BFBS here "is almost entirely in the Gurkha's native language of Nepalese, and includes programming produced and presented by the Gurkhas' wives".[1007]

There were initiatives, too, from unexpected external sources. Since 1993, there had been a series of RSL licences issued for sporting events; football, motor

1004 *Short Term Restricted Licences; Annual Report 1999*, Radio Authority.
1005 *Short Term Restricted Licences; Annual Report 2001*, Radio Authority.
1006 *The Future of Radio*, Ofcom 17 April 2007 para 6.14.
1007 *Short Term Restricted Licences; Annual Report 2000*, Radio Authority.

racing, rugby and horse racing. These included news and information about the event, and sometimes commentary where rights allowed. On 24 January 1997, Soo Williams and I met Peter Downie, who for some months had been bombarding us with demands that his new RefLink service should be licensed by the Authority. This involved transmitting on very low power, to spectators at a major rugby match, the wireless system linking the referee with his touch judges. After some initial broadcasts on AM, which were unsatisfactory, the Authority agreed in May 2001 to issue very low powered FM licences for broadcasts which were designed to be received only within the sports stadium.[1008] Once Ofcom was able to make decisions more readily across the whole of the spectrum, it became possible to offer licences called 'audio distribution systems' (ADS-RSL), deploying frequencies outside the broadcast bands. The referee's comments are now readily audible within radio commentaries, and sometimes as a separate audio channel alongside digital television broadcasts of sports matches, particularly rugby.

Ofcom wisely took the RSL regime and its style from the Radio Authority largely unchanged, and to protect it while attention was inevitably drawn to community radio. Ofcom's converged scope meant that it could deploy more forms of licensing and dissemination. In Northern Ireland, the practice had developed whereby church services, both Catholic and Protestant, were relayed on citizens band frequencies (CB) to parishioners who were too ill or infirm to attend them in person. This was extra-legal rather than illegal, and Ofcom found a way to legitimise it when it was deregulating CB radio. It ran an extended trial of what it called a Community Audio Distribution System (CADS), thus carefully distinguishing it from broadcasting and the panoply of statutory rules. The trials, between 2004 and 2006, were held in Northern Ireland and also in West Yorkshire. Ofcom was concerned that the opportunity should not be limited to the religious circumstances of Northern Ireland, and had received representations that CB or mobile phones might also be used for the Muslim muezzin's call to prayer. A long-term CADS scheme followed, to "meet a particular need that cannot readily be met through any other existing licence".[1009]

Access Radio

All this had followed from that afternoon in 1990, when David Vick and Bob Macey sat down to work out what to do in the new Radio Authority with those awkward 'special events licences'. The crowning consequences were to be the Access Radio experiment, and then full third-tier community radio. Relations between the Radio Authority and the community radio lobby, represented by the Community Radio Association (CRA), has been notably bad throughout most of the first two decades of ILR.[1010] One effect of the success of RSLs, however, was that it gradually became more usual for CRA members to be operators of RSLs than partisans of the old school, and contacts between the two sides improved a little as a consequence, at an operational level at least.

My first public speech after joining the Radio Authority was to the CRA's annual conference in Sheffield on 15 September 1995, and I had been briefed to

1008 *RA Paper* 60(01), 29 May 2001.
1009 *Community Audio Distribution Systems* policy statement, Ofcom, 12 September 2006.
1010 See chapter 12.

expect an awkward, even hostile reception. Actually, the response was courteous, cool, doctrinaire and sceptical. What I said very much reflected the thoughts of Authority staff and members at the time. I reviewed the Authority's work in licensing smaller-scale stations, praised the exponential growth in Restricted Service Licences, took a side swipe at the BBC's failure to provide small-scale services despite the public licence fee, and was gently but not completely dismissive of the lobbying by the CRA for a third tier of not-for-profit licences. The lobby had been knocked back too often, now, and I was sure that the Major government would show no interest in it. The issue seemed firmly on the back burner.

The CRA was soon renamed the Community Media Association (CMA), as its members started to look for outlets in television, since they lacked the political support to bring community radio back to the foreground of policy discussion. From 1997, with the new Blair government, there were some MPs keen to raise the point – but few who persisted beyond the initial explanation of the difficulties of definition, funding, and the risk of the acquisition of commercial licences 'through the back door' which was the Authority's standard riposte. As late as October 1999, the Authority was rejecting a proposal from the CMA to offer experimental community radio services on FM.

However, it was becoming increasingly apparent to the new executives and new members of the Authority that consolidation of ownership of ILR licences, and associated networking, was diluting the localness of the commercial sector. Soundings from DCMS officials, and from ministers, during 1999 suggested that they would welcome a move from the Radio Authority to re-open the community radio issue. As the government began to develop its own broadcasting policy, the likelihood of a fundamental re-organising of broadcasting regulation offered the opportunity to re-think the Authority's policy. Despite continuing scepticism of at least some of the senior staff of the Authority, when the Authority members met at their annual conference in November 1999 they were presented for consideration with a draft submission to the DCMS and DTI, which envisaged "taking the radical step of establishing a new third tier of non-commercial private radio".[1011]

The momentum towards finally realising community radio began at an Authority conference in the plush setting of Lainston House Hotel near Winchester, a venue which would have confirmed all the worst suspicions of the radical left of the seventies. The paper set out what I hoped the Authority would sign up to and recommend to the government. "The Authority's vision is of a new approach to harnessing the individuality and potential of non-commercial radio, and of using radio to assist in the broader aspects of education, social inclusion and social experimentation. Access Radio is not designed to be a publicly funded competitor to small-scale commercial radio, still less to be a way of unsuccessful applicants for small-scale licences finding a 'back door' onto the air ... we do not entertain any doctrinaire approach to self-styled 'community radio'. Access Radio is to be a third tier of services, fundamentally different in nature and scope from existing BBC and ILR stations ... The purpose is to enable public access to radio in a new and imaginative way."[1012] The most immediate proposal was that "once the direction of government policy becomes more clearly known, the Authority

1011 *Regulating Communications: approaching convergence in the Information Age*, July 1998, Cm 4022.
1012 *Radio Regulation in the 21st Century.* Radio Authority submission to DTI/DCMS, June 2000.

would propose to initiate a range of pilot experiments to cover as many aspects as possible of the proposed Access Radio sector".

The use of the phrase 'Access Radio' was conscious attempt to try from the outset to distinguish the proposed third tier from the baggage of 'community' radio's past. The proposals made explicit the notion that it was legitimate to deploy local radio for social purposes not just to maximise audiences, which seemed bizarre to much of the commercial radio sector. The Authority's proposals also addressed the issue of the impact on smaller stations, although noting that "small-scale commercial operators should be given the opportunity ... to convert to the Access Radio sector, thus helping to ensure the survival of the Radio Authority's aspirations in licensing services with a clear 'community' ethos which nevertheless find full commercial viability elusive". This was a pretty heroic assumption, in which the Authority was quietly acknowledging the precarious nature of some of the services it had licensed, where its own community aspirations had been dominant over commercial considerations.

The Authority members took up this new approach with enthusiasm, recommending to the government in June 2000 an experiment with Access Radio, with a view to establishing the practicality and desirability of a full-scale third tier. The proposals were detailed and specific, and they accurately prefigure the shape which community radio was to take in the parliamentary orders which followed the unspecific provision for community radio in the eventual Communications Act.[1013] Indeed, of all the Radio Authority's submissions on the way in which the regulation of the independent radio sector might be structured in the new decade, only the one for Access Radio came about as proposed.

Richard Hooper spoke to the Annual Festival of the CMA in Birmingham in October 2000, and floated the idea for an Access Radio Fund with the delegates. This was not to be a source of continuous finance, but was intended to provide "seed corn, start up, experimentation money."[1014] The CMA's reaction was rather graceless. Hooper's private notes observe that he "got a bit of a battering at the end of questions",[1015] but the show was now firmly on the road. The white paper, published in December 2000, acknowledged the success of RSLs and specifically sought "views on whether the benefits of community radio would justify greater public intervention".[1016]

To take the process forward, stimulate debate and boost the prospects of its proposals being adopted by the government, and with the co-operation of DCMS officials, the Authority organised an Access Radio seminar in February 2001. One hundred and sixty people attended, and although the commercial sector was relatively under-represented, there was consensus that "further expansion of not-for-profit radio was desirable, probably with a new structure ... that this sector should include non-profit distributing services, providing wide public access ... [and] a general wish, not least in government, that any new sector should link in closely with other initiatives for social inclusion and regeneration".[1017] There remained many issues to resolve, notably funding, and who should administer any

1013 Communications Act 2003, section 262
1014 Richard Hooper at Community Media Association Annual festival, Birmingham 7 October 2000
1015 Richard Hooper in an internal memorandum, "CMA thoughts 8 October 2000"
1016 *A New Future for Communications*, Cm 5010, December 2000, p 40
1017 RA Paper 18(01) 1 March 2001.

Radio Fund. There was also going to be a need to protect the smallest ILRs if some degree of advertising and sponsorship was to be allowed to Access Radio stations, and frequency availability for free-to-air services was going to be a headache. Nevertheless, the seminar confirmed that there was a fierce hunger among would-be community radio providers which greatly outweighed in volume the nervousness of the commercial sector. Ralph Bernard argued, in a speech otherwise notable for its vitriol towards the Radio Authority, that such a new tier of radio "would serve communities and special interest groups whose needs are unmet by the BBC and commercial stations, and would provide opportunities for those who cannot find a niche in mainstream broadcasting ... some communities of interest will really benefit from having their own radio service".[1018]

There was little doubt what the government's reaction was going to be, and it vindicated the Authority's careful cultivation of officials and ministers. Broadcasting minister Janet Anderson, visiting an RSL in Manchester on 2 March, indicated that the government "broadly support the principle of Access Radio and we welcome the Radio Authority's proposals" with the DCMS press release carrying the singular headline "Broadcasting Minister gives warm signals for Community Radio".[1019] On the same day, the Authority issued confirmed proposals, which recommended that Access Radio should be supported by mixed funding, allowing a measure of commercial money. Two weeks later Janet Anderson wrote to Richard Hooper confirming agreement to the pilot experiments which "should proceed as quickly as possible".[1020]

The commercial radio companies found themselves caught in a dilemma. They could scarcely oppose Access Radio outright, given the strong backing it was receiving from the government, although many disliked the idea intensely. The groups of smaller stations, and those individual small-scale stations which remained for the time being unconsolidated, saw a threat to their existence – which was already mostly rather precarious – if advertising, sponsorship and audiences were siphoned off to Access Radio stations, which enjoyed some measure of public subsidy. The Radio Authority had suggested two safeguards: first, that Access Radio stations would only be licensed having due regard for the local broadcasting scene, and by implication not up against struggling ILRs; and second, that small ILRs which wished to do so might convert to become Access Radio stations. The latter point had been taken up with some enthusiasm on behalf of small-scale Scottish stations at the Access Radio seminar, although in the event no station chose to convert.

CRCA determined to damn the idea with faint praise and look for safeguards, while awaiting an opportunity to ambush the plans at a later stage. Paul Brown wrote to secretary of state Chris Smith explaining that while "we understand the benefits of small-scale radio for training and social inclusion purposes" plenty of this was done already by existing ILR stations.[1021] He asserted that allowing Access Radio "to capture local commercial radio stations' revenue and effectively erode small-scale stations' income ... will seriously erode the service already available, and loved, by its listeners". To a degree CRCA's position was caused by the

1018 Ralph Bernard at Access Radio Seminar, 12 February 2001.
1019 DCMS Press Release, 2 March 2001.
1020 Letter from Janet Anderson to Richard Hooper, 17 March 2001.
1021 Letter from Paul Brown to Chris Smith, 21 March 2001.

growing awareness that ILR was now *commercial* radio, and was abandoning just those elements in its local involvement which were to be the chief staple of Access Radio.

The nine principles for the pilots were designed to ensure that the experiment replicated as far as possible the approach, patterns and structures which we then anticipated would govern any permanent third tier. I set them out in a speech to the Celtic Radio and Television Festival in Truro on 31 March 2001 (to an audience numbering barely a dozen people): "First, the pilots should be operated as not-for-profit services, in defined neighbourhoods, with clear public service content remits. Second, these experiments should, as far as possible, contain examples of the types of socially-regenerative and educational links which offer so much potential, and of training and development of local community capacity. Third, the pilots should cover as wide a range as is practical of the different types of locality – urban and rural, socially successful and socially disadvantaged, and reflecting the diversity of the Home Countries – without becoming so numerous as to get out of hand. Fourth, at least some of the services should be aimed at communities of interest within localities, not just all-embracing neighbourhood services, with the intention of establishing their role in serving minority groups and sustaining minority linguistic cultures. Fifth, the pilots should, in so far as the current legislation allows, experiment with a range of funding models, both to examine their implications for Access Radio and to inform Ofcom as to how it should take account of the local broadcasting ecology in deciding how to protect existing small-scale services from unsustainable levels of competition. Sixth, the regulations and administrative regime for the pilots should be modelled upon what we anticipate will be the eventual Ofcom arrangements, so that these can be tested and 'de-bugged' before they become a national standard. Seventh, the licences for the pilot will have to be for a fixed term … . Eighth, the licensing of Access Radio pilots must not interfere with the continued award of RSLs … . And ninth, the pilots themselves and the administrative and licensing regime must be monitored and researched formally, in order to provide a sound basis for their evaluation and for the proposals for the permanent arrangements which will follow."[1022]

The concept of 'social gain' had appeared in the original Radio Authority submission to the government. Once tested in the Access pilots, it was enshrined largely unchanged in the parliamentary order which enabled the start of community radio. "The achievement, in respect of individuals or groups of individuals in the community that the service is intended to serve, or in respect of other members of the public, of the following objectives: (a) the provision of sound broadcasting services to individuals who are otherwise underserved by such services, (b) the facilitation of discussion and the expression of opinion, (c) the provision (whether by means of programmes included in the service or otherwise) of education or training to individuals not employed by the person providing the service, and (d) the better understanding of the particular community and the strengthening of links within it."[1023]

Letters of intent to run the pilot stations were invited on 24 May 2001. By the closing date of 29 June nearly 200 had been received, and needed to be sifted at

1022 Tony Stoller at the Celtic Radio and Television Festival, Truro, 31 March 2001.
1023 Community Radio Order 2004. SI 1944.

top speed if the services were to get on air in time to inform the debates around the broadcasting bill. The Authority established a sub-committee of three members – Mark Adair as chair, Sheila Hewitt and Thomas Prag – and met for the first time on 20 June 2001. Thomas Prag was later to take over the chair, and his experience with small-scale radio in Scotland, and his involvement with the Scottish community radio scene generally, made him especially well fitted to be a champion for the new approach. With Soo Williams and me in support, the sub-committee worked hard to meet the Radio Authority's stated intention of testing as many as possible of the variables for a future nationwide community sector. On 8 August, 15 groups were invited to submit full applications. The stations were urban and rural, across the four home nations, religious and ethnic, mainstream and marginal, on FM and on AM, and on a full-time and a three-month basis. The eventual 16 services, drawn from 193 valid applications, encapsulated the nature and range of community radio aspirations as they had evolved over the previous 15 years.

The Authority had set out to manage the pilots and their evaluation with maximum transparency. With support from the Gulbenkian Foundation, it appointed an independent evaluator, Professor Anthony Everitt. His report, *New Voices, an evaluation of 15 Access Radio projects* is the central source for detailed information about the experiment. He summarised the different approaches of the stations as succinctly as possible for a group which was deliberately highly diverse.

"*New Style Radio* in Birmingham regards broadcasting as a valuable social tool for the development of African-Caribbean people. *Bradford Community Broadcasting* aims to serve all those living in a complex multi-cultural city. *Radio Regen* in Manchester created *ALL FM* and *Wythenshawe FM*, both of which target disadvantaged communities in the city. *Sound Radio* in Hackney sees itself as a 'local world service'. *Forest of Dean Radio* promotes community development in a rural area. *Takeover Radio* in Leicester enables children to run their own radio station, with minimum adult supervision. *Cross Rhythms* began by focusing on the Christian community of Stoke-on-Trent with a diet of community information and contemporary Christian music, but the Access Radio experience has led it to widen its approach; it now defines itself as a station serving the whole community with a Christian motivation. This is similar to the policy of *Shine FM* in Banbridge, County Down, another Christian radio project, which speaks to the community at large and promotes social reconciliation. *Angel Radio* in Havant broadcasts to people over sixty; *Awaz FM* in Glasgow sees itself as a much needed channel of communication between Glasgow's Asian community and the public and voluntary sectors. *Desi Radio* wishes to reconcile the different religious and social strands of Panjabi culture in Southall. *Northern Visions* places the arts and creative expression at the service of all communities in Belfast. *Resonance FM* on London's South Bank defines its community as artists, and broadcasts contemporary music and radio art. Two projects are alliances between different interest groups; first, the Asian Women's Project and the Karimia Institute which came together to run *Radio Faza* in Nottingham and, secondly, *GTFM*, a partnership between the residents' association of a housing estate in Pontypridd and the University of Glamorgan."[1024]

1024 Professor Anthony Everitt, *New Voices – an evaluation of 15 Access Radio Projects*, Radio Authority, January 2003, p 5.

Three stations – in Bradford, Stoke and Havant – went on air at the end of February 2002. Richard Hooper's hyperbole seemed justified: "This is a ground-breaking moment. These three stations, and the others that will soon follow them onto the airwaves, represent a pioneering experiment in not-for-profit radio. If successful, it could lead to a new tier of stations across the UK, finding new ways of harnessing the medium to serve local communities."[1025] Fifteen of the 16 stations had launched by mid-August 2002, with initial one-year licences. The last, Shine FM, designed to be the shortest duration of the experiments, at three months, followed in September. They all found success, in varying degrees, and impressed visitors and observers by their combination of purpose and enthusiasm, recalling perhaps the smaller of the first ILR stations. Remarkably, all but two of the Access Radio 'pilot' stations were still broadcasting at the end of 2008. One other chose instead to pursue community television options, while Sound Radio in Hackney fell victim to the re-development of its premises. The others continued for the experimental period, then for a 'grace and favour' extension while the political wheels ground out policy, and were subsequently licensed within Ofcom's community radio regime.

At least one station was internationally ground-breaking. When I visited Desi Radio in Southall, I was told it was the first 24-hour Panjabi radio station in the world. I met young Punjabi women who – arriving in England as very young brides – had never before been involved in any enterprise outside their homes. Their culture allowed them to work in the radio station, however, since it was run by the community, notably the formidable Amarjit Khera. As we arrived for a tour, there were these newest and most improbable recruits to the radio sector burning music tracks onto CDs to create a unique music library. Even the BBC inclined towards support. BBC media correspondent Nick Higham saw three of them as "just the kind of community stations which a small band of idealists and lobbyists have been lobbying for for years".[1026] When I met the BBC's director of radio Jenny Abramsky and director of nations and regions Pat Loughrey they offered more than cautious support, even contemplating giving spare studio equipment to the new stations, and they were both – typically – as good as their word.

Anthony Everitt produced an interim evaluation in October 2002, to contribute to government policy-making, which was advancing quickly. He noted the level of interest and demand, and argued that the existing licences should be extended as soon as the second reading of the communications bill made that constitutionally possible. The government duly extended the scheme on 11 December. Everitt's final report, *New Voices*, was published on 19 March 2003, and recommended that Access Radio should become a permanent third tier of not-for-profit radio for communities in the UK. He wanted such stations to have available to them professional expertise in administration, fund-raising and community liaison, and to that end he endorsed the Authority's proposal for a separate Radio Fund. Everitt proposed that stations should normally be allowed to raise up to half their funding from advertising and/or sponsorship. He advocated safeguards to ensure that the new Access stations should not compete directly with ILR stations, either in terms of programming or commercially, and he also advised that close

1025 Radio Authority News Release 1 March 2002.
1026 *Radio for the People* Nick Higham in *Ariel* 10 September 2001.

controls would be needed to oversee funding and fund-raising. The detail was valuable for the future, but at that moment it was the crucial overall judgement which mattered most. Everitt reported mostly favourably on the way in which the experiment had been managed, and then rang the bells with his conclusion. "This is not radio simply for the people, but by the people … I have little doubt that, if it is introduced, Access Radio will be one of the most important cultural developments in this country for many years."[1027]

There were to be two ambushes during the progress of the communications bill, and one failure of good faith. The first ambush, from CRCA, sought to limit the amount of revenue which an Access Radio station could derive from commercial sources. The Communications Act was passed in July 2003, but required a parliamentary order to give substance to its brief coverage of third-tier radio. A draft order was produced in February 2004, and the CMA decided its job was done. Not so the commercial companies. Led by the chair of CRCA, John Eatwell, they lobbied hard for concessions. The final order gave them plenty. All the new stations were to be limited to taking a maximum 50 per cent of their revenue from advertising or sponsorship. No community station would be permitted where it might overlap the coverage area of an existing small scale ILR station (defined as having a population coverage below 50,000). For the almost 60 ILR stations covering between 50,000 and 150,000 people, any community station would be permitted no commercial revenue at all. At the end of 2008, there were 17 community stations so restricted, including in Canterbury, Cambridge and a station serving elderly people on the Isle of Wight. Before licensing any community station, Ofcom must have regard to the likely impact on existing ILR services, and can (and does) limit the commercial revenues permitted to the not-for-profit service. A station for the army's Catterick Garrison is held to 25 per cent or below, and one for Knowslely College to no more than 15 per cent.

A second minor ambush came from the CMA, led by the chair of the All-Party Community Radio Group, Ian Stewart MP. He had always wanted to go back to the name 'community radio', perhaps showing his own roots in the notions of the past, and succeeded at a third reading in the Commons in persuading secretary of state Tessa Jowell on 14 July to change the name of the sector from 'access' back to 'community' radio, despite advice from her officials to the contrary.[1028] It seemed to be the last flourish of the old left, although the risk remains of long-discarded baggage coming back to clutter the new system.

However, in a final display of weakness amounting to bad faith, the government – despite its encouragement for the new third tier – managed in the end to find a mere £500,000 for the Radio Fund. This was a signally inadequate underpinning for what Anthony Everitt thought was potentially "the most important cultural development for many years". Moreover, a government amendment in the House of Lords on 8 July 2003[1029] extended the purposes of the Fund to local television as well. The value of the Fund had not by 2008 been increased, so its true value has been steadily eroded, despite the notable success which has attended the start of Community Radio. With the restriction on commercial funding, this

1027 Everitt, op. cit. p 3.
1028 *Hansard*, House of Commons, 14 July 2003, col 87.
1029 *Hansard*, Lords, col 243, 8 July 2003.

leaves most community stations unreasonably reliant on individual philanthropy, local fund raising and grants from social agencies. The onrushing recession will put ever more pressure on those sources, at the very time when the community 'glue' of such radio stations will be most needed. The failure to lend sufficient practical support to the new sector, while taking the credit for its introduction, and yet offering concessions to vested interests, was an unattractive aspect of government in those middle years. It was therefore reassuring that the interim report on *Digital Britain* at the start of 2009 proposed to re-visit the limits on commercial funding for community stations.[1030]

This was radio as social action once again, mirroring in a new time the aspirations of *independent* radio in the seventies and eighties. It was to be implemented and overseen by the new regulator, Ofcom, well supported by Radio Authority staff transferring across. Ofcom's current chief executive, Ed Richards, who had been Tony Blair's adviser through the passage of the communications bill, is in no doubt of the significance of the new sector: "a huge success in my view, a fabulous success. We have already offered licences to almost 200 stations, and could license twice as much of it. Stations will come and go, but it's a community, voluntary, creative phenomenon, and a soaraway success."[1031]

The story of Access Radio demonstrates the ability of regulators to innovate, against vested interests. Ofcom acknowledges that the existence of a widespread third tier of radio lets ILR off the hook to some degree, in terms of their own responsibilities for this type of content and involvement with the voluntary sector, but Richards is robust on this, arguing "so what if it does, given the very different place commercial radio finds itself in today". Thus, while ILR may have become *commercial* radio, there is a new opportunity to deploy the remarkable scope and capacity of *independent* radio to achieve local social good.

1030 *Digital Britain; the Interim Report*, Cm 7548, January 2009.
1031 Ed Richards in an interview with the author, 8 April 2008.

Chapter 24

Breaking the mould

2000 – 2003

The white paper, *A New Future for Communications*, was published in December 2000.[1032] It was issued jointly by the DTI and DCMS, on behalf of the two secretaries of state, Stephen Byers and Chris Smith. The previous month, in the USA, George W Bush had needed the help of a politicised US Supreme Court to filch the presidential election from a lack-lustre Al Gore, but in Britain New Labour's stock was still riding high. This was a new UK in many ways: one where the security for the wedding of Michael Douglas and Catherine Zeta Jones was run on behalf of *Hello!* magazine; yet a country where ten-year-old Damilola Taylor could bleed to death after being attacked in Peckham, without causing anyone to agonise over much about the state of the nation. Some things did not change; as the year ended, the first significant widespread snowfall in Britain for seven years brought the nation to a full stop.

For commercial radio, the white paper proposed two main changes: a formal, final shift from social to market priorities, and converged regulation. The white paper denied there was to be a third theme, the opening up of ownership of UK commercial broadcasting to international companies, but that happened too. Those three issues dominate the formal side of the history of commercial radio throughout the three years from December 2000 until December 2003, often to the exclusion of all else. Almost as a by-blow, the proposals flagged a new independent platform, the establishment at long last of community radio.[1033] Meanwhile, normal business continued throughout these years. Programming was broadcast, advertisements and sponsorship were (or, increasingly were not) sold, audiences listened and the whole caboodle of popular radio continued. But the mould of *independent* radio was finally being broken.

The first paradigm shift set in train by the white paper was to give primacy to the creation of a "dynamic market" in broadcasting and telecommunications, which was "fundamental to securing choice, quality of service and value for consumers".[1034] It was also to provide "a key input to the UK's international competitiveness". Broadcasting was to be regulated according to principles more

1032 *A New Future for Communications*, DTI and DCMS, December 2000, Cm 5010 [The 'white paper'].
1033 See chapter 23.
1034 *A New Future for Communications*, op. cit.

usually applied to telecommunications and the spectrum market. The government's thinking was almost wholly conditioned by the extent of technological convergence between the delivery of television and telephony, coming back together more than a century after Marconi and Edison. "Developing and sustaining a dynamic market is one of the government's key objectives for helping to sustain and develop this important industry [*note the singular noun*]. Competition is vital to dynamic markets."

This was unprecedented in British broadcasting, and marks the triumph of market philosophy in the media sector. Never before, since the route was laid out by the Sykes Committee and the Crawford Committee, had competition been the first and foremost issue for media regulation. For all the chapters of the white paper which follow on diversity, access and quality, the document was making clear that the official expectation of broadcasting had changed.

The second major proposal was for wholly new and converged regulatory arrangements for radio, television, telecommunications and spectrum management. The oversight of commercial radio, and the putative new third tier of not-for-profit radio, would be assumed by the Office of Communications (Ofcom). That would replace – crucially, not merge together – the five pre-existing regulators in this sector: the Radiocommunications Agency, once a department of the Department of Trade and Industry; the Office of Telecommunications (Oftel); the Broadcasting Standards Commission (itself an amalgam of the Broadcasting Standards Council and the Broadcasting Complaints Commission); the Independent Television Commission (ITC); and the Radio Authority. Government policy was strongly influenced by the potential commercial significance of the public resource of the electromagnetic spectrum for the growing possibilities offered by wireless communication. "We need a regulatory body with the vision to see across those converging industries, to understand the complex dynamics of competition in both content and the communications networks which carry services."[1035] Ofcom was charged with creating, first and foremost, a strong industry with international potential, by ensuring dynamic markets in the communications sector.

The Radio Authority continued to argue that radio would simply be swamped amid the huge numbers and the big battalions of television, telecommunications and spectrum. "There is a danger that small-scale and distinctive media – such as radio – will fail to receive adequate attention, and will suffer from the imposition of inappropriate regulation based on the needs of the larger media ... Independent radio's experience of converged regulation with television by the Independent Broadcasting Authority is discouraging in this respect, since it is generally accepted that at times radio became something of a 'Friday afternoon job'".[1036] This was unjust to the IBA, at least for most of the duration of its stewardship, but the phrase stuck. 'Friday afternoon job' was talked about in the parliamentary debates on the communications bill, and was often quoted back to me as I lobbied parliamentarians and legislators. However, a good slogan was not enough. Given that the radio companies broadly welcomed Ofcom as a chance to get rid of the interventions of the Radio Authority, the regulator seemed to be indulging in self-interested special

1035 Ibid., p 11.
1036 *Radio Regulation for the Twenty-first Century*, Radio Authority, June 2000.

pleading, just like the IBA a dozen years earlier. The white paper made it clear that this was not going to work, and that the Radio Authority would continue only until the regulation of commercial radio was taken over by Ofcom.

Despite being supposedly a catch-all measure, the new regulator of broadcasting and telecommunications would not oversee the BBC, to any significant degree. For radio, the involvement was even less than for television. Ofcom was to be required to conduct an early 'public service broadcasting' review, but as so often in those times 'broadcasting' was used as a synonym for television. Efforts to extend that to include BBC radio were wholly unsuccessful, given the fabled closeness between the DCMS and BBC radio.

The commercial radio groups welcomed both major shifts in government policy. They were excited that the way in which the white paper envisaged giving expression to the priority for dynamic markets was to overhaul and ease the ownership restrictions on ILR. The government wanted to "consider the possibility of devising a simpler, fairer regime for radio ownership to replace the current points system, or revoking the scheme completely".[1037] Here indeed was a sight of the promised land for the companies, and lobbying effort from CRCA was to focus above all on squeezing out the maximum relaxation on ownership rules.

For both radio and television, a third shift was firmly denied by the white paper. "We will retain the disqualifications on grounds of nationality ... Our current restrictions on foreign (non-EC and EEA) ownership of media interests are reflected across Europe, and indeed beyond ... We believe these restrictions play an important role in ensuring that European consumers continue to receive high quality European content."[1038] However, it was axiomatic for the Blair government in those years that open markets and open borders allowing inward investment were an unalloyed virtue. In the event, driven by Number Ten, the restrictions on foreign ownership were to be removed by the Communications Act. Taken together with much greater consolidation, the aim was to allow a very small number of UK broadcasting conglomerates to be major players across the world, not least to ensure the continued health of the domestic commercial television and radio industries.

Judging from where commercial radio and ITV found themselves by late 2009, such did not seem to be the consequence of the resulting consolidation. For the shareholders and executives of the radio companies in 2000, however, it offered the prospect of wealth as a consequence of being bought out for multi-millions by either their UK rivals, or the acquisitive giants of the US or Australia. In commercial radio, this prospect confirmed the groups in their wish above all to be seen to be profitable, in order to maximise their attractiveness to international predators, or to strengthen their own position as potential buyers. That in turn made them ever more reluctant to invest in programming – their 'product' – unless it could offer very short-term returns.

The white paper had urged discussion about the new ownership rules, and the Radio Authority and CRCA initially worked together accordingly. In June 2000, Neil Romain and Tim Schoonmaker had developed the notion of replacing the points systems for analogue ILR with a simpler principle agreed by both organisa-

1037 White paper, op. cit., p 41.
1038 Ibid. p 44.

tions. "We recommend that the new legislation should establish ... the broad aim that there shall not be fewer than three separate owners of commercial local radio services in any one locality." This was the basis for what became known as the '3 + 1' rule, which would allow consolidation only to the point in larger areas where there were at least three substantial owners. That underpinned an agreement between the Authority and CRCA in June 2001. CRCA announced at its annual congress that "proposals by the Radio Authority, supported by the Commercial Radio Companies Association, for a new ownership regime have been submitted to the Department for Culture, Media and Sport, and the Department for Trade and Industry". The jointly-worded press release confirmed that "the Radio Authority has developed its proposals for new ownership rules for radio, jointly with the Commercial Radio Companies Association, to be recommended to the government for the new communications bill. These offer further liberalisation nationally, whilst at the same time ensuring local plurality and diversity." Paul Brown for CRCA endorsed that. "Commercial radio [has] worked hard with the Radio Authority to arrive at local ownership proposals that would allow it to deliver new and more diverse services for listeners, while growing its business and ensuring a variety of separate local voices."[1039]

However, that seemingly done deal was not to be the end of the matter. Labour had been returned in the general election of 8 June 2001 with another huge majority, and Chris Smith had been replaced as secretary of state at DCMS by Tessa Jowell, who had tricky the task of enacting her predecessor's design. Among her political contemporaries was John Eatwell, economic adviser to Neil Kinnock from 1985 to 1992 and now a life peer, who had become chair of CRCA in 2000. Sensing that the government might be shifted further, and irrespective of the agreement trumpeted to Chris Smith the previous June, CRCA had flagged with DCMS by May that "further investigations were taking place".[1040] By June it was arguing that "it is over a year since we were strongly encouraged to seek consensus with our regulator in the matter of local radio ownership. Since then, there has been significant change in the broadcasting and telecommunications sector ... radio has, in contrast, not moved on."[1041] CRCA now argued that it would be enough to rely on competition rules, and a wider plurality scheme ensuring three local media owners in each market. With newspapers and television existing everywhere, that would let commercial radio off the hook entirely.

The argument raged on through 2002, into the pre-legislative scrutiny committee for the communications bill, and on to the fuller parliamentary debates and committees. CRCA commissioned a complex and theoretical report from Oliver and Ohlbaum to give legitimacy to its case, but this was really a naked matter of self-interest. The baldness of CRCA reneging on their previous agreement disturbed Tessa Jowell, and after a dinner on 30 October Eatwell felt obliged to write to her the following day. "You clearly remain concerned that our position has moved since we first jointly recommended 3+1 with the Radio Authority eighteen months ago. I explained that our view had had to change because the market within

1039 *Radio Authority News Release* 83/01, 28 June 2001.
1040 Letter from Paul Brown, CRCA, to Tessa Jowell, DCMS, received by DCMS on 31 May 2001.
1041 Letter from Alison Winter, CRCA, to Stuart Brand, DCMS, 6 June 2002.

which we will be operating will be so radically different from that which we had envisaged when devising the scheme."[1042] He wrote again on 7 November, inviting Jowell to re-open the argument in her upcoming speech at the Westminster Media Forum on 14 November. If she announced that the proposals remained unchanged, while welcoming the consultation he would "express regret and re-state CRCA's view that this will have a negative effect on listener choice, localness, and will do harm to the commercial radio industry".[1043] If on the other hand, Jowell offered the prospect of further easing of the rules, his public appreciation would know no bounds: "I will respond positively to any relaxation, congratulating the government. Naturally, the greater the deregulation the warmer my response will be ... I will lose no opportunity to commend the manner in which the government has moved to ... ensuring a stronger future for the local commercial radio industry." This exchange has about it a whiff of the novels of Trollope, John Major's favourite author. However, in the ascendancy of Tony Blair, this it appears is how business was actually done.

Tessa Jowell announced on 14 November that the government would look at the matter again. It was eventually resolved by the direct intervention of Ed Richards, who was managing the communications bill for Number Ten, and had taken a good deal of time to hear the arguments. 3 + 1 became 2 + 1, and the way was opened for huge consolidation within commercial radio. Richards said subsequently that "I wasn't aware of the deal that got you from 4 +1 to 3 + 1, so you could say that [CRCA] were a bit clever on that, as they certainly didn't behave as if there was a deal".[1044] Government policy was founded upon "a reliance on competition rules with a very clear insurance policy of 2 + 1. In most major cities that would mean you would be above 2 + 1, but you had a floor below which you couldn't go."

There were to be similar problems in resolving the limits on the ownership of digital multiplexes. Eatwell wrote to Jowell on 8 April 2003, as debate on the communications bill was nearing its end, that "we are becoming increasingly concerned about the proposed new rules for ownership of digital multiplexes and digital sound services", adding, with no apparent sense of irony, that "we aim to move forward in a helpful and consensual way".[1045] Uncertain of an agreement between the Radio Authority and CRCA actually holding, in the light of the analogue arguments, DCMS convened a Digital Radio Ownership Working Group. The Communications Act had by now received the royal assent, setting a vesting day for Ofcom of 23 December 2003, but radio ownership details were to be confirmed in a subsequent parliamentary order.

The Working Group met on 31 July. The Radio Authority might by this stage have been supposed to be something of a 'lame duck', with only a few months to go before being replaced by Ofcom. However, with the intellectual fire-power of David Vick, and the skill of Neil Romain in understanding how to deliver workable ownership systems, it was still a force to be reckoned with. Joint proposals between the Authority and CRCA emerged on 30 September, but were duly being qualified

1042 Letter from John Eatwell, CRCA, to Tessa Jowell, 31 October 2002.
1043 Letter from John Eatwell to Tessa Jowell, 7 November 2002.
1044 Ed Richards in an interview with the author, 8 April 2008.
1045 Letter from John Eatwell to Tessa Jowell, 8 April 2003.

in practice by CRCA as early as 2 October.[1046] CRCA, by this stage, had become pretty touchy. There was a further spat over 'grandfathering' existing ownership within the new rules, before the order went forward with a fudged outcome. CRCA accused the now-almost-extinct Authority of "post-rationalisation", and vaunted its own efforts "to be reasonable, keep things simple".[1047] Despite eventual agreement, CRCA typically hoped "that there may still be opportunities for change when Ofcom regularly inspects ownership rules over the course of its lifetime".[1048]

The companies had plenty of other things to worry about by this stage. Advertising and sponsorship revenue growth, buoyant through to the end of the century, hit a brick wall. In 2001, national advertising revenue fell by 11.4 per cent, and local revenue by 2 per cent, giving total revenues of £594.6m, down 7.7 per cent.[1049] Display revenues across the whole UK marketplace fell that year, so radio held its share steady at 6.3 per cent. However, given the continued addition of new stations – and the increasing costs associated with digital radio, which was offering no possibility of an early return – the picture was starting to look worrying. UK display revenue as a whole showed slight growth in 2002, but radio national revenue fell again, by 0.6 per cent, and the year was only rescued by stronger performances from local advertising and sponsorship sales, yielding an overall increase of 2.5 per cent. Given the falling away in commercial radio's share of radio listening from its true peak in the spring of 1998, these were signs of an industry in need of some urgent self-analysis.

The companies' solution was to cut their costs, not increase their investment in content. This led to more networking, fewer staff, shared management and reduced localness. Even the flow of youngsters into the ILR stations, which had been the very lifeblood of the system, was now drying up. Will Baynes, secretary of the Broadcast Journalism Training Council, was reported as saying that "one of my jobs is to place students on our recognised courses into work experience positions. But stations which took people last year, when I call them this year, say they no longer have anyone available to supervise a placement. With so many commercial stations coming under one, management jobs are being cut to increase revenues and, unfortunately, the first to go are in the newsroom".[1050]

Commercial radio's most bankable national personality was Chris Tarrant, who was steadily allowed by Capital more and more time away from presenting its flagship breakfast show. Neil Fox, who – by all measures except his attraction to City share-traders – was among the best breakfast talent around at that time, offers a harsh but clear judgement on it all. "The figures went down and the management ... [believed] ... that what they had to do at all costs was just to keep Tarrant." [1051] It was true that Capital's share price would fall whenever it was rumoured that Tarrant might be leaving, but the new arrangements broke breakfast continuity.

Chris Tarrant himself was no happier with where commercial radio was at, especially in its openness or otherwise to new talent. Collecting the Gold Award

1046 Email from Lisa Kerr, CRCA, to DCMS 2 October 2003.
1047 Email from Paul Brown to Helen Williams, DCMS, 22 October 2003.
1048 Letter from Paul Brown to Tessa Jowell, 10 December 2003.
1049 RAB/Ofcom, quarterly revenues.
1050 Quoted in the *Guardian*, 20 May 2000.
1051 Neil Fox in an interview with the author, 25 September 2008.

at the Sony Radio Awards in April 2001, he told the audience at London's Grosvenor House that the industry was at a "crossroads. When the next Kenny Everett walks into a radio station, for God's sake do not ask him to pre-record all his links for the next fortnight and put them into a computer ... Do not take spontaneity away".[1052] However, the die was cast, and as revenues continued to disappoint anything which saved costs was fair game, whatever it did to the radio product. Cost-reduction had been a feature of what happened when stations consolidated their ownership and then networked their programming, but now the radio groups saw it as a matter of survival. In a way it was, but not quite as they understood it; their short-term fixes caused permanent medium-term damage to their industry.

Things improved a bit in revenue terms in 2003, with national advertising notably more encouraging towards the end of that year. The four quarters from October 2003 were each to see double-digit percentage increases in national revenues. Yet this was a false dawn. From April 2005, national revenue was to fall for 10 consecutive quarters, only showing a 1 per cent increase for the last three months of 2007.[1053]

The Buggles had predicted in song in 1979 that video would kill radio, but they got it wrong. Radio survived television, even music video, by showing its inherent adaptability. What jeopardised its future now was not the competition, but the failure to change in time to respond to it. Radio companies were inclined to blame a combination of the internet, MP3 music downloads, and the restrictive regulatory regime, and to cut costs still further. The immediate impact was on group profits. Capital issued a profits warning on 22 March 2001, anticipating a 10 per cent fall due to declining advertising revenues. A statement from Scottish Radio Holdings (SRH) in late March 2001 warned that declining revenue would hit its half-year results in May. Things got worse. Capital issued a second profits warning in May, anticipating as much as a 25 per cent reversal. Emap followed suit in July. When GWR had announced in May a profit increase of 8.1 per cent for the previous year, it noted the downturn in advertising, but an optimistic Henry Meakin, the departing group chairman, reflected that "our forward order book is traditionally short-term, but we are confident that when advertising expenditure increases GWR will be in a strong position to benefit".[1054] Meakin's optimism had been misplaced. After he was succeeded as chairman by Ralph Bernard, GWR in November announced a 70 per cent decline in pre-tax profits, down to just £2.8m, and the company was looking urgently to divest overseas interests to service its high gearing.

There may not have been as much money to be made from selling advertising, but if you had a radio licence for sale you could still almost name your price. The radio groups could barely wait to begin their shopping spree, ahead of the new relaxed ownership regime, and many owners decided to cash in. Radio licences started changing hands for huge multiples of earnings, beyond anything prudence could dictate. SRH had bid for Border Television, eyeing its regional radio holdings, but its bid of £120m was trumped by Capital's £151m in April 2000. SRH

1052 Sony Radio Awards, 30 April 2001, reported in *Radio Magazine*, 1 May 2001.
1053 RAB/Ofcom, quarterly revenues.
1054 *Radio Magazine,* 23 May 2001.

Chief Executive, Richard Findlay, was as realistic as usual, but went unheard: "Capital's increased offer is a very high one and more than we are prepared to pay".[1055] Capital then promised to sell Border's television company to Granada. The television major could not acquire it under the ownership rules which applied at the time, so Capital agreed to hold on to the television licence – effectively to 'warehouse' it – until the rules changed.

GWR similarly acquired a raft of licences from the Daily Mail and General Trust (DMGT) in the middle of 2000, even though the UK stations within these also took it over its 'points' ceiling. DMG was taking a much closer interest in how GWR was getting on. It took another seat on the GWR board, to be occupied by Peter Williams, the director of DMGT's new media business. The deal had also involved GWR with DMG's Australian radio interests, and there was increased evidence of Australian radio practices feeding back into GWR's approach to its UK network.

The immediate regulatory concern was the breach of the statutory ownership limits. GWR therefore proposed to move 12 of its 17 Classic Gold licences to Classic Gold Digital Ltd (CGDL), a new company to be owned 80 per cent by UBC Media Group and 20 per cent by GWR. However, the agreement proposed between GWR and CGDL saw GWR selling the advertising, having involvement in sponsorship and promotions, and supplying most of the programming output from its studios in Dunstable, but under UBC's 'control'.[1056] There was also a 'put' option to permit GWR to 'dispose' of any other AM licences in a similar way. This new arrangement did not permit the Authority to determine that GWR still had 'control' of the twelve Classic Gold licences in the legal sense of the Broadcasting Acts, despite continuing to hold the other five in the network, and at its October meeting it duly sanctioned the disposal as passing 'control' to UBC. The result was to bring GWR within the ownership points limit, with 14.58 per cent of the total points universe[1057], and to bring the statutory regime into further disrepute.

The Radio Authority had also to consider the position of Vibe, the regional dance station in East Anglia. As part of its deal with DMGT, GWR had acquired 49.99 per cent of Eastern Counties, which owned East of England regional station Vibe FM, and the Authority determined that GWR did indeed have *de facto* control of Eastern Counties, for the purposes of points calculations. The Radio Authority was not the only regulator interested in the issue. The Office of Fair Trading also reviewed the purchases, looking at the extent to which GWR's acquisition of Vibe FM in East Anglia added to its existing ILR holdings there, and might require a Competition Commission investigation. In this instance, they decided to allow the transaction.[1058] GWR was not to do so well later. In September 2002, GWR and SRH jointly agreed to buy Vibe 101, a regional licence covering Bristol and South Wales, from Chrysalis. The OFT was not satisfied this time about the competition implications, and referred the deal to the Competition Commission (CC) for a full investigation. The adjudication, when it came the following May, rejected the deal. The CC found that "there are no benefits to this merger that would

1055 *Radio Magazine*, 20 April 2000.
1056 *RA Paper* 76(00) 1 September 2000.
1057 *RA Paper* 92(00), 29 September 2000.
1058 Acquisition by GWR Group plc of the radio interests of the *Daily Mail* and General Trust plc, adjudication no. 00154/C, 18 October 2000.

sufficiently offset the adverse effects identified".[1059] GWR consequently sold its 49 per cent stake in the acquisition to SRH.[1060]

In July 2000 Capital bought the central Scotland regional station, Beat 106, for £33.75m, an extraordinary price for a station which had only been on air for eight months. Up against the might of SRH, the station had managed just over a 5 per cent share of radio listening in its first RAJAR.[1061] Kelvin MacKenzie's Forever Broadcasting brought the small Bolton & Bury station, Tower FM, for £3.5m. Acquisition fever ruled, fuelled by improbable share prices. In December the Scottish Media Group, now owners of Virgin radio, increased its stake in SRH, which had been left exposed by its refusal to pay over the odds for radio licences. It had diversified with some success into radio, newspapers and outdoor media in Ireland, but it found itself at the centre of an increasingly acrimonious takeover battle. Once the radio rules relaxed with the arrival of the new communications legislation, SRH was to be snapped up by Emap in August 2005.

The trend started by GWR for networking was soon followed by the other groups. At its April 2001 meeting, the Radio Authority had two such requests to consider. Emap had already networked some of the output of all of its AM stations, operating under the Magic brand: in Manchester, Liverpool, Preston, Leeds, Sheffield, Hull, Teesside and Newcastle. In March it had asked to reduce the locally produced and presented hours on all these stations from 17 or 18 down to just to seven. Emap argued that it currently used automation extensively, and networking would serve the listener better. It argued that "our investment in Magic is immense" and that it was putting forward "reasonable proposals [compared with] other AM stations across the country run by other major groups".[1062] That was the bind in which the Authority now found itself. Having agreed to GWR's series of networking proposals, it could hardly turn down similar, better-resourced proposals. The Authority agreed to the request, "because the character of these music-led services will not substantially alter, and *because this is consistent with the Authority's established approach to part-networking on AM* [my italics]".[1063]

At the same meeting there was a request from Radio Investments to "move towards some standardisation of format"[1064] for its portfolio of rather individualistic small stations. Again, the Authority was effectively bound by its own precedents. It had only just agreed to eight hours per day on Mercury's FM stations, and that had set the rules – or sold the pass, depending on your point of view – for small stations. However, this time there were 23 stations involved, most of which had won their licences on the basis of providing distinct local services, albeit with scant resources. The Authority deferred the decision, but the outcome could never

1059 The completed acquisition by Scottish Radio Holdings plc and GWR Group plc of Galaxy Radio Wales and the West Limited, adjudication ME 1550/02, 2 May 2003.

1060 These two cases existed on the boundary between the radio regulator and the competition regulator. The creation of Ofcom was intended to provide clarity, by bringing together all media and communications oversight. Competition responsibilities were to be exercised first by Ofcom; and appeals from its decisions to go directly to the Competition Appeals Tribunal (CAT), not to the Competition Commission (CC). Yet where matters of pricing are involved, the CAT then has to refer the matter back to the CC; and issues relating to Ofcom's regulation of the communications sector are subject to a regulatory review by the CC. Institutional neatness can be elusive.

1061 *RAJAR*, Q1 2000.

1062 Letter from Phil Roberts, Emap Performance, to Martin Campbell, Radio Authority 16 March 2001.

1063 *RA Paper* 34(01), 26 March 2001.

1064 *RA Paper* 35(01), 29 March 2001.

really be in doubt. After some minor changes, the networking was agreed in June.[1065]

Kelvin MacKenzie continued to manage the system with more skill than most of his competitors. Desperate to get live football coverage on his re-branded TalkSPORT national station, and entirely priced out of the national radio market by BBC Five Live, he came up with the ruse of splitting his two AM frequencies, 1053 and 1089, which covered different parts of the UK. If he could do that, then he could bid against other owners for local commentary rights, and have live football to offer. There was plenty of logic in this, allowing listeners in localities to hear teams they would be more likely to be interested in, and enhancing competition. The original notes of guidance when INR3 was advertised back in 1994 had specifically envisaged a limited amount of what it called "regional programming opt-outs ... restricted to a maximum of two hours per day",[1066] which seemed to have almost been designed for football match commentary. Agreement was given to such evening broadcasts for London in 2000, although they were not then broadcast. MacKenzie returned in May 2001 to ask for a Scottish split, too, agreed in July on the grounds that "Scotland and Northern Ireland are ... special cases because of their devolved status and wholly separate football structure".[1067] This 'salami-slicing' served TalkSPORT well, giving it live football commentary, an important audience attraction. It hardly needs saying that other commercial groups were most unhappy at facing further competition for commentary rights, while the BBC – for whom MacKenzie and his station were *bêtes noires* – was incandescent.

There were still a few more issues in programming regulation during these last years of the old regime, even though most of the heritage companies had long since abandoned any of the ambitious programming which might push the boundaries. Just before midnight on 18 January 2002, Jon Holmes on Virgin Radio was playing 'swear word hangman' in which callers guess the letters of a sexually explicit phrase that uses swear words. The broadcast involved a nine-year old child taking part by phone in the live on-air competition. Even in the context of alternative comedy, for which Holmes has a justifiable reputation, this was hugely offensive.

I recall how the Radio Authority members, meeting on 14 March to adjudicate and decide on sanctions, were for a while unsure what decision to reach. The contribution from Mark Adair, the thoughtful but often reserved member for Northern Ireland, was brief but decisive. He pointed out that the key issue was the unacceptable exposure of the child. That was exactly what the members and staff thought it was, and the maximum fine for a local station of £75,000 was levied on Virgin. Richard Hooper made the Authority's position clear. "I and my eight colleagues on the Radio Authority are fully aware that attitudes amongst adults, about programmes by and for adults, towards what constitutes indecency and offence have changed markedly over recent years. However, where children are involved, the Authority will use the full range of its powers to preserve clear standards of what is unacceptable, in order to protect children."[1068]

1065 RA Minutes 6(01) of a meeting on 6 June 2001.
1066 Radio Authority, *Notes of Guidance for INR3*, November 1993.
1067 *Radio Authority News Release* 101/01, 30 July 2001.
1068 *Radio Authority News Release* 34/02, 19 March 2002.

The year before, in May 2001, Sunrise Radio owner, Avtar Lit, stood for parliament in the May 2001 general election. On 20 March, Sunrise Radio twice carried extracts from an interview with Lit. One of them concerned his views about the political dissatisfaction of people living in Ealing, Southall, and the other outlined his policy proposals for the constituency. Whatever else an owner could do with his ILR station, self-promotion of his political ambitions was, and remains, strictly outlawed. Sunrise was fined £10,000, and was asked "for assurances that suitable action will be undertaken to ensure future compliance with the Broadcasting Act and its Codes".[1069]

Castle Rock FM was fined £1,000 for signal over-deviation in September 2001. That apart, the seven yellow cards in that year and the six in 2002, covered all but the major issues. What the Radio Authority's June 2000 paper had dubbed "negative content regulation"[1070] had relatively little impact on commercial radio, excepting the isolated high-profile issues. As for "positive content regulation" – essentially policing the adherence to formats – the companies by then expected Ofcom's regime to be much lighter and more permissive, so there seemed little point in tweaking the tail of the old lion. Nevertheless, the animal still had enough spirit to bite Chiltern for failing to produce logging tapes for output on two successive days, leading to a fine of £5,000 in February 2003. Asian Sound broke the rules yet again, and in June was fined £3,000 for permitting undue prominence of commercial terms in programmes, and for carrying an advertisement of a political nature.

Whatever the political and corporate activity, radio remained fun for those working in it, even if much of the imagination went into events which were for promotional rather than broadcast purposes. Even those were not without their legal problems. *Party in the Park* was a public, open-air concert which put the promoting station very visibly into the centre of local life. When Capital sought to register that as its own trademark in 2002, it was challenged by Emap in a case which went as far as the Patent Office's *inter partes* adjudicator. Capital Radio won the legal argument, on the grounds that its own visual branding distinguished it from anything done by Emap stations, and was awarded costs of £1,350.[1071] This was how commercial radio now chose to conduct itself.

Capital Radio in London had picked up the *Party in the Park* idea in 1998 with great success, and it became an established and hugely popular feature of London life at the start of the new century, raising money for the Prince's Trust charity. These were major pop occasions. The 2000 event, for example, featured Bon Jovi, Lionel Richie, Backstreet Boys and Elton John. The 2002 concert attracted over 100,000 visitors to see some of the biggest names in world pop music at the time, including Shakira, Wyclef Jean, Westlife and Blue. This format, copied across the UK according to the resources available to local stations, was a new way of relating to younger listeners. When the group announced that the events were to be discontinued, it seemed to be a significant admission of growing weakness. For sure, Bob Geldof's renewed plans for a *Live Aid* concert in 2005 represented

1069 *Radio Authority News Release* 73/01, 11 June 2001.
1070 *Radio regulation for the 21st Century,* op. cit. p 20 ff.
1071 Patent Office, Trade Mark *inter partes* decision, 0/412/02, 14 October, 2002.

impossible competition, but the decision not to bring *Party in the Park* back in 2006 was another sign of the end of the Capital era.

BRMB, like most major commercial stations was endlessly seeking new promotional stunts, and eager to copy ideas from the US. However, it was caught cold by a competition it ran in Birmingham in August 2001. The contest, *The Coolest Seat in Town*, involved listeners seeing how long they could sit on a block of dry ice, with an average temperature of minus 79 degrees centigrade, in an attempt to win tickets and backstage passes for *Party in the Park* in Birmingham. The *Guardian* takes up the story of the final four contestants. "After an hour of gritting their teeth, they were numb and crying in agony. They were taken to hospital and three were transferred to a specialist burns unit at Selly Oak hospital in the city. One woman has 18 per cent burns on her legs, thighs and lower back. The youngest contestant, a 12-year-old boy, has burned buttocks."[1072] The company was eventually fined £15,000 for breaching section three of the Health and Safety at Work Act. The headline writers had a field day; the *Guardian* closed its files with a story headlined "Frosty response to dry ice stunt".[1073]

Then Richard Park in March 2001 left his job as group programme director of Capital. Park is the guru of commercial contemporary hits radio in the UK, and his absence was to be felt acutely by Capital, as it faced increasingly skilled competition for listeners in London. Fourteen years previously, Park's unique touch had seminally created the Capital sound anew, and that of ILR as well. It was his loss, just as much as the protracted departure of Chris Tarrant, which accelerated Capital's decline, leading to its eventual absorption by GWR into the new GCap group in the spring of 2005.

There was still plenty of pep in the Radio Authority's licensing activity as the new century began. This had been the characteristic feature of the regulator's time between 1991 and 2003, and it was determined to complete – so far as possible – the licensing of ILR services to utilise the available spectrum. It is a measure of that achievement that the number of ILR licences in issue at the end of 2000, when the future closure of the Radio Authority was announced, stood at 248; by the time it closed its doors in December 2003, the total was 272. David Witherow recalled that "we were determined to keep going right to the end. The last licence, quite a large one for Glasgow, was awarded at the November meeting, and there was still plenty to do at the December one."[1074]

Seven new ILR licences were awarded in 2001: Kendal & Windermere (Lakeland FM); Omagh & Eniskillen (Q101.2 FM); Pembrokeshire (Haven FM); Rugby (Rugby FM); South & West Yorkshire (Real Radio); Warminster (3TR FM) and West Midlands (Saga FM). The continued development of the Guardian Media Group's Real Radio group, and the arrival of Saga Radio, showed the continuing level of interest in the larger, regional licences. Re-licensing continued throughout 2001, with 15 existing ILR licences re-advertised, all being awarded to the incumbents.

During 2002, six new analogue ILR licences were awarded: Reading (New City FM); Chester (Chester FM); Mid Ulster (Mid FM); East Midlands (Saga

1072 *Guardian*, 25 August 2001.
1073 *Guardian*, 15 January 2003.
1074 David Witherow in an interview with the author, 1 October 2008.

FM); Skye & Lochalsh (Cuillins FM); and Worthing (Splash FM). Saga Radio's acquisition of a second regional licence in the East Midlands offered this new and diverse entrant the possibility of critical mass, although it denied it to Jazz FM which was accordingly disappointed.

A further 13 ILR licences were re-awarded in 2002 after re-advertisement, including Premier Christian Radio, which, after its long agony over the accidental statutory prohibition of any fast-track approach for a religiously-owned station, found itself unopposed.[1075] One of the re-advertised licences, however, did not go to the incumbent, Liberty Radio, but was awarded instead to Club Asia. The licence had originally been awarded in July 1995 to Viva Radio, and targeted entirely at women by women. It was re-launched as Liberty Radio in December 1996, and limped on until it was acquired in summer 2000 by Universal Difusao, a Lisbon-based offshoot of the Universal Church of the Kingdom of God. As Viva, it had attracted huge headlines, but neither audiences nor revenue. As Liberty, it was always hindered by poor AM frequencies. The Authority's reasoning for the non-award focussed on the appeal of the Club Asia application. "Members were impressed with Club Asia's proposals for a service appealing to an under-served young Asian community in Greater London and its extensive research, which made a compelling case for such a service."[1076] BBC media correspondent Nick Higham's assessment was more robust and direct. "It's not really a surprise that London's Liberty Radio is about to lose its licence. The AM radio station is the least successful in Britain. It accounts for just 0.1 per cent of all radio listening in London, the lowest share of any station ... Liberty was not a success when it was launched as Viva, a station for women. It was not a success when it was taken over and renamed by Harrods boss Mohamed Al Fayed, who lavished £7m on it in recent years."[1077]

As it entered its last year of operation, 2003, the Radio Authority had advertised, but not yet awarded, new ILR licences for a further seven areas. It had also identified publicly a further 12 areas for future new licences, on the assumption that Ofcom would wish to complete the programme of bringing ILR, wherever possible, to the whole of the UK. To the Authority, this seemed a necessary step in ensuring the continuation of 'business as usual' during the handover of responsibilities. However, the new regulator – feeling that there had to be a better way of doing it – decided to pause and re-examine the licensing process before embarking on the Radio Authority's legacy of a continuing working list of new areas. As a consequence, the impetus which new licensing always gave to ILR was lost at a critical time in the gathering decline of the sector.

In the first quarter of 2003 the Authority advertised its last four new analogue local licences. No new licences were advertised after March. The Radio Authority, diligent to the last, noted that it "could not be confident of being in a position to reach an award decision in advance of the date when licensing would become the responsibility of Ofcom".[1078] These four local licences were subsequently awarded, and along with a further seven licences which had been advertised in

1075 See chapter 19.
1076 *Radio Authority News Release*, 139/02, 7 November 2002.
1077 Nick Higham, *BBC News*, 19 November 2002.
1078 *Radio Authority Annual Report* 31 December 2003. The report was not published in printed form, but is available online from Ofcom.

2002, brought to 11 the total number of new analogue licences awarded during the year. The new local licences were for: Barnsley (Dearne FM); Buxton (High Peak Radio); Carmarthenshire (Radio Carmarthenshire); Gairloch & Loch Ewe (Two Lochs Radio); Glasgow (Saga Radio); Helensburgh (Castle Rock FM); Maidstone (CTR); North Norfolk (North Norfolk Radio); West Lothian (River FM); West Midlands (Kerrang!); and Yeovil (Ivel FM).

The Glasgow award, made by the Radio Authority at its meeting on 6 November, was the last occasion on which a group of people all driven the principles of *independent* radio exercised this central function. Their statutory brief bore no reference to competition, except in the sense of competition for listeners' attention, nor was it directed at creating a dynamic industry or market. After yet another long and close debate among the Authority members over the merits of 13 applications for the licence, the winning application was that from Saga Radio. Saga was about the last of those companies capable of winning larger-scale licences which offered more than just a version of CHR radio, and had a target audience which extended beyond younger consumers. The final licensing press release has an elegiac footnote. "Throughout its thirteen year history, the Authority has been committed to facilitating the provision of high quality services which offer a wide listening choice. With the award of this new licence, there is a total of 272 analogue ILR licences in issue across the UK, as well as three national analogue licences. The Authority also awarded a digital multiplex licence today, bringing the total of local digital multiplex licences across the UK to 46, in addition to the one and only commercial national multiplex."[1079]

While the radio industry waited to find out what Ofcom's regulation would be like, and wrestled with a slowing down in advertising, the Radio Authority quietly but determinedly moved to its close. I left as chief executive in June 2003 to join Ofcom; Richard Hooper departed as chair the following month, to become deputy chair of the new regulator, and was joined on that board by Sara Nathan. David Witherow – typically putting duty before personal comfort – agreed to become executive chair of the Radio Authority for its final months. David Vick, diligent and indispensible to the last, became chief operating officer. Of the directors, only Martin Campbell and Mark Thomas took their radio expertise to Ofcom, but many other staff transferred, leaving only nine redundancies, none of them forced. Witherow paid proper tribute to the winding up process, and the huge amount of work involved, in his final annual report. "Inevitably and rightly, a great deal of effort during the year was expended on Ofcom matters: in contributing to the final stages of the communications bill, in helping to get Ofcom up and running, in the processes by which staff would be selected to move over to the new body, in ensuring an orderly transfer of duties and responsibilities, and, ultimately, of closing down and 'switching off the lights'."[1080]

The Authority commissioned a 'memory book'. Leading media journalist Maggie Brown interviewed staff and members, to produce a compilation of what it felt like to be a part of the Radio Authority. It stands as a memorial of what was, for those who worked there, a very special experience. Amid the – mostly rose-tinted – comments, senior programming officer Mike Phillips catches the general

1079 *Radio Authority News Release* 115/03, 6 December 2003.
1080 *Radio Authority Annual Report* 2003, op. cit.

mood. "I've had a great thirteen years working at the Authority. It's good so many new stations have started and DAB has been introduced ... It may sound like a cliché, but there really has been a family atmosphere here ... For a radio enthusiast this has been a fabulous place to work."[1081] It is not a bad epitaph, both for the institution and for *independent* radio, too.

1081 *Radio Authority Memory Book*, private publication.

V. Postscripts

Chapter 25

Epilogue

The challenges facing *commercial* radio in the years after 2003 seemed even more daunting than those overcome by *independent* radio in the seventies, when there was the energy which came from creating something new. The top line figures tell only part of the story, but they are stark enough. Commercial radio's share of all radio listening in the UK had tumbled, settling down 17 per cent from its peak in the nineties. Revenue, adjusted for inflation, had never come close to matching the figure achieved in 2000, and was down almost a quarter by 2006 – and that before the impact of the credit crunch in 2008 and impending deep recession.

Expand those indicative figures and the picture is one of bad times for commercial radio, when the rest of the economy was still on the upswing which followed the dot-com bust. Revenues of £641m in 2004 fell to £582m in 2006, and rose only slightly to £598m in 2007.[1082] The next year they fell again, to just £560m, the lowest real figure for ten years. Programme sponsorship was playing a larger and larger part, representing typically between a fifth and a quarter of all broadcast revenues, as radio spot advertising lost its hold on the media buyers and their clients. Commercial radio's listening share sank into a new trough by the end of 2008, at just 42.2 per cent.[1083] Over 30 million people were still tuning in each week, but for much less time while BBC average hours stayed constant.

The failure of the underlying psychology was especially distressing. Commercial radio had suffered death by a thousand cuts. The long-term impact of networking programming had undermined the local relevance of the output. The lack of variety produced a mostly bland product which held its listeners by inertia; nothing must be broadcast which might cause people to tune away, and nothing surprising was going to be broadcast which might attract new custom. The supply of news from news hubs, rather than from a genuine local newsroom, accentuated this mind-set. The automation of output confirmed it. The decision by the larger groups to do away with local managing directors for individual stations denied their companies a local presence to sustain essential commercial interests. The huge amount of debt incurred during consolidation would have been challenging in normal times; in the circumstances of late 2008 it looked unsustainable. BBC radio still flourished, but commercial radio was widely perceived of as being in serious

1082 RAB/Ofcom.
1083 *RAJAR*, Q4, 2008.

trouble, and did little to contradict that. As new media platforms arrived, the commercial music radio model was left exposed, and the various pronouncements made by successful predators about the dire prospects they had inherited served only to confirm the general impression of a medium which had lost its relevance.

The radio companies had high hopes of the new regulatory arrangements, although it had taken the government a year longer than originally intended to get the communications bill written, through the process of pre-legislative scrutiny, and then into parliament for debate. That had required the enactment of the Ofcom Act the previous year, which allowed the Ofcom board to be appointed, and initial planning to be funded. David Currie[1084] as chair had selected a board which was to include two members of the outgoing Radio Authority; Richard Hooper, who was to be deputy chair of Ofcom and chair the Content Board; and Sara Nathan, who was a member of both boards. In a last spark from the embers of the Radio Authority fire, Sara Nathan and Richard Hooper were involved in a spat with the radio industry at the Westminster Media Forum in April 2003, over a poor response by local stations in Peterborough and Harlow to a weather emergency, a criticism which made GWR's Steve Orchard "incandescent".[1085]

The Ofcom Act also made possible the appointment of a chief executive in January 2003, Stephen Carter, most recently managing director of NTL.[1086] As a strikingly young chief executive of J Walter Thompson until the autumn of 2000, Carter had been one of those in the advertising agency business sympathetic to radio. He therefore arrived with an instinctive sympathy for radio's commercial aspirations, and an understanding of what the companies hoped for. His evaluation of the regulatory regime which he was succeeding, however, is typically shrewd. "The Radio Authority was proper, well-run, took itself seriously, but was not strategic enough or radical enough. It was more interested in the social than the commercial".[1087] Apart from the 'not radical' tag, the Radio Authority would not have disagreed. Carter judged that commercial radio was "dominated, squeezed by ITV and the BBC in almost everything, and existed within a very fragile space". His prescription, creating larger units through consolidation in order to give radio commercial critical mass, was entirely at one with the market liberalism of the times.

The timing of the substantive Communications Act had been dictated by the European telecommunications directives, and received the Royal Assent at the end of July 2003. Ofcom was to be vested with its full powers on 23 December 2003, a very short space of time in which to take over duties from what were termed the 'legacy regulators' for many months. There had been a significant last-minute argument in the House of Lords over the statutory list of duties. David Puttnam had chaired the Joint Scrutiny Committee which had examined the draft bill. In June 2003, supported by Tom McNally, Elspeth Howe, Melvyn Bragg, Peta Buscombe and other peers with significant interest and history in broadcasting

1084 Lord Currie of Marylebone, a distinguished economist and Dean of the Cass Business School, was a working peer, taking the Labour whip, which he resigned as soon as he was appointed. It is a measure of the high esteem in which he is held, both personally and politically, that his appointment raised no significant partisan complaint.

1085 E mail from Sara Nathan to the author, 6 October 2008.

1086 After a spell with the Brunswick Group after Ofcom, and then as strategy chief for the Prime Minister, he became Lord Carter of Barnes in October 2008, and Minister of Communications.

1087 Stephen Carter in an interview with the author, 11 September 2008.

matters, he tried to replace the neutral tone of the list of duties with a clear steer towards the primacy of citizenship issues.

These peers had been not a little upset by Ofcom's formulation of its role as "to further the interests of citizen-consumers through a regulatory regime which, where appropriate, encourages competition".[1088] This seemed to them to indicate that it would be more concerned with market than with citizenship issues. David Currie was determined not to be hamstrung. Speaking for only the second time in the whole of the parliamentary process over this bill, he offered the reassurance that all would be well. "The dual concept of the citizen/consumer has been at the heart of the appointments to the Ofcom board, in the selection of its senior management, and in the nascent processes that we are putting in place. The Ofcom board reflects a wide and appropriate range of experience, to regulate in the interest of both the consumer and the citizen ... These issues are fundamental to the role of the citizen: plurality, impartiality, high quality, diversity, and effective support for democratic discourse are critical outputs of our society."[1089]

The amendment was defeated, and the spat passed, but from their perspective the peers were right to be uncertain. Ofcom was to be first and foremost an economic regulator, relying – as its brief required – on well-ordered markets to deliver benefits to telecoms users and television and radio audiences, with little time for the old-fashioned ideas of *independent* broadcasting. That was reinforced since there was no longer to be a main board of independent members making decisions based on staff advice, as in the old broadcasting regulators, but an integrated group of executives and non-executives, all determinedly market-aware.

Ofcom had decided early on that it would absorb radio matters into its general structures. There was to be no 'radio silo', meeting the radio companies' wish for "a genuinely integrated decision-making body ... and not a silo mentality for any sector".[1090] The companies hoped to flourish within a converged regulator that would have no time to micro-manage their businesses, and one which was welcoming to consolidation; the regulator in turn would try to be "more hands-off" in the regulation of radio, to "draw back to another level" in its contacts with the stations.[1091]Giving the Guardian Media Lecture at the Radio Festival in Birmingham in July 2003, David Currie had told an eager audience of the major radio companies that "there are no over-large players in ILR" and that he "expected to see consolidation".[1092] He set out clearly Ofcom's approach; it would "want to have comprehensive, accurate, research-based data about these developments and their effect upon the radio market as a whole". Data first, decisions second.

The absence of a separate radio function proved to be a disadvantage to a changing and struggling industry. Stephen Carter observes that "when we set up Ofcom, in order to avoid putting the regulation of radio into a ghetto, we distributed it broadly around the organisation. Ironically, it might have been to the industry's benefit to have had more senior attention earlier. It started to receive that with the appointment of Peter Davies as director of radio and multi-media".[1093] Davies took up that appointment in January 2005, and even then "it

1088 *Foundation and Framework*, Ofcom, autumn 2003.
1089 *Hansard*, Lords, 23 June 2003, cols.17,18.
1090 *Radio Magazine*, 2 May 2002.
1091 Interview with Neil Stock, 11 December 2008.
1092 *Guardian Media Lecture*, Lord Currie of Marylebone, 7 July 2003.

was an uphill struggle to get some people in Ofcom to take radio seriously".[1094] The initial arrangements had risked being that 'Friday afternoon job' after all, just adding radio to potentially more interesting television duties. By the time that was changed, the external climate for radio had got a lot more difficult.

Ofcom's approach to the regulation of commercial radio differed fundamentally from that of its predecessors, being deductive where they had been inductive. The concept of regulation by market research and analysis led it to seek top-down solutions. For the IBA and the Radio Authority, services grew from their local areas; for Ofcom, there needed to be a structure of national/regional/small-scale overlaid onto the inconvenient plethora of pre-existing stations. That made digital radio critically important to Ofcom's vision, since there was no other credible way in which the national tier of commercial radio could be expanded. Just as the radio industry had shifted from bottom-up to top-down management of its business and its programming output, so the regulator was now to follow suit.

In the course of 2003, although it had still to assume its full statutory functions, Ofcom had worked to permit the merger of Carlton and Granada, producing something close to a single ITV company. The Radio Authority had been instinctively nervous about consolidation; Ofcom welcomed it warmly. By 2004, it started to get its wish in radio. To the considerable astonishment of industry commentators, who had seen Capital as the major predator in the expected raft of takeovers, Capital and GWR announced on 29 September 2004 that they were to 'merge', and effectively GWR was to take over Capital. Chief executives David Mansfield of Capital and Ralph Bernard of GWR announced that they planned to run the new company jointly. "If Ralph and I can't work together, the business won't work", Mansfield asserted.[1095] He was to depart from the consolidated radio company on 19 September 2005, barely a year after that statement.

The new company was branded GCap. When the OFT announced that it was not going to refer the merger to the Competition Commission, it flashed the green light for future takeovers in the sector. The second major takeover – and in this instance there was no attempt to disguise the nature of the deal – was Emap's acquisition of Scottish Radio Holdings for £391m, announced on 22 June 2005. This too was swiftly approved by the OFT on 8 August, although Ofcom was obliged to seek some re-arrangement of the new company's digital holdings to stay within the statutory ownership rules. And so, before the end of 2005, these two groups – GCap and Emap – effectively dominated commercial radio. Down from the top table, the Guardian Media Group flourished under John Myers' shrewd stewardship; UTV took over TalkSport's parent, the Wireless Group; the Local Radio Company acquired Radio Investments; and the *Times of India*, along with the radio investors Absolute, paid £53m to buy Virgin Radio from the Scottish Media Group, which had laid out £225m for it eight years earlier. The consolidation which the radio companies had been working towards since the Heathrow Conference in 1984 had at last been achieved.

Wisely is it said, 'be careful what you wish for'. Ralph Bernard left GCap on

1093 Stephen Carter in an interview with the author, 9 January 2009.
1094 Peter Davies in an interview with the author, 24 November 2008.
1095 *Guardian*, 29 September 2004.

23 November 2008, depriving digital radio of having its leading man at the centre of the industry. His successor promptly put the boot into DAB in no uncertain terms.[1096] In the event, neither major conglomerate was to survive long. Emap sold its radio operations to German publisher H Bauer in December 2007. GCap was taken over by the privately-owned Global in April 2008, in a highly-geared takeover hugely dependent on cash flow to service the debt incurred. Bauer, a private company which owns 166 magazines across three continents, was new to radio. The new owner of Global was Ashley Tabor, about whom *The Times* had said that his multimillionaire father "effectively bought him GCap Media and Chrysalis Radio for a combined £545 million".[1097]

Commercial radio's two biggest groups were therefore privately owned. That need not be a bad thing. The *Economist* and the *Guardian* both enjoy the protection of trust status. Channel Four is similarly outside the normal commercial pattern. In that context, the continuing presence of the Guardian Media Group in radio offers some encouragement. However, the abrupt withdrawal of Channel Four from its DAB undertakings was a body blow. It was looking as though the fall of *independent* radio had not after all produced a *commercial* model with any prospect of flourishing in the new media world.

Having paused to ponder the weaknesses of the Radio Authority's approach to licensing new stations, Ofcom announced proposals for consultation in February 2004. To its disappointment, the legislation was clear, requiring Ofcom to award licences for FM commercial radio services in open competition based on statutory criteria. The new Ofcom approach would feature "a simpler, clearer process to apply for a licence, including clear guidance on how the statutory criteria relating to a particular licence will apply; a timetable for the licences to be offered over the year ahead; more information available to applicants when a licence is offered, including an analysis of the relevant market; a reduction in the amount of information that applicants need to send to Ofcom, with a better focus on providing the information which matters".[1098] The whole process marked an overdue shift from paper to electronic applications, and applicants would also be invited to write their own 'format'.

It was a fairly modest re-invention. In May 2004, Ofcom was able to announce a renewed programme of licensing, and would hope to advertise some 30 new FM licences, with nine being advertised that year.[1099] In practice, it largely followed the Radio Authority's working list, and then moved to those areas for which its predecessor had identified frequency options. The first advertisements appeared in June, for Edinburgh and Blackburn, and the licences were awarded in December to The Wireless Group's all-speech Dunedin FM, and the locally-owned Blackburn Broadcasting Company respectively. The gap in licence awards created by Ofcom's re-evaluation was 13 months, from the Glasgow award in November 2003 to that of Edinburgh in December 2004, but it was a critical time to have lost the impetus which new licences always provided to the industry.

Licensing decisions were made by a committee which included executive staff and members of the Ofcom board and content board. However, apart from the

1096 See chapter 20
1097 *The Times*, 17 September 2008.
1098 Ofcom News Release, 2 February 2004.
1099 Ofcom News Release, 12 May 2004.

awards being made more swiftly, the outcomes seemed to outsiders no less capricious – and no more popular within the industry – than those made by the Radio Authority. It was becoming clear that it was the statutory regime, far more than the prejudices of the decision-makers, which was dictating the outcomes. Thirty-seven new FM stations were licensed between 2004 and 2007, ending with a regional licence for South Wales, with one final new licence – for North and Mid Wales – awarded in December 2008, very much a nod to political expectations in Wales.

New DAB multiplexes were also advertised, aimed at "filling the gaps as spectrum became available",[1100] in descending order of population coverage. By 2007, the localities left were too small to be viable, especially given the growing problems for DAB. Thirteen multiplex licences had been being awarded up to that date. Initially, digital radio seemed to many of the new regulators to match other strands in Ofcom's vision of the future of communications, so much so that they missed some of the rather obvious and uncomfortable realities, and the significant differences between radio and television. Kip Meek – who was along with Ed Richards Ofcom's senior partner – asserted at his first Radio Festival appearance, in Birmingham on 13 July 2004, that "we're aiming for 2010 to 2012 in digital switchover in TV and I don't think it will be that long thereafter [for radio]".[1101] In truth, DAB in its UK form was well suited to deliver national radio channels, and larger regional stations. However, it was inappropriate for the large numbers of smaller stations (other than perhaps on L band) which were the heart of ILR, and hopeless for Ofcom's new radio responsibility, community radio.

As already related,[1102] Ofcom's enthusiasm to get ahead fast with DAB, and offer a second national multiplex, brought it into head-to-head confrontation with Digital One, owned by GWR and then GCap, in 2004. When DAB suddenly shifted from being an asset to an expensive liability for the companies, a couple of years later, they found to their horror that they were unable to exit from digital without surrendering their analogue licences too, since those licences had enjoyed automatic renewal on the strict basis of a continuing digital commitment. The business seemed locked into a descending spiral, to which it responded once more by demanding regulatory relaxation over its few remaining localness obligations. Did anyone yearn for the resilient old *independent* radio model? If so, it was now as irrecoverable as Lyonesse.

Against this background, the upbeat interim report of the government's Digital Radio Working Group, under UK Digital chair Barry Cox, seemed to be startlingly unrealistic. It generated headlines in June 2008 by its proposals to 'migrate' all analogue radio services to digital by 2020. "We believe that the government should set an aspirational timetable for migration."[1103] The subsequent full report, published in December 2008, brought the possible date forward to "at least 2017".[1104] It drew from the *Guardian* the apposite comment that "there's some old advice to the effect that if you're in a hole, stop digging. This

1100 Neil Stock interview, op. cit.
1101 *Media Guardian* website, 13 July 2004.
1102 See chapter 20.
1103 *Interim Report of the Digital Radio Working Group*, June 2008.
1104 *Final Report of the Digital Radio Working Group*, 19 December 2008.

unfortunately does not suit the UK's digital radio industry, or, it seems, the government-appointed Digital Radio Working Group."[1105]

The group, and subsequently the government, may have been unduly influenced by the commercial desperation of the radio sector and its backers. It seemed disproportionately concerned with those who had purchased the 7m DAB sets, rather than with the owners of 150–200m analogue sets who would be disenfranchised by switch-over. It paid more attention to the interests of the failing commercial radio conglomerates, rather than the much longer tradition of smaller, local independent radio companies. The Working Party demonstrated little understanding of radio history. When the BBC had tried to replace Radio Four on long wave – of all the supposedly anachronistic platforms – it provoked a march of Middle England onto Broadcasting House.[1106] Radio listening habits change slowly. Just because a medium can be received through a particular technology does not mean that listeners will lightly agree to abandon their beloved old sets.

At the end of 2008, the proponents of DAB had still to find solutions for six crucial issues. Power usage – even with the new generation of microchips – made DAB radio sets hugely less portable than analogue, as well as vastly more expensive. Deploying DAB rather than the new DAB Plus, with improved coding, made the transmission efficiency much less than optimal. There was no early prospect of finding transmission space for the many smaller ILR stations, far less of finding a business model to make their migration to digital feasible. Extensive factory-fitting of DAB radios into cars seemed as far away as ever, and the manufacture of new DAB car radios had virtually dried up since there was so little demand; given the importance of in-car listening, this was a huge lacuna. As the BBC had demonstrated successfully, it was content which drove digital take-up, and the commercial sector had largely lost its ability (or indeed desire) to produce compelling new programming. And the challenge of accommodating the burgeoning new community radio sector – the one undisputed success for independent radio while the commercial sector languished – simply could not be met by the transmission arrangements which were then envisaged.

Given that there was little prospect of solving any of these six issues quickly, and that each of them was a potential deal-breaker on its own, early digital radio switch-over seemed preposterous, only to be contemplated by a government rushing to DAB judgement, just as John Major's did back in 1996. Radio realists, reading in disbelief the DRWG reports, could perhaps take some comfort from DAB pioneer and booster, Mark Thomas, when he said "someone's always saying they're going to turn off analogue radio in *n* years time, but it never happens".[1107]

Indeed – to stray briefly into current affairs – at the start of 2009, the government published an interim strategy document, *Digital Britain*,[1108] which proposed a much more measured approach than its Working Party, no longer flirting with the improbable notion of early analogue switch-off. However, the full *Digital Britain* report, published on 16 June 2009, brought the putative switchover date forward to 2015,[1109] and the Digital Britain bill published on 19 November

1105 *Guardian*, 19 December 2008.
1106 See chapter 12.
1107 Mark Thomas in an interview with the author, 1 July 2008.
1108 *Digital Britain; the Interim Report*, Cm 7548, January 2009.
1109 *Digital Britain. Final Report*, Cm 7650, June 2009, p 93.

was designed to permit major stations to switch when 50 per cent of all radio listening is to digital platforms. It left unspecified what would be the fate of the remaining half of all radio listening, or of those many radio stations which would be stuck on analogue. Such *dirigisme* may encourage listeners to abandon their beloved analogue radio sets. On the other hand, the history of radio suggested that where radio developments are conceived in the interests of providers rather than listeners – which seemed to be the case with the 2009 initiative – they do not succeed. The failure of attempts to structure radio in the thirties, and then again in the sixties, to suit the powers-that-be and the monopoly broadcaster against the wishes of the mass of radio listeners, show that such an approach is not sustainable in the medium term. As the first decade of the new century came towards its close, the significant risk – unacknowledged, unmeasured and perhaps even unappreciated at the time – was that attempting to force the digital radio issue too soon, amid a fierce recession, chiefly to rescue the ailing commercial radio industry, could have the effect of causing irreversible damage to the radio medium as a whole.

Programming regulation within Ofcom had up to the end of 2008 largely eschewed structural remedies, or the seeking of quality by proxy. It accepted the companies' arguments that it should regulate 'outputs' (that is to say what was broadcast) not 'inputs' (which were the structures, systems and rules conditioning broadcasting). It had therefore largely accepted automation, and had permitted extensive – but not wholly unrestricted – co-siting of stations, news hubs and untrammelled name-changes. It tried to hold a firm line only on formats, and by this means to protect localness. There were risks in this approach from the outset. By focusing only on outputs, Ofcom lost the chance to ameliorate behaviour before issues became critical. The initial 'anti-silo' approach to radio regulation also meant that a large number of non-specialists were involved, and they might thus seek to regulate what they did not like to hear, rather than address the underlying issues of what should be broadcast.

The dangers were illustrated by an escalating series of compliance issues. Galaxy Manchester was fined just £2,500 on 16 October 2003 for broadcasting a pre-recorded phone call in which "a male caller advocated physical harm to women".[1110] Key 103 FM in Manchester was then fined £125,000 on 24 November 2005, for a range of breaches: offensive jokes and comments about the death of Iraq hostage Ken Bigley, offensive references to and treatment of Muslims, alleged incitement to racial hatred and a racist comment, together with undue prominence for a presenter's political views, all in the same six-week period.[1111] Notwithstanding a requirement that it transmit the details of the regulator's findings, it seems to have got off fairly lightly for such a catalogue of failings from such an established station. Kiss FM in London was fined a total of £175,000 in total for a series of offensive and unfair broadcasts between April and November that year.[1112]

Then, on 26 June 2008, came the big one. Amid a flurry of issues with broadcast competitions, which affected the BBC and ITV, 30 of GCap's stations were fined a total of £1.1m for fixing phone-in competitions so that listeners continued to pay for texts in the vain hope of winning. Ofcom's adjudication is

1110 Ofcom Content Sanctions Committee, 16 October 2003.
1111 Ofcom Content Sanctions Committee, 23 November 2004.
1112 Ofcom Content Sanctions Committee, 5 July and 15 November 2005.

stark. "GCap had submitted information to Ofcom ... that was ambiguous, both in terms of the nature and the extent of the unfair conduct. Ofcom considered it wholly inadequate that GCap had demonstrated an unwillingness to disclose for several months the specific details and seniority of those responsible for the unfair conduct. This was the first case of its kind in which the behaviour of the licensee (or as in this case, the parent company acting on behalf of the licensees) had effectively hindered Ofcom's investigation. GCap's attempts on behalf of itself and the licensees to remedy the consequences of the breaches were viewed by Ofcom as entirely inadequate. Further, GCap's conduct during Ofcom's investigation, and in particular, its lack of full disclosure of the facts of the case, was a matter of significant concern."[1113]

Ofcom took up its responsibilities to introduce community radio with considerable enthusiasm, even though these were miles away from its general approach to creating dynamic commercial markets in broadcasting. Opening the first round of licensing of community stations in September 2004, it received 194 applications and had awarded 107 licences by the end of the round in May 2006. The second round of licensing was conducted on a region-by-region basis, and by the end of 2007, 155 community radio licences were in issue, with more to come. Although the demand for trial restricted services licences (RSLs) ended with the completion of new FM and digital licensing, RSLs continued to be popular and valued. Nearly 500 short-term RSL licences were issued in each of 2004, 2005, 2006 and 2007. At the end of 2007, there were also 96 long term RSLs licensed across the country.

By the end of 2007, Ofcom had conducted two major reviews of the radio industry. *Radio - Preparing for the Future* appeared in two sections: at the end of 2004[1114]; and in October 2005.[1115] A second study, *The Future of Radio*, was issued in April 2007.[1116] with a follow-up statement in November.[1117] Ofcom continued to "think about radio a lot", but felt "hamstrung" by the past: "the world has changed but we are still stuck with the same legislation".[1118] Ed Richards, who became Ofcom's chief executive in October 2006, continued to assert at the end of 2007 – just as he had done while piloting through the communications bill, which loosened the binding on ILR – that "our research shows that localness is still important to listeners, and we believe that this should be protected ... We are confident that our revised proposals strike the right balance between easing financial pressures faced by industry, and safeguarding the interests of listeners".[1119] However, events in 2008 and 2009 were moving with uncomfortable swiftness for commercial radio, mostly downwards. As the financial and economic crisis which followed the credit crunch gathered momentum, it was becoming hard to see how commercial radio was going to escape intact, and unclear how far Ofcom would feel able to protect localness, the last remaining element of *independent* radio.

1113 Ofcom Content Sanctions Committee, 8 February 2007.
1114 *Radio Review - Preparing for the future*, Ofcom, 15 December 2004.
1115 *Radio - Preparing for the Future Phase 2: Implementing the Framework*, Ofcom, 19 October 2005.
1116 *The Future of Radio*, Ofcom, 17 April 2007.
1117 *The Future of Radio – the next phase*, Ofcom, 22 November 2007.
1118 Neil Stock interview, op. cit.
1119 Ofcom News Release, 22 November 2007.

All of this has happened so recently that for this book it is still, effectively, in the realm of current affairs. History provides the perspective. Britain had been denied commercial radio at the start of the medium in the twenties, because it would not have fitted with the politico-social norms which those who managed British society expected. When commercial output was transmitted from the continent in the thirties, it proved highly popular, only to be snuffed out by war and then airbrushed out of the official memory. The efforts to introduce commercial radio in the sixties found expression in the offshore pirate services, which were sunk by the government, leaving an apparent choice between commercial and state radio. An unexpected general election result in 1970 challenged that dichotomy, and produced a bold experiment, adopting a version of the successful model from television 15 years previously.

Independent radio was to be public service radio, funded by advertising. It was to be demotic and autochthonous, and it took listeners by storm. That popular success broke the party-political divide. Independent radio made progress – despite difficult economic circumstances – by maintaining the settlement of the seventies, and marrying commercial funding with popular, public service programming and local ambition. By the mid-eighties, however, commercial ambition began to be dominant, and in keeping with the changing zeitgeist, the old settlement was overturned in favour of a new concept of local and national *commercial* radio.

Commercial radio only gradually left behind the qualities of its independent past. It showed the genuine possibilities for a soundly-based radio medium in the long boom from the mid-nineties, and by the end of the century it looked to be set fair. However, by then the inherent tension between the old independent attitudes and the new commercial demands was already undermining it, driven by the demands of the City and its shareholders. Stations had moved in the early nineties from being owned typically by local consortia to having single owners, themselves usually public companies. The consolidated radio groups set out to cut their costs, which meant removing the remaining elements of local public service output and the resources to make original and local programming. Commercial radio narrowed its demographic focus to what seemed to be the economically advantageous audience groups, and ceased to offer all the sounds of their lives to a wide range of listeners. Instead, it had bet its shirt on being music radio, and failed to anticipate the impact which new media, such as MP3 and the internet, would have on its business.

To the concerned observer at the end of 2009, it appears that commercial radio now has such a narrow base that it will be a stiff task for it to survive in a meaningful form the shock of this deepest of post-war recessions. For the radio enthusiast, hope lies chiefly in the historical evidence of radio's seemingly indestructible ability to re-invent itself. There remain a good number of people, either independent or strong-minded within groups, who still understand how to do that. What is now needed is evidence that the industry as a whole will manage the trick.

Of the three stations which played the most major parts in the history of independent radio, Capital Radio, for a time the most substantial commercial radio station in the world, and essential to the entire fabric of its city, is now not even the market-leader in London. Radio Clyde, once dubbed the 'jewel in the crown' of independent radio, is pretty much just one more station within a consolidated group. In March 2009, Independent Radio News (IRN) – long since sundered from LBC – was replaced by Sky News as virtually the sole supplier of national and international news to the UK's commercial radio stations.

This history began on the outskirts of Reading in the seventies. If you were to drive along Radio Way in Calcott now, you would be hard pressed to find any traces of the semi-rural outskirts of an old-fashioned market and industrial town. It is now a dormitory for London workers going east, and the British version of Silicon Valley going west, with not much left of its quietly distinctive character. Small-scale Radio 210 became the larger Radio Two Ten, and since March 2009 has been 'Heart', part of a huge group which has changed most of its inherited FM station names to composite brands, and which networks the majority of its output. The station, which once claimed to offer all the sounds of your life – diverse music, news, features, information, sport, discussions, documentaries, and on-air community access, all intensely local – now invites you to "listen to the best music, get the latest showbiz gossip and catch-up with all the action from the shows".[1120]

The house of *independent* radio has fallen; its *commercial* radio heirs are largely indifferent to their legacy; the old music has died. These days, if you want to listen to the sounds of your life in a single accessible local medium, you will mostly search in vain ... except in the echoes of the past.

1120 www.heart.co.uk

Acknowledgements, sources and bibliography

So many people have given me generous assistance in researching and writing this book. John Libbey has been a most helpful and accommodating publisher, and Elizabeth Kay a sympathetic and assiduous editor. Ofcom also has pride of place. Stephen Carter and then Ed Richards have been most encouraging in making possible both the research and the eventual publication of this history. More than that, Ofcom has allowed me access to the full archives of the ITA, IBA and Radio Authority. The daunting task of actually getting to grips with this mass of material was made manageable by the kindness of all those in the Ofcom Secretariat; Elaine McDowall and Caroline Sims most of all. The Knowledge Centre at Ofcom is a fine source of information, and both Julia Fraser and Jan Kacperek have been outstandingly helpful and flexible. I owe a huge debt to the IT team in Riverside House, particularly to John Brennan, who never for a moment let me know how unskilled – and no doubt annoying – I am when confronted with a lap-top computer.

Emma Wray, as well as undertaking a huge amount of the detailed research through the Ofcom files, has also been an invaluable guide and constructive critic. Professor Sean Street has been an inspiration and a rock throughout. The Centre for Broadcasting History Research at Bournemouth University (CBHR) has helped me as far as it is possible to overcome my weaknesses as an academic historian. My thanks go to the many academic and library staff, students and researchers at Bournemouth who have provided information and encouragement. As part of its growing and impressive broadcast archive, CBHR now houses the ITA and IBA archives. It also has a full set of press cuttings taken by the legacy television regulator, which are an object lesson in how to preserve such material – the librarians of the IBA and ITC deserve every researcher's thanks and admiration.

All my former colleagues both at the IBA and at the much-missed Radio Authority have once again given me support and kindness. I hope I have done justice to the project which we shared. Pre-eminently, John Thompson has been fulsome and helpful to someone who has been messing around in the roots of his legacy. Mark Thomas has helped me specifically on technical matters, David Witherow extensively on DAB, Martin Campbell on programming, Neil Romain on finance, Soo Williams on RSLs and community radio, and Neil Stock on licensing. Eve Salomon's sympathetic sharpness has lost nothing since 2003, while

David Vick's perceptiveness, memory and capacity for detail seem if anything to have increased, which I would not have thought possible.

People throughout the radio industry have been generous with their time and memories, and my many interviewees have all been helpful and positive. Specific quotes are sourced in the text. Some of those who spoke with me have asked not to be named, so to avoid speculation I must content myself by simply thanking everyone who spoke to or corresponded with me. That is rather like asking the DJ when you phone in to dedicate the next song to "everyone who knows me"; but you know who you are, and I am most grateful. Several radio people have also allowed me freely to mine their own files, notably Ralph Bernard, Richard Findlay, Jimmy Gordon, Brian West, David Witherow and John Whitney. Special thanks are due to Deanna Hallett of Hallett Arendt, who has provided most of the audience research data used in this book, and much good advice and judgement about the history of radio. Fran Nevrkla and his colleagues at PPL/VPL have generously provided material about the tribunals. Paul Carter's creative camera has provided a defining image of independent radio for the front cover, helped by the patience of Charlie Antrobus who modelled as our DJ.

Turning to the bibliography, the pre-eminent work on the early days of UK radio is of course Asa Briggs' monumental *History of Broadcasting in the United Kingdom* (Oxford University Press, in five volumes published between 1961 and 1995), but its perspective is firmly through the lens of the BBC, even when covering the years after 1955. Sean Street's seminal work on commercial radio broadcasting to Britain from Europe, between 1922 and 1945, *Crossing the Ether* (John Libbey Publishing, 2006), is not just a useful corrective to Briggs but also a major piece of broadcasting social scholarship in its own right. Unexpectedly, Quaker philanthropist B Seebohm Rowntree also has material of relevance to the inter-war radio period, *Poverty and Progress* (Longmans Green & Co, 1941, now available from the Joseph Rowntree Foundation). The administrative history of ITV is contained in the six volumes of *Independent Television in Britain*, begun by Bernard Sendall, taken forward by Jeremy Potter and completed by Paul Bonner and Lesley Aston (Macmillan, between 1982 and 1998). The offshore pirate radio ships found their chronicler in DJ and radio manager Keith Skues, in his *Pop Went the Pirates* (Lambs Meadows Publications, Sheffield, 1994).

In the era of independent radio, there are very few books specifically about the medium. It is noticeable how many academic *tours d'horizon* of broadcasting omit non-BBC radio almost entirely, although Andrew Crisell includes chapters on ILR in *Understanding Radio* (Routledge, 1994) and in his *Introductory History of British Broadcasting* (Routledge, 2002). Two books from Sean Street give full space to private UK radio: *A Concise History of British Radio, 1922–2002* (Kelly Publications, Tiverton, 2005); and his *Historical Dictionary of British Radio* (Scarecrow Press, Oxford, 2006). Important also is Street's review of programme sharing in *The Hidden History of Commercial Radio*, published in *Aural History – Essays on Recorded Sound* (British Library, National Sound Archive 2001). Mike Baron's *Independent Radio; the Story of Commercial Radio in the UK* (Terence Dalton Limited, Suffolk, 1975) has a title that gets over the problem of nomenclature, but sadly stops almost as ILR began. Stephen Barnard's fine *On the Radio: Music Radio in Britain* (Open University Press, 1989) is notably strong on ILR as well as BBC popular radio, but it too ends in 1987, before commercial radio truly arrived in the UK. Meg Carter's book, *Independent Radio; the First 30 Years* (Radio Authority 2003), written for the regulator, is the only attempt so far to cover the whole span. Tim Crook's study, *International Radio Journalism*, Routledge 1998, has a useful chapter on the history and development of independent radio journalism in Britain. Michael Freegard and Jack Blake's *The Decisions of the UK Performing Right and Copyright Tribunal* (Butterworths, London 1997) will satisfy any appetite for yet more information on copyright matters. It sets out in detail all the decided cases, includes the relevant legislation and regulations, and has a useful review of the state of copyright law in regard to these issues.

Written histories of individual ILR stations are almost non-existent. Kathy Barham's touching *Radio City, the Heart of Liverpool* (lulu.com 2006) is a rare exception. The Local Radio Workshop's polemic against Capital Radio, *Public Radio; Private Profit* (Comedia Publishing, 1983) is an interesting anachronism. Anthony Everitt's evaluation of the Access Radio experiment, *New Voices* (Radio Authority, January 2003) is perceptive and remains relevant. Ofcom's two major consultations about radio have started with publications containing extensive data: *Radio – Preparing for the Future* (Ofcom December 2004); and *The Future of Radio* (Ofcom, April 2007). Capital Radio released a 25-year anniversary DVD in 1998, and Radio Trent produced a nostalgic 30-year anniversary DVD, *The Castle Gate Years*, in 2005. Many stations have broadcast compilation programmes, notably Capital's *Now we are Five* in 1978, and Radio Victory's (or more properly Chris Carnegy's) *Farewell to Victoryland* in 1986. Among the competition, BBC Radio Four has attracted a major historical work in David Hendy's award-winning *Life on Air* (Oxford University Press, 2007). The history of Radio One is told by former controller Johnny Beerling in his *Radio 1; the inside scene* (Trafford Publishing, 2008). The Third Programme and Radio 3 are the subject of *The Envy of the World: Fifty Years of the Third Programme and Radio Three* by Humphrey Carpenter (Weidenfeld and Nicholson, 1996). The stories of two relevant rival television platforms are splendidly told by Matthew Horsman in *Sky High; the Amazing Story of BSkyB* (Orion Business Books, 1997) and Maggie Brown in *A Licence to be Different; the Story of Channel 4* (British Film Institute, 2007).

Annex A

Radio advertising and sponsorship revenue 1972–2008

	Net advertising revenue £000 (Years to 30 September)
1975	6,167
1976	11,529
1977	17,999
1978	24,075
1979	33,204
1980	44,358
1981	41,545
1982	52,161
1983	61,467
1984	61,464
1985	67,418
1986	66,403
1987	83,906
1988	113,173
1989	123,552
1990	124,638
1991	112,322
	Gross advertising and sponsorship revenues £000 (Years to 31 December)
1992	141,000
1993	178,300
1994	220,100
1995	270,700
1996	308,800
1997	354,100
1998	418,100
1999	464,400
2000	594,600
2001	549,000
2002	562,700
2003	604,000
2004	641,000
2005	613,400
2006	581,700
2007	598,200
2008	560,200

Notes:

1. Until 1990, the IBA recorded annual net advertising revenue totals as at the end of September each year. The figures from 1991, by the Radio Advertising Bureau, are annual gross totals for advertising and sponsorship income, and include agency commission. They are correspondingly inflated compared with the pre-1991 figures.

2. From 2003 onwards, the RAB's published figures are for moving annual totals (MAT) each quarter; this table uses the final quarter of each year for comparison.

Annex B

Independent radio licences in issue 1972–2008

	ILR licences	Short term RSLs	Local digital multiplexes	Community licences
AM and FM franchised together as one ILR licence				
1973	3			
1974	9			
1975	16			
1976	19			
1977	19			
1978	19			
1979	19			
1980	26			
1981	33			
1982	38			
1983	43			
1984	50			
1985	51			
1986	51			
1987	51			
1988	51			
1989	60 includes 'incremental' licences on single waveband			
1900	79			
AM and FM awarded as separate licences				
1991	130	178		
1992	143	241		
1993	154	188		
1994	166	262		
1995	172	318		
1996	177	324		
1997	205	350		
1998	226	344		
1999	242	393		
2000	248	464	9	
2001	255	423	24	
				Access Radio pilot services
2002	261	450	32	15
2003	272	492	36	14
2004	268	488	46	14
				Community Radio licences from March 05
2005	276	498	46	62
2006	288	475	46	107
2007	297	496	46	155
2008	299	438	46 plus 16 awarded but not yet issued	191

Sources: Independent Broadcasting Authority and Radio Authority archives; Ofcom information office

Annex C

Radio audiences 1972–2008

	Independent/commercial radio listeners 15+ '000s	Independent/commercial radio share %	
	JICRAR		
1977	13553	28.5	
1978	14043	32.0	
Autumn 1980	14318	33.9	
Autumn 1981	14233	31.6	
Spring 1982	17189	33.0	
Spring 1984	15066	27.9	
1986	17430	28.7	
1987	16946	28.9	
1988	17332	29.8	
1990	22255	35.0	
1991	22884	37.9	
	RAJAR		BBC share %
1992	24731	37.7	58.4
1993	26480	42.8	54.9
1994	28034	49.0	48.6
1995	30091	49.7	47.2
1996	27885	48.3	49.6
1997	28408	49.5	47.9
1998	28454	49.3	48.5
1999	30862	46.7	51.3
2000	30871	46 .0	51.7
2001	31877	44.6	53.4
2002	31946	45.5	52.5
2003	31539	45.3	52.9
2004	31176	44.2	54.0
2005	30888	42.8	55.1
2006	31346	43.2	54.4
2007	30716	42.4	55.4
2008	31210	42.2	55.7

Notes:

1. From 1986, the figures given in this table are for the final quarter of each year.
2. A new RAJAR specification applied from 1999, and a further revision from June 2007.

Sources:

Hallett Arendt/JICRAR; RAJAR/RSL/Hallett Arendt; www.rajar.co.uk

Index

Abramsky, Jenny 23, 92, 284, 287-8, 383
Access radio 4, 175, 269, 317-325
AIRC (Association of Independent Radio Contractors) 62-4, 77, 85-7, 92-4, 101-2, 118-9, 125, 128-9, 132, 136, 139-140
  Heathrow conference 142-153, 159, 163, 174-5
  copyright 181-196, 203, 205, 209, 216-7, 227, 230, 232, 247, 249
Amnesty International 263-5
Annan Report 3, 36, 68, 82, 102, 105-7, 109, 155-6
Associated Newspapers 62, 85, 120, 214, 237, 253
Atlantic 252, 217
Attenborough, Sir Richard (Lord Attenborough) 44, 48, 64-5, 87, 127-8, 232
Audience research (see also JICRAR, RAJAR) 15, 58, 62-3, 67, 78, 84-95, 304
Automation 171, 257, 272-4, 312, 334, 343, 350
Bannister, Matthew 244, 295-6
Barton, Michael 22, 108
Bate, Terry 19, 44, 63, 78, 179
BBC - radio policy issues affecting Independent Radio 2, 13n, 14-23, 32, 41, 49, 51, 57, 64-7, 75-6, 84-9, 92-5, 97, 105-8, 129, 131-2, 143, 152, 154-5, 158-9, 163-4, 168-72, 179, 206, 215, 232, 244, 246, 250, 270,295-7, 304, 311, 323, 328, 343-4
BBC music copyright 181-91, 194-6
BBC and digital radio 275-277, 281, 284, 288, 290-1, 349
BBC Radio One 22, 24, 27-8, 57, 80, 97-8, 140, 164-5, 168, 200, 217, 244, 295-6
BBC Radio Two 4, 22, 57, 94, 215, 217, 296-7
BBC Radio Three (Third Programme) 16, 214-5
BBC Radio Four 57, 111, 169, 215, 219, 230, 237, 244, 296, 349
BBC Five Live 94, 216, 219, 288, 290, 296, 335
BBC local radio 22, 35, 51, 53, 57, 68-9, 71, 80, 88, 97, 106-8, 119-20, 150, 154-5, 248
BBC - privatisation of Radio One/Two 30, 164, 247
Beacon Radio 73-5, 79, 102, 118, 125, 133, 166, 200, 254, 256
Benn, Tony 20, 66-7
Bernard, Ralph 79, 118, 143, 212-4, 253, 255, 269-270, 277, 282-3, 289-90, 312, 320, 332, 346-7
Betton, Michael 265-6, 273
Birch, Philip 19, 39, 44, 60, 102, 119
Birt, John 217, 242, 281, 295-6
Blackmore, Tim 56, 98
Blackwell, Eddie 63, 78, 118-9
Blair, Tony – broadcasting issues 293, 299, 310, 312, 328
Bottomley, Virginia 278-9, 293
Bradford, John 70, 118-9, 143-4, 150
Braham, Charles 61, 119
Branson, Sir Richard 217-9, 240
Breakfast television 107, 128-9, 143
Brittan, Leon 117, 134, 146-7, 158-60
BRMB 46-7, 51, 60, 62-4, 69, 73, 79, 81-2, 88, 100, 132-3, 166-7, 172, 184, 200, 224, 337
Brown, Paul 63, 118, 126, 160, 172-3, 195, 202, 207, 224, 248-9, 254-5, 263, 268, 273, 288, 321, 329
Bukht, Michael 48, 61, 98, 110, 135-6, 212, 214-5
Cable Authority 143, 168, 179
Campbell, Martin 269, 272, 303, 305, 339
Capital Radio 4, 24, 45-6, 48-9, 55-9, 61-5, 69, 73-4, 78-9, 81-2, 85-9, 98, 100-1, 105, 109-10, 119-20, 122, 125-6, 128, 132-3, 135-7, 138-140, 143, 146, 157, 164-5, 167, 160, 174, 176, 193-5, 200-1, 212, 214-5, 218, 224, 226-7, 235-6, 240-2, 245, 252-3, 255, 266, 268, 280, 285-6, 293, 295, 297, 299, 308, 331-4, 336-7, 346, 352
Carnegy, Chris 123, 127, 231, 272
Carter, Stephen 289, 344-5
CBC (Cardiff Broadcasting Company) 109, 116, 124, 142, 156-7
Centre Radio 116-7, 124
Chalfont, Alun (Lord Chalfont, Alun Gwynne Jones) 201-5, 211, 213-4, 216, 218, 228, 238, 240-2, 248, 262
Channel Four 4, 105-6, 122n, 127-9, 143, 164, 291, 347
Chataway, Christopher 28-30, 32, 34-6, 39-40, 52-3, 148, 166, 184, 200, 227
Choice FM 224, 235, 239, 258
Classic FM 5, 169, 194, 211-6, 229, 232, 235, 244-5, 252-3, 256, 280, 305
Colville, Sir John (Jock) 1, 2, 72
Community radio 4, 109, 116-7, 127, 145, 147-8, 152, 154-161, 168, 172-4, 180, 203, 208, 223, 225, 313-325, 326, 348-9, 351
Consolidation 4, 47, 199, 204, 226, 256-7, 299, 303, 318, 328-30, 343-6
Copyright 42, 62-3, 77, 97, 134, 148-9, 151, 181-196, 211, 227
CRA (Community Radio Association) 157, 161, 317-8
CRCA (Commercial Radio Companies Association) 63, 93, 181, 273, 287, 304, 311, 320, 324, 328-31
Crown Communications 148, 166, 200, 227, 236
Currie, David (Lord Currie of Marylebone) 344-5
DAB (Digital Audio Broadcasting) 93, 270, 275-291, 299, 302, 309, 315, 347-9
Daubney, Chris 49-50, 67
DCMS (Department for Culture, Media and Sport) 271, 293, 312, 318-320, 326, 328-331
Death on the Rock 119, 148n, 153, 177-9
DNH (Department for National Heritage) 228, 278, 280
Downtown Radio 72, 144
Dubs, Alf 264
East of England licence award 250, 267-9, 298
Eatwell, John (Lord Eatwell) 94, 312, 324, 329-30

EBU (European Broadcasting Union) 140, 202, 276-7
Emap 4, 226, 245, 248, 252, 258, 261-3, 265, 274, 285, 290, 309, 332, 334, 336, 346-7
Eyre, Richard 149, 274, 295
Ffitch, George 110, 143, 146
Findlay, Richard 70, 118-9, 129, 144, 146-152, 205, 333
Forgan, Liz 217, 244, 275, 295-6
Fox, Neil 141, 164-5, 233, 331
Francis, Stuart 216, 227, 230
Fraser, Sir Robert 30, 31n
Frost, Sir David 143, 218, 240
Garnett, Cecilia 63, 118, 188
GCap 4, 257, 290, 306, 337, 346-8, 351
Gibbings, Sir Peter 222, 248, 251-2, 254, 260, 262, 267, 280, 298, 301-3, 307
Gillard, Frank 20, 22, 154-5
Global Radio 4, 290, 347
Gordon, Jimmy (Lord Gordon of Strathblane) 39, 60, 75, 85, 119, 131, 144, 148, 188, 191-6, 230, 233, 258, 293
Gorst, John 20-1, 29, 39-40, 55, 58
Grant, John 201-2, 204, 237, 241, 248, 266
Green, Hughie 20, 28, 30, 44, 65
Guardian Media Group 248, 251, 261-2, 285, 346-7
GWR 2, 4, 71, 143, 200, 212-5, 226-7, 237, 245, 251-6, 258, 268-70, 273, 277, 280, 282-3, 285-6, 289, 294, 305, 332-4 337, 346, 348
Hallett, Deanna 85-7, 94, 212
Heart FM 229, 241, 353
Heath, Edward - broadcasting issues 28, 32
Heathrow Conference 4, 70, 75n, 115, 117, 119, 125-6, 142-153, 162, 164, 166, 174, 180, 191, 230, 274, 297, 346
Home Office 106, 108, 116, 127, 144, 147-8, 152, 156, 158-60, 163, 170, 172-4, 177, 201, 204, 206, 313
Hooper, Richard 201-2, 204, 228, 238, 248, 274, 302-3, 311, 319-20, 335, 339, 344
Howard, Quentin 276-7, 282
Hurd, Douglas 147-8, 159-60, 172-3, 175, 178
Hussell, Bob 70
Hussey, Duke (Marmaduke) (Lord Hussey) 143, 150
Hutton, Bill 61-2, 64, 79-80
IBA (Independent Broadcasting Authority) 30n, 32-4, 36-9, then passim to 189, 201-2, 205-7, 223, 225, 229, 231-2, 236, 243, 246, 248, 256, 263, 311, 314, 327-8, 346
Incremental licences 89, 160-1, 173-6, 203, 206-7, 212, 221, 224-9, 250
IRN (Independent Radio News) 36, 48-9, 54-5, 57, 62, 64-7, 79, 97-8, 100-1, 120, 123, 125, 133, 137-8, 140, 143, 163, 166, 216, 227, 236, 297
ITA (Independent Television Authority) 28-39, 41-2, 49-52, 178, 184, 187, 209
ITV (independent television) 16-7, 20-1, 24, 30-2, 34-5, 37-8, 42, 45, 47, 49, 52, 61, 70, 76-8, 85, 90, 95, 96, 103-4, 106, 119, 121, 127, 129, 131, 135-8, 143, 163, 178-9, 208, 228, 288, 328, 344, 346, 350
Jazz FM 175, 201, 213, 224, 229-30, 234, 236, 239, 285, 338
JICRAR (Joint Industry Committee for Radio Audience Research) 80, 85-9, 92-3, 95, 126, 128, 236, Annex C
Jowell, Tessa 324, 329-30
Judicial review 256, 238, 248, 252, 260-3, 265-6, 270-1, 273, 290
Kennedy, Bob (Robert) 51, 103, 124, 212
Kiss FM 160, 176, 201, 226, 234, 236, 239, 294, 350
Laser 558 143
LBC (London Broadcasting Company) 45-9, 51, 54-59, 61-6, 69, 73, 75, 79-80, 85, 87-8, 89, 97-8, 100-1, 105, 109-10, 121, 124-6, 128, 137, 139-40, 143, 148, 163, 166, 174, 176, 200-1, 219, 227-8
LBC licence loss 234, 236-9, 241, 243, 248, 303
Leicester Sound 124, 253-4, 258
Licensing process 40, 42, 45, 109, 125, 180, 202, 206-8, 231, 234-243, 249-53, 257, 276, 280-1, 284-6, 293-4, 297-8, 300-2, 304, 314, 317, 337-9, 347-8, 351, Annex B
Lincs FM (also Trax FM) 221, 235-6, 265-6
Littler, Dame Shirley 108, 174, 178
Live music 2, 36, 82, 132, 134, 136, 181-3, 186-7, 196
Lloyd, David 281, 303
Local Advisory Committees 32, 51, 107, 202, 209
Local Radio Association 20-1, 29, 35, 43, 57, 70, 184n
Local Radio Workshop 101, 154n, 156
London Greek Radio 160, 176, 225, 257
Lucas, Christopher 51, 70, 118
MacArthur, Douglas 232, 293
MacDonald, Bill 37, 61, 119, 151n, 172, 191
MacKenzie, Kelvin 94, 269n, 282, 285, 303-4, 334-5
Maker, David 61, 211-3, 224
Mansfield, David 143, 268, 280, 295, 297, 308, 346
Manx Radio 20, 183-5, 193
Meaningful speech 103-4, 120
Mellor, David 152, 203-5, 213, 228, 238, 278
Moir, Jim 4, 296
Moriarty, Michael 201, 241, 242n, 246, 248, 251, 267, 303
MU (Musicians Union) 33-5, 39, 42, 63, 137, 181-2, 184-5, 187, 189, 191-4, 196
National Broadcasting School 120, 135
Networking 4, 29, 138, 171, 254-7, 266, 272, 318, 331, 334-5, 343
Newsnight 222, 266-9, 274, 294, 295n, 311
Norrington, John 201, 228, 240, 264, 267-9, 303
NTL (National Transcommunications Limited) 162, 179, 206, 229, 282, 344
Ocean Sound 126-7, 170, 199
Ofcom 5-6, 161, 169n, 206, 257-8, 269, 274, 286, 289-90, 310-1, 314, 316-7, 321, 324-5, 327-8, 330-1, 334n, 338-9, 344-8, 350-1
Orchard, Steve 2, 255-6, 273, 344
Oxygen FM 250, 306-8
Oyston, Owen 167, 205, 261-2, 309

Park, Richard 24, 60, 164, 337
Parliamentary broadcasting 65-7, 133, 135, 137, 236
Peacock Committee 152, 159-60, 163-4, 166-8
Piccadilly Radio 28, 47, 51-2, 60, 63-4, 69, 78, 82, 87-8, 99-100, 121-2, 132-4, 165-7, 170, 200, 205, 258, 350 (as Key FM)
Pilkington Committee 17, 155
Pinnell, David 60, 64, 119, 144n, 172, 184
Pirate Radio 18-24, 27-9, 32-3, 44, 50, 57, 60-1, 71, 82, 84, 111, 130, 143, 145, 147-8, 157-8, 171, 175-6, 184, 221, 258, 313, 352
Plugge, Leonard 14-5, 46n
Plymouth Sound 57, 63-4, 69-70, 73, 97, 119, 200, 226
PPL (Phonographic Performance Limited) 34, 170, 181-196, 211, 227
Premier Radio 342, 270-1, 274, 290, 338
Programme sharing 4, 137-141
Promise of performance 94, 208, 223-4, 226, 230, 240, 257-8, 269, 306 (Formats 305-6)
PRS (Performing Right Society) 182-7, 189-90, 193, 195
RAB Radio Advertising Bureau) 230, 232, 244, 247, 293
RACC (Radio Advertising Copy Clearance Centre) 249
Radio 210 (also Two Ten FM) 1-2, 5, 52, 72, 82, 91, 99, 118, 120, 139, 166, 200, 225, 298, 353
Radio Authority 4, 32n, 91-2, 94, 140, 148, 153, then passim to 340, 344, 346, 348
Radio Caroline 19-21, 44, 176, 184
Radio City 57, 61, 69, 79-82, 87-8, 107, 122-3, 139, 165, 200, 226, 261
Radio Clyde 38-9, 47, 51, 57, 60, 62, 64, 66-7, 69-70, 75, 78-9, 82, 87-8, 99-100, 105, 118, 121, 132-3, 137, 165-6, 170, 200, 212, 224, 227, 258-9, 352
Radio Hallam 37, 57, 61, 63, 69, 79, 82, 87-8, 97, 118-9, 139, 200, 235, 265, 308
Radio London 19, 21, 27, 39, 44, 71, 184
Radio Luxembourg 14-5, 17-9, 21-2, 130, 214, 217
Radio Trent 64, 71, 73, 79, 82, 101, 124, 166, 200, 254
Radio Victory 69, 71, 118, 123, 125-7, 170, 301
RAJAR (Radio Industry Joint Audience Research) 87n, 89, 92-5, 217, 232, 236, 244, 293, 296, 304, 334, Annex C
RCC (Radio Consultative Committee) 74-5, 77, 86, 102-4, 120, 138, 163
Regional licences 213, 218, 229-30, 245-6, 248, 251-2, 258, 281, 284-6, 294, 300-1, 332, 337, 348
RSLs (Restricted service licences) 160, 265, 294, 304, 309, 313-317, 320-1, 351, Annex B
Reynolds, Gillian 61, 74, 81, 98, 131, 139, 219, 221
Robinson, Neil 61, 195, 232, 258
Romain, Neil 201-2, 204, 211, 282, 305, 328, 330
Sally licences 266, 300-1
Secondary rental 4, 33, 38, 43, 67, 100, 109, 115, 120-2, 124-5, 131-41, 143, 162, 186, 256
Selkirk Communications 45, 47-8, 61-2, 64-5, 73, 126-7
Shadow Radio Authority 91-2, 202-9, 239
Sharkey, Feargal 111, 303
Showtime 212-3, 219
Smith, Chris 268, 271, 293, 320, 326, 329
Smith, Terry 61, 119, 129, 177, 226
Sony Radio Awards 140, 157, 332
Spectrum Radio 176, 228
Split frequencies and simulcasting 29, 91, 140, 149, 163, 168, 170, 176, 237, 242, 290, 305
Standard Broadcasting 48, 65, 73, 121, 166
Starks, Michael 103-4, 137
Stewart, Marshall 62, 80, 98
Stiby, Robert 45-6, 65, 74, 126, 268
Stoller, Tony (including first person reportage) 51-2, 64, 71, 72, 75-6, 86-7, 89, 94, 101, 104, 118, 120, 137n, 139, 159n, 188, 202, 209, 225n, 239, 248-9, 254, 265, 267-71, 274, 280, 283-4, 308, 311-2, 317-8, 321, 323, 327, 335, 339
Street, Sean 6, 14-5, 40
Sunrise Radio 175, 223-4, 235, 239-40, 336
Sunset Radio 175, 225, 228, 235
Talk Radio 94, 218-20, 282, 303-4
Talksport (later TalkSPORT) 94, 304, 305, 335
Tarrant, Chris 331-2, 337
Thatcher, Margaret - broadcasting issues 67, 80, 119, 178-9, 191
Thomas, Mark 202, 246, 275, 277-8, 281, 339, 349
Thompson, John 31, 36-40, 42-3, 45-6, 49-51, 55, 58, 63-4, 70, 74, 102-3, 107-8, 118, 131-2, 134, 138, 148-9, 152-3, 162, 171-2, 184-8, 196
Thomson, George (Lord Thomson of Monifieth) 91, 118, 146-50, 158, 177-8
TWC (Trans World Corporation) 205, 248, 252, 261-2, 265, 274
NUJ (National Union of Journalists) 58, 62, 64, 79, 98, 101, 151
UCB (United Christian Broadcasters) 271
Vick, David 6, 90-3, 148, 173-5, 201, 203, 207-8, 211, 222, 229-30, 239, 249-50, 280-1, 300, 314, 317, 330, 339
Virgin Radio 211, 214, 216-20, 240-3, 246, 248, 252-3, 257, 282, 290, 297, 308, 315, 334-5, 346
Walters, Colin 60, 99, 140, 166-7, 170
West, Brian 136, 146, 159, 174, 191-2, 194, 247
Whitney, John 17, 20-1, 30, 43, 46, 48-9, 56, 62-4, 118-9, 129, 136, 146-9, 152-3, 162-3, 173, 177, 188, 212
Wildman, Francis 266
Wilson, Harold - broadcasting issues 28, 54, 65
Witherow, David 276, 278-9, 303, 337, 339
Wren Orchestra 79, 98, 136-7
Xfm 239, 241-2, 293, 295, 307-8
Yellow card 306, 308, 336
Young, Sir Brian 35-6, 58, 67, 70, 75-6, 102, 118, 188
105-108 MHz band 168, 218-9, 246-7, 249, 294

Note on titles: The titles used are those which chiefly applied when people were playing a part in this history; subsequent honours etc. have not been shown in this index or in the main text.